# Scheduling Theory
and its Applications

# Scheduling Theory and its Applications

*Edited by*

**Philippe Chrétienne**
*Université Pierre et Marie Curie, France*

**Edward G. Coffman, Jr.**
*AT&T Bell Labs, USA*

**Jan Karel Lenstra**
*Technische Universiteit Eindhoven, The Netherlands*

*and*

**Zhen Liu**
*INRIA, France*

JOHN WILEY & SONS
Chichester · New York · Brisbane · Toronto · Singapore

Copyright © 1995 by John Wiley & Sons Ltd,
Baffins Lane, Chichester,
West Sussex PO19 1UD, England

*National*     01243 779777
*International*   (+44) 1243 779777

All rights reserved.

No part of this book may be reproduced by any means,
or transmitted, or translated into a machine language
without the written permission of the publisher.

*Other Wiley Editorial Offices*
John Wiley & Sons, Inc., 605 Third Avenue,
New York, NY 10158-0012, USA

Jacaranda Wiley Ltd, 33 Park Road, Milton,
Queensland 4064, Australia

John Wiley & Sons (Canada) Ltd, 22 Worcester Road,
Rexdale, Ontario M9W 1L1, Canada

John Wiley & Sons (SEA) Pte Ltd, 37 Jalan Pemimpin #05-04,
Block B, Union Industrial Building, Singapore 2057

**British Library Cataloguing in Publication Data**

A catalogue record for this book is available from the British Library

ISBN 0 471 94059 3

Typeset in 10/12pt Times by Techset Composition Ltd., Salisbury, Wiltshire.
Printed and bound in Great Britain by Biddles Ltd, Guildford and King's Lynn

# Contents

| | | |
|---|---|---|
| **Preface** | | ix |
| **List of Contributors** | | xiii |
| **Chapter 1** | **Computing Near-Optimal Schedules** | **1** |
| | *J. K. Lenstra and D. B. Shmoys* | |
| | 1.1 Introduction | 1 |
| | 1.2 A general principle via a motivating example | 2 |
| | 1.3 Multiprocessor scheduling | 5 |
| | 1.4 Shop scheduling | 9 |
| | 1.5 A concluding remark | 12 |
| | Acknowledgments | 13 |
| | References | 13 |
| **Chapter 2** | **Recent Asymptotic Results in the Probabilistic Analysis of Schedule Makespans** | **15** |
| | *E. G. Coffman, Jr. and W. Whitt* | |
| | 2.1 Introduction | 15 |
| | 2.2 The on-line greedy policy | 18 |
| | 2.3 Off-line policies | 22 |
| | 2.4 Policy-free error asymptotics | 24 |
| | 2.5 Flow shops | 26 |
| | 2.6 Final remarks | 28 |
| | References | 29 |
| **Chapter 3** | **A Tutorial in Stochastic Scheduling** | **33** ✓ |
| | *G. Weiss* | |
| | 3.1 Introduction | 33 |
| | 3.2 Single-machine nonpreemptive scheduling | 37 |
| | 3.3 Single-machine preemptive scheduling | 40 |
| | 3.4 Machines in parallel | 42 |
| | 3.5 Machines in series | 48 |
| | 3.6 Queueing networks with a single server | 52 |
| | 3.7 Restless bandits | 56 |
| | 3.8 Fluid approximations | 58 |
| | 3.9 Brownian approximations | 59 |
| | Acknowledgments | 59 |
| | References | 59 |

## Chapter 4   Scheduling with Communication Delays: A Survey    65
*P. Chrétienne and C. Picouleau*

| | | |
|---|---|---|
| 4.1 | Introduction | 65 |
| 4.2 | Problem definition and notation | 66 |
| 4.3 | Duplication and an unlimited number of processors | 67 |
| 4.4 | No duplication and an unlimited number of processors | 73 |
| 4.5 | No duplication and a limited number of processors | 85 |
| 4.6 | More general problems | 87 |
| 4.7 | Conclusions | 88 |
| | References | 89 |

## Chapter 5   Profile Scheduling by List Algorithms    91
*Z. Liu and E. Sanlaville*

| | | |
|---|---|---|
| 5.1 | Introduction | 91 |
| 5.2 | Notation | 92 |
| 5.3 | Nonpreemptive profile scheduling of UET tasks | 94 |
| 5.4 | Preemptive profile scheduling | 100 |
| 5.5 | Stochastic profile scheduling | 104 |
| | Acknowledgment | 108 |
| | References | 109 |

## Chapter 6   Efficient Algorithms and a Software Tool for Scheduling Parallel Computation    111
*A. Gerasoulis and T. Yang*

| | | |
|---|---|---|
| 6.1 | Introduction | 111 |
| 6.2 | Program partitioning and data dependence | 112 |
| 6.3 | Granularity and the impact of partitioning on scheduling | 118 |
| 6.4 | Scheduling algorithms for MIMD architecture | 124 |
| 6.5 | The PYRROS software tool | 135 |
| 6.6 | Conclusions | 140 |
| | Acknowledgments | 141 |
| | References | 142 |

## Chapter 7   Scheduling Parallel Programs Assuming Prelocation    145
*V. J. Rayward-Smith, F. Warren Burton and G. J. Janacek*

| | | |
|---|---|---|
| 7.1 | Introduction | 146 |
| 7.2 | Model, assumptions and terminology | 147 |
| 7.3 | Worst case scheduling | 148 |
| 7.4 | Optimal scheduling | 151 |
| 7.5 | Level scheduling | 153 |
| 7.6 | Balancing the allocation function | 155 |
| 7.7 | Using a random allocation function | 159 |
| 7.8 | Conclusions | 163 |
| | Acknowledgments | 164 |
| | References | 164 |

## Chapter 8   Real-Time Scheduling of Periodic Tasks    167
*C. Chaouiya, S. Lefebvre-Barbaroux and A. Jean-Marie*

| | | |
|---|---|---|
| 8.1 | Introduction | 167 |

|     |     |     |
| --- | --- | --- |
| 8.2 | A single-server system with periodic arrivals and preemption | 170 |
| 8.3 | A single server under hard real-time constraints | 176 |
| 8.4 | A communication network with real-time constraints | 182 |
|     | Appendix: illustration of the proof of Theorem 8.6 | 188 |
|     | References | 190 |

## Chapter 9  Cyclic Scheduling on Parallel Processors: An Overview   193
*C. Hanen and A. Munier*

|     |     |     |
| --- | --- | --- |
| 9.1 | Introduction | 193 |
| 9.2 | The basic cyclic scheduling problem | 195 |
| 9.3 | Dynamic scheduling | 198 |
| 9.4 | BCS with resource constraints | 208 |
| 9.5 | Properties of periodic schedules | 212 |
| 9.6 | Algorithms for periodic scheduling on parallel processors | 217 |
| 9.7 | Other models | 221 |
| 9.8 | Conclusions | 223 |
|     | References | 224 |

## Chapter 10  Some Graph Coloring Models for Cyclic Scheduling   227
*D. de Werra and P. Solot*

|     |     |     |
| --- | --- | --- |
| 10.1 | Introduction | 227 |
| 10.2 | Cyclic scheduling | 229 |
| 10.3 | Equitable edge colorings and scheduling | 232 |
| 10.4 | Cyclic compact schedules | 236 |
| 10.5 | Conclusions and additional questions | 238 |
|      | References | 239 |

## Chapter 11  Transforming Cyclic Scheduling Problems into Acyclic Ones   241
*F. Gasperoni and U. Schwiegelshohn*

|     |     |     |
| --- | --- | --- |
| 11.1 | Introduction | 241 |
| 11.2 | Machine model | 242 |
| 11.3 | Tasks and cyclic tasks | 245 |
| 11.4 | Periodic schedules | 247 |
| 11.5 | Core schedules | 250 |
| 11.6 | Main result | 252 |
| 11.7 | Applications and conclusions | 256 |
|      | Acknowledgments | 257 |
|      | References | 257 |

## Chapter 12  Recent Advances in Branch-and-Bound Procedures for Resource-Constrained Project Scheduling Problems   259
*W. S. Herroelen and E. L. Demeulemeester*

|     |     |     |
| --- | --- | --- |
| 12.1 | Introduction | 259 |
| 12.2 | Branch-and-bound for the RCPSP | 263 |
| 12.3 | Extension of the DH procedure to the GRCPSP | 270 |
| 12.4 | Extension of the DH procedure to the PRCPSP | 272 |
| 12.5 | Conclusions | 274 |
|      | References | 274 |

## Chapter 13 The Job Shop Scheduling Problem: A Concise Survey and Some Recent Developments 277
### E. Pinson

13.1 Introduction 277
13.2 Problem statement and complexity results 278
13.3 Optimization algorithms 280
13.4 Approximation algorithms 289
13.5 Conclusions 290
Acknowledgments 291
References 291

## Chapter 14 Application of Majorization to Control Problems in Queueing Systems 295
### D. Towsley

14.1 Introduction 295
14.2 Stochastic ordering and majorization 296
14.3 Optimality of the smallest remaining processing time (SRPT) policy 299
14.4 Extremal properties of queueing disciplines 303
14.5 Applications to routing 306
14.6 Summary 309
Acknowledgment 310
References 310

## Chapter 15 Scheduling and Interchangeability in Tandem Queues 313
### R. Weber

15.1 Fundamentals 313
15.2 The two machine flow shop 316
15.3 Interchangeability 319
15.4 Literature 322
References 324

## Chapter 16 Dynamic Routing and Sequencing in Open Queueing Networks 327
### C. N. Laws

16.1 Introduction 327
16.2 Heavy traffic analysis 329
16.3 Routing and sequencing policies 333
References 335

## Chapter 17 Bandit Processes: Control, Analysis and Characterization 337
### H. Kaspi, A. Mandelbaum and R. J. Vanderbei

17.1 Introduction 337
17.2 Prior research 338
17.3 The multi-armed bandit model and its solution 339
17.4 Open problems and future directions of research 340
References 346

**Author Index** 349

**Subject Index** 359

# Preface

This book collects tutorials, surveys, and articles with original results from a week-long professional meeting on scheduling theory held in France in 1992. The meeting combined the purposes of a summer school with those of a standard scientific conference, in a setting that promoted the close interactions of a workshop. The organization of the school took an unusually broad view of scheduling theory, as can be seen from the table of contents. To better understand the aims of the meeting and to give the setting for the material presented in this book, a quick review of the history of scheduling will be helpful.

The first results of modern scheduling theory appeared some 40 years ago; the field came into its own in the 1960s, grew sharply in the late 1960s and 1970s, and has retained its momentum ever since, with strong support from academia, industry and government. In the 1960s the early theory focussed on industrial applications, and was developed largely by researchers in management science, industrial engineering and operations research. There is still a lively interest in classical flow shop and job shop models incorporating new features that reflect modern industrial processes. The 1980s saw the beginning of fundamentally new directions, with developments in flexible manufacturing systems. The process of automation and the technological advances in robotics will continue to spawn new research in this application of scheduling theory.

The early modeling approaches were those of deterministic scheduling theory and queueing theory, which are both still much in use today. In the former the problem data are assumed known in advance, and the objective is to compute schedules that optimize a given performance metric or, in those cases where this computation is too time-consuming, to compute schedules with performance metrics taking values near the optimum. Job running times and sequencing constraints are examples of problem data; latest job finishing time and average job finishing time are examples of performance metrics.

In classical queueing models the customers or jobs to be scheduled arrive and depart at unpredictable times. Probability laws describing interarrival and service times are given, so a typical problem has been to determine the probability distributions governing customer waiting times under various scheduling policies.

While both the deterministic queueing models are concerned with the design and analysis of scheduling policies, their obvious differences, which extend to mathematical foundations, account for the fact that the two approaches have been pursued by and large as two independent fields of research, with very few researchers working in both fields.

Since the late 1960s, computer science and electrical engineering have had a major impact on scheduling theory, with computer science embracing the combinatorial theory in its general theory of algorithms. A rapid expansion of computer science as an academic discipline in the late 1960s and 1970s coincided roughly with the accelerated growth of scheduling theory during that period. The influence of computer science and electrical engineering has taken several forms: results in complexity theory led to deeper understanding of the inherent difficulty of scheduling problems, and their relation to cognate partitioning and graph problems such as coloring and traveling salesman problems; in the analysis of simple approximation algorithms, techniques were developed for obtaining bounds on performance relative to optimal algorithms; and the computer itself became a major application of the theory, rich in new problems. This source of applications has evolved with technique; for example, scheduling theory has devised new models, or put a special emphasis on existing models, in studies of multiprocessor systems, distributed systems and computer networks. The network application has come to include virtually all of the major advances in computer and communication systems of the past 20 years. For example, relatively recently the theory has had to deal with the challenging problems of scheduling processor communications on the interconnection networks of massively parallel computers, and, on a completely different level, the congestion control problems of high-bandwidth multimedia communication systems.

Two distinctly new components of scheduling theory became active in the 1970s and 1980s. The first was stochastic scheduling, which shares important problems with the theory of bandit processes, stochastic control theory and Markov decision theory. In typical problems of this type probability distributions govern the job running times or sequencing constraints, and the objective is to determine a scheduling policy that optimizes a given performance metric in some stochastic sense, for example in expectation or in distribution. In cases where these objectives have proven intractable, simpler but still interesting questions have been successfully answered, for example those concerning convexity properties and stochastic comparisons.

The second component was the probabilistic analysis of scheduling algorithms, where again probability distributions were defined on problem instances. Typical of this research are the purely analytical problems of finding the expected or average-case behavior under given scheduling algorithms. Because of the difficulty of the problems, a characteristic of this approach has been the use of asymptotic probabilistic analysis; this technique exploits the greater regularity of performance metrics for large numbers of jobs, and is entirely analogous in queueing theory to the study of stationary processes with no concern for the transient dependence on initial states. The tools in this approach rely heavily on the classical limit laws, and the inequalities and bounds of applied probability.

Finally, in this discussion of methodology, we should mention the classical empirical or experimental approach. The design of more efficient (e.g. parallel) algorithms and the increases in computer speed and storage capacity have improved dramatically the effectiveness of these techniques, particularly in the evaluation of the more elaborate scheduling algorithm, such as those using heuristic search and branch-and-bound methods, that trade off computing times with the accuracy of solutions. Although originally more in the province of the engineer than the scheduling theorist, simulation has become a useful tool in stochastic analysis; it is often used to identify schedule properties that can be proved, or, failing that, supply key ideas in the proofs of other results.

This brief historical sketch has shown the great diversity in scheduling theory, in terms of academic disciplines, application areas, fundamental approaches and mathematical skills. Thus meetings such as the one reported here are especially effective in coping with the fragmentation in the field. Acknowledging that it may be too much to ask that a researcher be equally conversant in all aspects of scheduling theory, this book presents tutorial surveys designed to make researchers aware of the progress in the various areas of specialization, and to see the possible influences that this progress may have on their own specialties. In addition, few disciplines are driven so much by a continually changing and expanding technology—a fact that gives scheduling a permanence and the excitement that attaches to the design and analysis of brand new systems. This book plays a vital role here as well, bringing the researcher up to date on the scheduling models and problems of many of the newest technologies in industry, commerce, and the computer and communication sciences.

We should like to take this opportunity to thank Dr Francois Baccelli for having initiated the summer school, and to COMETT, $C^3$, IBP and ARCHIPEL for having sponsored the workshop. We are very grateful to INRIA, in particular to Mrs Catherine Juncker, Patricia Perez and Dany Sergeant, for their efficient administrative work and local organization. This workshop would not have been such a success without the excellent tutorials, the motivating technical sessions and active participation of all the audience. We thank all the participants. Finally, we would like to extend our gratitude to Mr and Mrs Simon, the hosts of the workshop, for having provided us with such a picturesque and convivial environment.

# List of Contributors

**Dr F. W. Burton**
*School of Computing Science*
*Simon Fraser University*
*Burnaby*
*BC*
*Canada V5A 1SJ*

**Dr C. Chaouiya**
*INRIA Centre Sophia Antipolis*
*2004 Route des Lucioles*
*BP 93*
*06902 Sophia Antipolis Cedex*
*France*

**Professor P. Chrétienne**
*Institut Blaise Pascal*
*Laboratoire LITP*
*Université Pierre et Marie Curie*
*4 Place Jussieu*
*75252 Paris Cedex 05*
*France*

**Dr E. G. Coffman, Jr.**
*AT&T Bell Laboratories*
*600 Mountain Avenue*
*Murray Hill*
*NJ 07974*
*USA*

**Dr D. de Werra**
*Swiss Federal Institute of Technology*
*CH-1015 Lausanne*
*Switzerland*

**Dr E. L. Demeulemeester**
*Department of Applied Economic Sciences*
*Katholieke Universiteit Leuven*
*Naamestraat 69*
*B-3000 Leuven*
*Belgium*

**Dr F. Gasperoni**
*Telecom Paris*
*ENST*
*46 Rue Barrault*
*75634 Paris Cedex 13*
*France*

**Dr A. Gerasoulis**
*Department of Computer Science*
*Rutgers University*
*New Brunswick*
*NJ 08903*
*USA*

**Dr C. Hanen**
*Laboratoire LITP*
*Case 168*
*Université Pierre et Marie Curie*
*4 Place Jussieu*
*75252 Paris Cedex 05*
*France*

**Dr W. S. Herroelen**
*Department of Applied Economic Sciences*
*Katholieke Universiteit Leuven*
*Naamestraat 69*
*B-3000 Leuven*
*Belgium*

# LIST OF CONTRIBUTORS

**Dr G. J. Janacek**
School of Mathematics
University of East Anglia
Norwich
NR4 7TJ
UK

**Dr A. Jean-Marie**
INRIA Centre Sophia Antipolis
2004 Route des Lucioles
BP 93
06902 Sophia Antipolis Cedex
France

**Dr H. Kaspi**
Faculty of Industrial Engineering and Management
Technion—Israel Institute of Technology
Haifa 32000
Israel

**Dr C. N. Laws**
Department of Statistics
University of Oxford
1 South Parks Road
Oxford
OX1 3TG
UK

**Dr S. Lefebvre-Barbaroux**
INRIA Centre Sophia Antipolis
2004 Route des Lucioles
BP 93
06902 Sophia Antipolis Cedex
France

**Professor J. K. Lenstra**
Department of Mathematics and Computing Science
Technische Universiteit Eindhoven
Postbus 513, 5600 MB Eindhoven
The Netherlands

**Professor Z. Liu**
INRIA Centre Sophia Antipolis
2004 Route des Lucioles
BP 93
06902 Sophia Antipolis Cedex
France

**Dr A. Mandelbaum**
Faculty of Industrial Engineering and Management
Technion—Israel Institute of Technology
Haifa 32000
Israel

**Dr A. Munier**
Laboratoire LITP
Case 168
Université Pierre et Marie Curie
4 Place Jussieu
75252 Paris Cedex 05
France

**Dr C. Picouleau**
Institut Blaise Pascal
Laboratoire LITP
Université Pierre et Marie Curie
4 Place Jussieu
75252 Paris Cedex 05
France

**Dr E. Pinson**
Institut de Mathématiques Appliquées
Université Catholique de l'Ouest
BP 808
49008 Angers Cedex 01
France

**Dr. V. J. Rayward-Smith**
School of Information Systems
University of East Anglia
Norwich
NR4 7TJ
UK

**Dr E. Sanlaville**
Laboratoire LITP
Case 168
Université Pierre et Marie Curie
4 Place Jussieu
75252 Paris Cedex 05
France

**Dr U. Schwiegelshohn**
Institute for Information Technology Systems
University of Dortmund
44221 Dortmund
Germany

**Professor D. Shmoys**
*School of OR & IE*
*E & TC Building*
*Cornell University*
*Ithaca*
*NY 14853*
*USA*

**Dr P. Solot**
*Ciba-Geigy AG*
*IS 7.2—Operations Research*
*R-1008.72.27*
*CH-4002 Basel*
*Switzerland*

**Dr D. Towsley**
*Department of Computer Science*
*University of Massachusetts*
*Amherst*
*MA 01003*
*USA*

**Dr R. J. Vanderbei**
*Department of Civil Engineering and*
*Operations Research*
*School of Engineering/Applied Science*
*Princeton University*
*Princeton*
*NJ 08544*
*USA*

**Dr R. Weber**
*Statistical Laboratory*
*University of Cambridge*
*16 Mill Lane*
*Cambridge*
*CB2 1SB*
*UK*

**Dr G. Weiss**
*School of ISyE*
*Georgia Tech*
*Atlanta*
*GA 30332*
*USA*

**Dr W. Whitt**
*AT&T Bell Laboratories*
*600 Mountain Avenue*
*Murray Hill*
*NJ 07974*
*USA*

**Dr T. Yang**
*Department of Computer Science*
*University of California*
*Santa Barbara*
*CA 93106*
*USA*

CHAPTER 1

# Computing Near-Optimal Schedules

**J. K. Lenstra**
*Department of Mathematics and Computing Science, Technische Universiteit, Eindhoven; CWI, Amsterdam*

**D. B. Shmoys**
*School of OR & IE, Cornell University*

## Abstract

We survey a number of results on computing near-optimal solutions for $\mathcal{NP}$-hard scheduling problems. For many $\mathcal{NP}$-hard optimization problems there are polynomial-time approximation algorithms for finding solutions that are provably quite close to the optimum, whereas for others no such algorithm is known. We concentrate on results that state that certain performance guarantees are unlikely to be attained, in the sense that if there is such a good algorithm, then $\mathcal{P} = \mathcal{NP}$. In particular, we survey results for multiprocessor scheduling and shop scheduling problems.

## 1.1 INTRODUCTION

Many scheduling problems are $\mathcal{NP}$-hard. Consequently, it is unlikely that any such problem has a polynomial-time algorithm that, given any instance, finds an optimal solution. The algorithm designer must then settle for less: either to relax the requirement that the algorithm be polynomial-time, as in the design of branch-and-bound algorithms, or to relax the requirement that the algorithm always deliver an optimal solution, as in the design of approximation algorithms. We shall focus on issues related to the latter approach.

When designing an approximation algorithm for an $\mathcal{NP}$-hard minimization problem, one can evaluate its performance in a variety of ways. Most intuitively, one can try the algorithm on several particular examples and see how well it works. This may also involve computing a lower bound on the optimal value, so as to be able to have some a posteriori guarantee on the quality of the solution found. However, this

*Scheduling Theory and its Applications* Edited by P. Chrétienne, E. G. Coffman, Jr., J. K. Lenstra and Z. Liu
© 1995 John Wiley & Sons Ltd

approach has the disadvantage that the algorithm, on some important set of data, might perform extremely badly. Thus, one also would like to use an approximation algorithm that comes with some a priori performance guarantee on the worst-case deviation from the optimal value.

The theory of $\mathcal{NP}$-completeness proves that a wide class of decision problems are equivalently hard, in the sense that if there is a polynomial-time algorithm for one $\mathcal{NP}$-complete problem, then there is a polynomial-time algorithm for any $\mathcal{NP}$-complete problem. However, the robustness of the notion of a polynomial-time algorithm blurs the distinction between the computational difficulty of optimization problems. For example, for the types of instances that arise in practice it is easy to solve very large instances of the knapsack problem to optimality, whereas for the job shop scheduling problem instances of rather modest size resist such attacks. If the purpose of a scientific theory is to explain phenomena that occur in practice, then, in this respect, computational complexity theory has not fully achieved its goal. Consequently, it is of interest to consider additional theoretical measures of the tractability of a computational problem.

If one considers the extent to which near-optimal solutions can be obtained in polynomial time for $\mathcal{NP}$-hard optimization problems, then optimization versions of $\mathcal{NP}$-complete problems no longer appear to be equivalent. For some problems, for any arbitrarily small error tolerance, there is a polynomial-time algorithm that delivers a solution of objective function value within that tolerance. On the other hand, for some problems one can show that finding a solution of value guaranteed to be within a certain factor of the optimum is itself $\mathcal{NP}$-hard. Thus, with respect to approximation algorithms with a priori performance guarantees, $\mathcal{NP}$-hard optimization problems vary in much the same way that they do with respect to computing optimal solutions in practice.

We shall survey a number of results in the area of performance guarantees of approximation algorithms for scheduling problems. In particular, we shall concentrate on so-called negative results: results that imply limits on the performance guarantees obtainable by polynomial-time algorithms. There are a number of scheduling models for which one can prove the hardness of obtaining near-optimal solutions. We shall discuss results for several multiprocessor scheduling models: a result of Lenstra and Rinnooy Kan (1978) for minimizing the length of a schedule on identical machines subject to precedence constraints, an extension of this result in which there are communication delays between machines due to Hoogeveen et al. (1995) and a result of Lenstra et al. (1990) for minimizing the length of a schedule on unrelated machines. We shall also discuss recent results of Williamson et al. (1994) for several shop scheduling problems.

## 1.2   A GENERAL PRINCIPLE VIA A MOTIVATING EXAMPLE

The main technique for proving that certain approximation algorithms are unlikely to exist is based on proving a related $\mathcal{NP}$-completeness result. Before presenting this relationship in its full generality, we first give a simple example that illustrates it.

In the *bin packing problem* the input consists of a set of $n$ items, where the $i$th item has size $a_i > 0$, $i = 1, \ldots, n$, which we wish to pack into bins of a given size $b > 0$. A subset of items fits in a bin if the sum of their sizes is at most $b$. We wish to find a packing of all items into bins, so as to use as few bins as possible. For an input $I$, let $OPT(I)$ denote the optimal number of bins.

Consider the special case of the bin packing problem for which $b = \frac{1}{2}\sum_{i=1}^{n} a_i$. In this case it is easy to see that $2 \leq OPT(I) \leq 3$. However, if we wish to decide if $OPT(I) = 2$, then we must decide if the items can be partitioned into two sets of equal total size. This latter problem is the *number partition problem*, which was one of the 21 problems proved to be $\mathcal{NP}$-complete by Karp (1972). Consequently, the problem of deciding if a given instance $I$ of the (general) bin packing problem has $OPT(I) \leq 2$ is $\mathcal{NP}$-complete.

We now show that such an $\mathcal{NP}$-completeness result has implications that limit the kind of performance guarantees that we can expect to prove for polynomial-time approximation algorithms. For an approximation algorithm $A$, let $A(I)$ denote the number of bins used in the packing computed for the instance $I$. Suppose that there is a polynomial-time algorithm $A$ for the bin-packing problem and a constant $\rho < \frac{3}{2}$ such that $A(I) \leq \rho OPT(I)$ for each instance $I$. We shall show that algorithm $A$ can be used to decide if $OPT(I) \leq 2$. Since this decision problem is $\mathcal{NP}$-complete, such an algorithm does not exist unless $\mathcal{P} = \mathcal{NP}$. For any instance $I$ with $OPT(I) \leq 2$, it generates a packing with $A(I) \leq \rho OPT(I) < \frac{3}{2} OPT(I) \leq 3$ bins; since it always uses an integral number of bins, $A(I) \leq 2$. On the other hand, if $OPT(I) \geq 3$, then $A(I)$ must also be at least 3. Thus the algorithm $A$ can be used to decide if $OPT(I) \leq 2$. We conclude that, unless $\mathcal{P} = \mathcal{NP}$, there does not exist a polynomial-time algorithm $A$ with such a performance guarantee.

Polynomial-time approximation algorithms with good a priori performance guarantees do exist for the bin packing problem. The First Fit Decreasing algorithm (FFD) first sorts the items in order of nonincreasing size, and iteratively packs the items one at a time, each time putting the next item in the first bin in which it fits. Johnson (1974) showed that $FFD(I) \leq \frac{11}{9} OPT(I) + 4$ for any instance $I$; in fact, Yue (1991) improved this result to be $FFD(I) \leq \frac{11}{9} OPT(I) + 1$. The latter result implies that $FFD(I) \leq \frac{3}{2} OPT(I)$ for each instance $I$, so that in terms of worst-case relative error, the algorithm FFD is the best possible, unless $\mathcal{P} = \mathcal{NP}$.

The results discussed above suggest that, for the bin packing problem, it is easier to obtain solutions with smaller relative error as the optimum value increases. This behavior is rather atypical of combinatorial optimization problems in which numbers constitute an essential part of the input. Consider the *traveling salesman problem*: given an $n \times n$ matrix $C = (c_{ij})$, where each $c_{ij}$ is a non-negative integer that specifies the distance from city $i$ to city $j$, find a tour of minimum total distance that visits each city exactly once. By multiplying the matrix by a constant $k$, we obtain an equivalent instance of the traveling salesman problem, which has optimal value exactly $k$ times as large. Informally, we shall say that a problem has the *scaling property* if, for any constant $k$, each instance $I$ can be transformed into an equivalent instance $I'$, where each feasible solution of $I$ can be mapped into a feasible solution for $I'$, of objective function value $k$ times as large. For any problem with the scaling property, if there is a polynomial-time algorithm with a performance guarantee $\rho OPT(I) + \beta$, then one can apply this algorithm to a suitably rescaled equivalent instance, and thereby obtain an algorithm with performance guarantee $\rho OPT(I)$.

The bin packing problem does not have the scaling property, and hence the difference between the two types of guarantees is quite significant. In fact, it is still possible that there is a polynomial-time algorithm $A$ for the bin packing problem such that $A(I) \leq OPT(I) + 1$ for each instance $I$. The best algorithm currently known for the bin packing problem, due to Karmarkar and Karp (1982), uses $OPT(I) + O([\log (OPT(I))]^2)$ bins for each instance $I$.

All of the scheduling problems that we shall consider do have the scaling property. Consequently, it is without loss of generality that we restrict attention to performance guarantees of the form $\rho OPT(I)$, where $OPT(I)$ now denotes the optimal value of the particular optimization problem at hand. We shall say that algorithm $A$ is a *$\rho$-approximation algorithm* if, for each instance $I$, $A(I)$ is within a factor of $\rho$ of the optimal value. We shall say that a family of algorithms $\{A_\rho : \rho > 1\}$ is a *polynomial approximation scheme* if, for each $\rho > 1$, $A_\rho$ is a polynomial-time $\rho$-approximation algorithm. We shall say that a family of algorithms $\{A_\rho : \rho > 1\}$ is a *fully polynomial approximation scheme* if, for each $\epsilon > 0$, $A_{1+\epsilon}$ is a $(1 + \epsilon)$-approximation algorithm and its running time is bounded by a polynomial function of the size of $I$ and $1/\epsilon$.

**Impossibility Theorem 1.1** *Consider a combinatorial optimization problem for which all feasible solutions have non-negative integer objective function value. Let $c$ be a fixed positive integer. Suppose that the problem of deciding if there exists a feasible solution of value at most $c$ is $\mathcal{NP}$-complete. Then, for any $\rho < (c+1)/c$, there does not exist a polynomial-time $\rho$-approximation algorithm $A$ unless $\mathcal{P} = \mathcal{NP}$.*

**Proof** The proof of this theorem is exactly analogous to the argument given above for the bin packing problem. If there exists such an algorithm $A$, then it can be used to decide if $OPT(I) \leq c$ in polynomial time. Suppose that $A$, is a polynomial-time $\rho$-approximation algorithm for this combinatorial optimization problem, for some $\rho < (c+1)/c$. If $OPT(I) \leq c$, then

$$A(I) < \frac{c+1}{c} OPT(I) \leq c + 1.$$

Since the algorithm finds a feasible solution, which thereby has an integer objective function value, $A(I)$ is at most $c$. If $OPT(I) \geq c + 1$, then $A(I)$ must also be at least $c + 1$. The algorithm $A$ correctly decides if $OPT(I) \leq c$, and hence $\mathcal{P} = \mathcal{NP}$.

In fact, an identical proof can be used in the case when $c = 0$. In this case, the impossibility theorem states that, for any $\rho > 1$, there does not exist a polynomial-time $\rho$-approximation algorithm unless $\mathscr{P} = \mathscr{N}\mathscr{P}$; furthermore, $\rho$ need not even be a constant for this claim to be true, but it can be an increasing function of the input size. For the traveling salesman problem, Karp (1972) proved that deciding if there is a tour of length at most 0 is $\mathscr{N}\mathscr{P}$-complete, which has this rather strong implication about the nonexistence of approximation algorithms for the traveling salesman problem (Sahni and Gonzalez 1976). However, there are polynomial-time approximation algorithms for this problem with good a priori performance guarantees if one restricts attention to the case when the distance matrix satisfies the triangle inequality, that is, $c_{ij} + c_{jk} \geq c_{ik}$, for each $i, j, k = 1, \ldots, n$ (Christofides 1976; Johnson and Papadimitriou 1985).

## 1.3 MULTIPROCESSOR SCHEDULING

We first consider the problem of scheduling $n$ independent jobs $J_1, \ldots, J_n$ on $m$ identical machines $M_1, \ldots, M_m$. Each job $J_j$, $j = 1, \ldots, n$, is to be processed by exactly one machine, and requires processing time $p_j$. A *schedule* is such an assignment of each job to a machine. For any schedule, the *load* of $M_i$ is the total processing requirement of the jobs assigned to it, and the *length* of the schedule is the load on the most heavily loaded machine. We wish to find a schedule of minimum length. Even if we restrict attention to the special case of this problem in which there are two machines, the problem of computing a minimum length schedule is $\mathscr{N}\mathscr{P}$-hard; as in the previous section, this contains the number partition problem as a special case.

The study of approximation algorithms for $\mathscr{N}\mathscr{P}$-hard problems can be viewed as starting with a result of Graham (1966), who analyzed a simple algorithm that finds a good schedule for this multiprocessor scheduling problem. Graham showed that if we list the jobs in any order, and whenever a machine becomes idle the next job from the list is assigned to it, then the length of the schedule produced is at most twice the optimum: in other words, it is a 2-approximation algorithm. Graham (1969) later refined this result, and showed that if the jobs are first sorted in order of non-increasing processing times, then the length of the schedule is at most $\frac{4}{3}$ times the optimum. Subsequently, there were a number of polynomial-time algorithms with improved performance guarantees. Hochbaum and Shmoys (1986) showed that, in fact, there is a polynomial approximation scheme for this problem. A simple modification of their result can be used to show that for any $\epsilon > 0$ there is a $(1 + \epsilon)$-approximation algorithm that runs in $c_\epsilon n$ time, where $c_\epsilon$ is a constant that depends double-exponentially on $1/\epsilon$. Even though there is a linear-time $\rho$-approximation algorithm for any $\rho > 1$, Garey and Johnson (1978) showed that there is no fully polynomial approximation scheme unless $\mathscr{P} = \mathscr{N}\mathscr{P}$.

We next consider two generalizations of this problem for which it is harder to obtain near-optimal solutions. In the first generalization we introduce a precedence relation $\prec$ among the jobs: if $J_i \prec J_k$, then job $J_j$ must have completed processing before $J_k$ can begin. In the second generalization the machines are no longer identical: if $J_j$ is assigned to $M_i$, then it requires processing time $p_{ij}$. The results currently known for these two models are similar, even though the techniques used are quite different. For both, there is a polynomial-time 2-approximation algorithm. For the first problem, deciding if there is a schedule of length at most 3 is $\mathcal{NP}$-complete, so that, for any $\rho < \frac{4}{3}$, there does not exist a polynomial-time $\rho$-approximation algorithm unless $\mathcal{P} = \mathcal{NP}$. For the second problem, the negative results hold even for a schedule of length at most 2 and hence for any performance ratio smaller than $\frac{3}{2}$.

We first consider the problem of multiprocessor scheduling on identical machines subject to precedence constraints. Graham (1966) showed that the following algorithm is a 2-approximation algorithm: the jobs are listed in any order that is consistent with the precedence constraints, and whenever a machine becomes idle, the next job on the list with all of its predecessor completed is assigned to that machine; if no such job exists, then the machine is left idle until the next machine completes a job.

Lenstra and Rinnooy Kan (1978) showed that, even if each job $J_j$ has processing requirement $p_j = 1, j = 1, \ldots, n$, deciding if there is a schedule of length 3 is $\mathcal{NP}$-complete. We shall give the proof of this result. One of the problems proved to be $\mathcal{NP}$-complete in the seminal paper of Cook (1971) was the CLIQUE problem: given a graph $G = (V, E)$ and an integer $k$, does there exist a subset $C \subseteq V$ of size $k$ such that $\{v, w\} \in E$ if $\{v, w\} \subseteq C$? We shall show that the CLIQUE problem can be reduced to this scheduling problem. Given an instance $(G, k)$ of the CLIQUE problem, let $\bar{k} = |V| - k$, $l = \frac{1}{2}k(k-1)$, and $\bar{l} = |E| - l$. We construct the corresponding instance of the scheduling problem as follows. There are $m = \max\{k, \bar{k} + l, \bar{l}\} + 1$ machines and $n = 3m$ jobs. We introduce a job $J_v$ for each $v \in V$ and a job $K_e$ for each $e \in E$, with $J_v \prec K_e$ whenever $v$ is an endpoint of $e$. We also need dummy jobs $X_x$ (for $x = 1, \ldots, m - k$), $Y_y$ (for $y = 1, \ldots, m - l - \bar{k}$) and $Z_z$ (for $z = 1, \ldots, m - \bar{l}$), with $X_x \prec Y_y \prec Z_z$ for all $x, y, z$.

We claim that the instance of the CLIQUE problem has a solution if and only if there exists a feasible schedule of length at most 3. The basic idea behind the reduction is the following. In any schedule of length 3 for the dummy jobs there is a certain pattern of idle machines that are available for the vertex and edge jobs. This pattern is chosen such that a complete feasible schedule of length 3 exists if and only if there is a clique of size $k$.

More precisely, suppose that a clique of size $k$ exists. We then schedule the $k$ jobs corresponding to the clique vertices and the $m - k$ jobs $X_x$ in the first time slot. In view of the precedence constraints, we schedule the $l$ jobs corresponding to the clique edges and the $m - l - \bar{k}$ jobs $Y_y$ in the second time slot; we also schedule the $\bar{k}$ remaining vertex jobs there. We finally schedule the $\bar{l}$ remaining edge jobs and the $m - \bar{l}$ jobs $Z_z$ in the third time slot. This is a feasible schedule of length 3.

Conversely, suppose that no clique of size $k$ exists, and yet there is a schedule of length 3. In any schedule of length 3 no machine can be idle at any time. As a result of the precedence constraints, in this schedule each of the jobs $X_x$ must be scheduled in the first time slot, each of the jobs $Y_y$ must be scheduled in the second time slot, and each of the jobs $Z_z$ must be scheduled in the third time slot. Thus exactly $k$

vertex jobs are processed in the first time slot, and they release at most $l - 1$ edge jobs for the second slot. However, the only other jobs that can be scheduled in the second slot are the $m - l - \bar{k}$ jobs $Y_y$ and the $\bar{k}$ remaining vertex jobs. Hence there must be an idle machine then, which is a contradiction.

Invoking the impossibility theorem, we obtain the following result.

**Theorem 1.2** *For any $\rho < \frac{4}{3}$, there is no polynomial-time $\rho$-approximation algorithm for the problem of finding a minimum-length multiprocessor schedule on identical machines subject to precedence constraints, unless $\mathscr{P} = \mathscr{NP}$.*

Furthermore, this theorem can be strengthened by observing that the problem has the scaling property: one can simply multiply every processing time by a suitably chosen integer $k$.

One might also be interested in approximation algorithms for the special case in which $p_j = 1$. The $\mathscr{NP}$-completeness result still implies that, for any $\rho < \frac{4}{3}$, no polynomial-time $\rho$-approximation algorithm exists unless $\mathscr{P} = \mathscr{NP}$, and yet it might be possible to devise a polynomial-time algorithm that always finds a schedule of length at most one more than the optimum. However, even though we cannot simply multiply the processing times by $k$, this special case does have a sort of scaling property. Given an instance $I$, construct $k$ copies $I_1, \ldots, I_k$ of this instance using disjoint sets of jobs; then introduce $k - 1$ new jobs, so that the $i$th new job is a successor of each job in $I_{i-1}$ and a predecessor of each job in $I_i$, for $i = 2, \ldots, k$. While this construction does not exactly multiply the value of any feasible solution by $k$, it is still sufficient to prove that if there exists a polynomial-time algorithm that always finds a solution of length at most $\rho OPT + \beta$, then there exists a polynomial-time algorithm that always finds a solution of length at most $\rho OPT$. Thus we see that whether or not a problem has a scaling property can be a rather subtle issue; nonetheless, it has important consequences in the types of performance guarantees that might be attainable.

In a related model $J_j \prec J_k$ implies the following stronger constraint: if $J_j$ and $J_k$ are processed on different machines, then not only must $J_k$ be processed after $J_j$ completes, but it must be processed at least $c_{jk}$ units afterwards. The special case in which each $p_j = 1$ and each $c_{jk} = 1$ was shown to be $\mathscr{NP}$-complete by Hoogeveen et al. (1994); more precisely, they showed that deciding if there is a schedule of length at most 4 is $\mathscr{NP}$-complete. The proof of this result is reminiscent of the proof given above: the main difference is that there also is a second job $\bar{J}_v$ for each vertex $v \in V$, where $J_v \prec \bar{J}_v$; this is used to ensure that, for any edge $e$ and one of its endpoints $v$, the jobs $J_v$ and $K_e$ are not scheduled in successive time slots. Consequently, for any $\rho < \frac{5}{4}$, no polynomial-time $\rho$-approximation algorithm exists for this problem unless $\mathscr{P} = \mathscr{NP}$. Furthermore, Picouleau (1991) gave a simple polynomial-time algorithm to decide if there is a schedule of length at most 3. On the positive side, Rayward-Smith (1987) has given a 3-approximation algorithm for the special case of this problem in which the communication delays and the processing times are all of unit length. It is surprising that both the algorithmic and the hardness results are weaker in this model than their analogues in the setting without communication delays.

Hoogeveen et al. (1994) also considered the following model: there are precedence constraints with communication delays as above, but there is no limit on the number of identical machines that may be used to construct the schedule. For the special case in which each $p_j = 1$ and each $c_{jk} = 1$, they gave a polynomial-time algorithm to decide if there is a schedule of length 5, and yet deciding if there is a schedule of length 6 is $\mathcal{NP}$-complete. Hence, for any $\rho < \frac{7}{6}$, no polynomial-time $\rho$-approximation algorithm exists unless $\mathcal{P} = \mathcal{NP}$. There has been essentially no work done on analyzing the performance guarantee of approximation algorithms for this model.

We next consider the case in which the machines need not be identical. Let $p_{ij}$ denote the time required to process job $J_j$ if it is assigned to machine $M_i$, $i = 1, \ldots, m, j = 1, \ldots, n$. We might simply have the case in which the machines run at different speeds, so that $p_{ij} = p_j/s_i$, where $s_i$ denotes the speed of $M_i$. The results known for this model are quite similar to those known for the case of identical machines. Gonzalez et al. (1977) have shown that if the jobs are sorted in order of nonincreasing processing times, and each successive job is assigned to the machine on which it would complete earliest, then the length of the schedule computed is at most twice the optimal. Hochbaum and Shmoys (1988) have given a polynomial approximation scheme.

The model in which the machines are completely unrelated, in that there is no restriction on the matrix of processing times $P = (p_{ij})$, is substantially harder. There is no simple rule to assign the jobs to machines one at a time that is known to produce schedules of length within any constant factor of the optimum. The best algorithm currently known is a 2-approximation algorithm, which is due to Lenstra et al. (1990). This algorithm first solves the linear relaxation of a natural integer programming formulation, and then carefully rounds the fractional values. On the other hand, Lenstra et al. (1990) also show that deciding whether there is a schedule of length 2 is $\mathcal{NP}$-complete, which implies that for any $\rho < \frac{3}{2}$ no polynomial-time $\rho$-approximation algorithm exists unless $\mathcal{P} = \mathcal{NP}$.

We first prove a weaker result: we show that deciding whether there is a schedule of length 3 is $\mathcal{NP}$-complete. The reduction is from the *3-dimensional matching problem*, which is defined as follows: given disjoint sets $A = \{a_1, \ldots, a_n\}$, $B = \{b_1, \ldots, b_n\}$, and $C = \{c_1, \ldots, c_n\}$, along with a family $F = \{T_1, \ldots, T_m\}$, where each $T_i$, $i = 1, \ldots, m$, is a triple of elements, one from each of $A$, $B$ and $C$, does there exist a subset $F' \subseteq F$ such that each element of $A \cup B \cup C$ is contained in exactly one triple of $F'$?

Given an instance of the 3-dimensional matching problem, we construct an instance of the scheduling problem with $m$ machines and $2n + m$ jobs. For each triple $T_i$, $i = 1, \ldots, m$, we introduce a machine $M_i$. For each element $e \in A \cup B \cup C$, we form a job $J_e$. We also need dummy jobs $X_x$, for $x = 1, \ldots, m - n$. (If $m < n$, then construct a trivial 'no' instance of the scheduling problem.) Each element job $J_e$ takes 1 unit of time on any machine $M_i$ for which $e \in T_i$, and takes 3 time units otherwise. Each dummy job $X_x$ takes 3 units of time on each machine.

It is quite simple to show that there is a schedule of length 3 if and only if there is a 3-dimensional matching in the original instance. Suppose that there is a matching $F'$. If $T_i = (a, b, c)$ is in $F'$, schedule the element jobs $J_a$, $J_b$, and $J_c$ on $M_i$. Since $F'$ is a matching, each element job is scheduled exactly once. Schedule the dummy jobs on the machines corresponding to the triples that are not in $F'$. This is a schedule of length 3. Conversely, suppose that there is such a schedule. Each of the dummy jobs is assigned by itself to some machine. Hence all of the $3n$ element jobs are assigned to a subset of $n$ machines, and each job $J_e$ is assigned to a machine on which it takes one unit of processing time. In other words, each machine $M_i$ is assigned $J_a$, $J_b$ and $J_c$, where $T_i = (a, b, c)$. Let $F'$ be the set of $n$ triples corresponding to the machines not processing dummy jobs. Since each element job $J_e$ is scheduled exactly once, $F'$ is a matching. This concludes the proof that deciding if there is a schedule of length 3 is $\mathcal{NP}$-complete.

We can refine the previous reduction to prove that deciding if there is a schedule of length 2 is $\mathcal{NP}$-complete. As above, for each triple $T_i$, we introduce a machine $M_i$. If $a_j \in T_i$, we say that machine $M_i$ is of *type j*. Let $t_j$ be the number of machines of type $j$. For each element $e \in B \cup C$, introduce a job $J_e$. Refining the previous construction, we need dummy jobs $X_{j,x}$, $j = 1, \ldots, n$, $x = 1, \ldots, t_j - 1$. (If some $t_j = 0$, then again construct a trivial 'no' instance of the scheduling problem.) Each element job $J_e$ takes 1 unit of time on each machine $M_i$ for which $e \in T_i$, and takes 2 units of time otherwise. Each dummy job $X_{j,x}$ takes 2 units of time on any machine of type $j$, but 3 time units on the other machines.

Suppose that there is a matching $F'$. For each $T_i \in F'$, schedule $J_b$ and $J_c$ on $M_i$, where $b$ and $c$ are the elements of $T_i$. For $j = 1, \ldots, n$, this leaves $t_j - 1$ machines of type $j$ that are still idle, and so we can schedule the $t_j - 1$ dummy jobs $X_{j,x}$, $x = 1, \ldots, t_j - 1$, on these machines. This is a schedule of length 2. Conversely, suppose that there is such a schedule. Each dummy job $X_{j,x}$ is scheduled on a machine of type $j$. Therefore there is exactly one machine of type $j$ that is not processing a dummy job, $j = 1, \ldots, n$. Since there are $2n$ element jobs processed by the remaining $n$ machines, each of these machines must be processing two element jobs in one time unit each. If the remaining machine of type $j$ is processing $J_b$ and $J_c$, then there is a triple $(a_j, b, c)$ corresponding to it. Let $F'$ be the set of $n$ triples corresponding to machines that are not processing dummy jobs. Since each element job is scheduled exactly once, and there is one machine of each type scheduling element jobs, $F'$ is a matching.

Invoking the impossibility theorem, we obtain the following result.

**Theorem 1.3** *For any $\rho < \frac{3}{2}$, there is no polynomial-time $\rho$-approximation algorithm for the problem of finding a minimum-length multiprocessor schedule on unrelated machines, unless $\mathcal{P} = \mathcal{NP}$.*

## 1.4 SHOP SCHEDULING

In a shop scheduling problem we are given a set of jobs $\mathcal{J} = \{J_1, \ldots, J_n\}$, a set of machines $\mathcal{M} = \{M_1, \ldots, M_m\}$ and a set of operations $\mathcal{O} = \{O_1, \ldots, O_t\}$; each operation $O_k \in \mathcal{O}$ belongs to a specific job $J_j \in \mathcal{J}$ and must be processed on a specific machine $M_i \in \mathcal{M}$ for a given amount of time $p_k$, which is a non-negative integer. At any time, at most one operation can be processed on each machine, and at most one operation of each job can be processed. We shall consider only nonpreemptive models: each operation must be processed to completion without interruption.

Shop models are further classified based on ordering restrictions for the operations of a job. In an *open shop* the operations of each job may be processed in any order. In a *job shop* the operations of each job must be processed in a given order specific to that job. A *flow shop* is the special case of a job shop in which each job has exactly one operation on each machine, and the order in which each job is processed by the machines is the same for all jobs. In all three models we define the *length* of a schedule as the time at which all operations are completed.

Gonzalez and Sahni (1976) have shown that the open shop problem is $\mathcal{NP}$-hard, even in the case when there are only 3 machines. Racsmány has observed that it is easy to devise a 2-approximation algorithm for the open shop problem (see Bárány and Fiala 1982). At any point in time, an operation is said to be *available* if it has not been processed yet, and no other operation of the same job is currently being processed. Consider the following algorithm: whenever a machine is idle, it starts processing any of its available operations; if no such operation exists, it remains idle until the next operation has been completed on some other machine. Suppose that $O_k$ is the last operation completed, where $O_k$ belongs to $J_j$ and must be processed by $M_i$. The length of the schedule is equal to the time that $M_i$ spends processing plus the time it is idle. Its processing time is clearly a lower bound on the length of an optimal schedule. Its idle time is no more than the total processing requirement of $J_j$, which is also a lower bound on the length of an optimal schedule. Hence the length of the schedule is at most twice the length of an optimal one.

Williamson *et al.* (1994) have shown that the problem of deciding whether there exists a schedule of length 4 is $\mathcal{NP}$-hard. This proof relies on a reduction from the $\mathcal{NP}$-complete problem MONOTONE-NOT-ALL-EQUAL-3SAT. In this problem we are given a set $U$ of Boolean variables, and a collection $C$ of clauses over $U$ such that for each clause $c \in C$, $|c| = 3$, and each clause contains only unnegated variables; we wish to decide if there a truth assignment for $U$ such that each clause in $C$ has at least one true variable and at least one false variable. (Equivalently, this is the problem of deciding if a given 3-uniform hypergraph has a 2-coloring.)

Although we shall not give the complete proof of this result, we give the reduction, and sketch the proof of its correctness. Suppose that we are given an instance of MONOTONE-NOT-ALL-EQUAL-3SAT with $U = \{x_1, \ldots, x_u\}$ and $C = \{c_1, \ldots, c_v\}$, in which each variable $x_i$ appears $t_i$ times. For notational convenience, we shall view the $k$th occurrence of $x_i$ as the variable $x_{ik}$. Furthermore, let $\sigma(x_{ik})$ denote the next occurrence of $x_i$, cyclically ordered; that is, $\sigma(x_{ik}) = x_{il}$, where $l = 1 + k \bmod t_i$. We transform this instance into the following instance of the open shop scheduling problem. For each variable $x_{ik}$, we construct two machines: $M_A(x_{ik})$ and $M_B(x_{ik})$. We construct three types of jobs:

(1) For each variable $x_{ik}$, we construct an *assignment job* $J_{ik}$ with operations $A(x_{ik})$ and $B(x_{ik})$, each of length 2, which are to be processed by $M_A(x_{ik})$ and $M_B(x_{ik})$, respectively.
(2) For each variable $x_{ik}$, we construct a *consistency job* $K_{ik}$ to ensure that its value is equal to the value of its next occurrence, $\sigma(x_{ik})$. It has two operations $\hat{B}(x_{ik})$ and $\hat{A}(x_{ik})$ of length 2 and 1 respectively, which must be processed by $M_B(x_{ik})$ and $M_A(\sigma(x_{ik}))$.
(3) For each clause $c = (x \vee y \vee z)$, we construct a *clause job* $L_c$ with three unit-length operations, $T(x)$, $T(y)$ and $T(z)$, to be processed on $M_A(x)$, $M_A(y)$ and $M_A(z)$ respectively.

The intuition behind the correctness of the reduction is that each assignment job will denote the truth assignment of an occurrence of a variable. Consider the assignment job $J_{ik}$; it has operations of length 2 on $M_A(x_{ik})$ and $M_B(x_{ik})$. In a schedule of length 4 one of these assignment operations must run on one machine from time 0 to 2 and the other operation must run on the other machine from time 2 to 4. Hence we can consider each assignment job as a switch, which can be set in one of two positions depending on whether the job runs first on $M_A(x_{ik})$ or $M_B(x_{ik})$. We shall say that $x_{ik}$ is true if the job runs first on $M_A(x_{ik})$, and false if it runs first on $M_B(x_{ik})$. The consistency job $K_{ik}$ prevents assignment jobs from being scheduled at the same time on machines $M_B(x_{ik})$ and $M_A(\sigma(x_{ik}))$, thus ensuring that the truth assignment will be consistent among all occurrences of the variables $x_{ik}$, $k = 1, \ldots, t_i$. Finally, given the assignment and consistency jobs, each clause job $L_c$ will not be able to have all of its operations scheduled on machines that correspond to true (or false) variables. This property will enforce the not-all-equal constraint for the clause $c$.

Invoking the impossibility theorem, we obtain the following result.

**Theorem 1.4** *For any $\rho < \frac{5}{4}$, there is no polynomial-time $\rho$-approximation algorithm for the problem of finding a minimum-length open shop schedule, unless $\mathscr{P} = \mathscr{NP}$.*

Williamson et al. (1994) have also given a polynomial-time algorithm to decide if there exists a schedule of length 3. This algorithm is based on a well-known connection between the open shop problem in which each operation is of unit length and the problem of edge coloring a bipartite graph. The latter problem can be solved in polynomial time (see e.g. Bondy and Murty 1976). In essence, for any bipartite graph $G = (V_1, V_2, E)$, we can view $V_1$ as the set of machines and $V_2$ as the set of jobs, where an edge $(M_i, J_j)$ reflects that there is a unit-length operation of $J_j$ to be processed by $M_i$. An edge coloring is an assignment of colors to edges so that no two edges with a common endpoint are assigned the same color. Each color class is a set of operations that can be scheduled simultaneously. Thus we see that the problem of using the minimum number of colors is equivalent to the problem of finding a minimum length open shop schedule. If we wish to decide if there is a schedule of length 3, then operations of length 0 and 3 can be scheduled trivially, and we are left with operations of length 1 and 2 only. This can be formulated as a bipartite edge coloring problem with additional constraints on the allowed colorings, which can then be reduced to a bipartite matching problem; see the paper for details.

The flow shop and job shop scheduling problems appear to be significantly harder, in that the performance guarantees known for any polynomial-time algorithm are much worse. It is easy to show that the length of any schedule in which some operation is always being processed is at most $m$ times the optimal length. Röck and Schmidt (1982) have given an aggregation algorithm for the flow shop problem, which relies on the fact that the 2-machine case can be solved in polynomial time; this can be shown to be a polynomial-time $\lceil \frac{1}{2} m \rceil$-approximation algorithm. More recently, Shmoys et al. (1994) have given a polynomial-time $O(\log^2 m\mu)$-approximation algorithm for the job shop problem, where $\mu$ denotes the maximum number of operations of a single job; this was subsequently improved by Schmidt et al. (1993) to yield a polynomial-time $O(\log^2 m\mu / \log \log m\mu)$-approximation algorithm.

Williamson et al. (1994) have shown that, for the flow shop problem, deciding if there is a schedule of length 4 is $\mathcal{NP}$-complete. Hence, for any $\rho < \frac{5}{4}$, there does not exist a polynomial-time $\rho$-approximation algorithm, unless $\mathcal{P} = \mathcal{NP}$. This clearly implies the analogous result for the job shop problem. Furthermore, they gave a polynomial-time algorithm for the job shop problem to decide if there is a schedule of length 3.

## 1.5 A CONCLUDING REMARK

For several scheduling problems, we have seen that there exists a polynomial-time $\rho$-approximation algorithm and that there is some $\delta > 0$ such that, for any $\rho' < \rho - \delta$, no $\rho'$-approximation algorithm exists unless $\mathcal{P} = \mathcal{NP}$. This leaves a troublesome gap between the positive and negative results. Furthermore, the technique for proving the negative results (that is, proving the intractability of the decision problem for a small constant) cannot be used to prove stronger negative results. The optimists among us would claim that this suggests that we must develop better approximation algorithms, and this may indeed be the truth. However, there is no reason to believe that stronger negative results cannot be proved by other techniques.

Recently, a new technique has been developed to prove limits on the extent to which near-optimal solutions can be found in polynomial time, assuming that $\mathcal{P} \neq \mathcal{NP}$. Feige et al. (1991) observed that approximation results are complete for randomized complexity classes, in the sense that in order to decide if a given input is a 'yes' input, we need only approximate the probability with which the input is accepted. Consequently, a randomized characterization of a deterministic complexity class due to Babai et al. (1991) was then invoked to prove negative results about obtaining near-optimal solutions for the maximum clique problem. This line of research culminated in a new characterization of $\mathcal{NP}$ (Arora and Safra 1992; Arora et al. 1992) that showed, for example, that the size of the maximum clique cannot be approximated within a factor of $|V|^\epsilon$, for some $\epsilon > 0$, and the maximum 3-dimensional matching cannot be approximated within a factor of $1 + \epsilon$, for some $\epsilon > 0$, unless $\mathcal{P} = \mathcal{NP}$. In both examples the value of $\epsilon$ is quite small, and so this approach cannot currently be used to bridge the gap mentioned above. It remains a challenging open problem to give a technique that can obtain such results.

## ACKNOWLEDGMENTS

The work of D. B. Shmoys was partially supported by NSF PYI Grant CCR-8896272 with matching support from UPS, Sun, Proctor & Gamble and DuPont, and by the National Science Foundation, the Air Force Office of Scientific Research and the Office of Naval Research, through NSF Grant DMS-8920550.

## REFERENCES

Arora S., Lund C., Motwani R., Sudan M. and Szegedy M. (1992) Proof verification and hardness of approximation problems. In *Proceedings of the 33rd IEEE Symposium on Foundations of Computer Science*, pp 14–23.

Arora S. and Safra S. (1992) Probabilistic checking of proofs. In *Proceedings of the 33rd IEEE Symposium on Foundations of Computer Science*, pp 2–13.

Babai L., Fortnow L. and Lund C. (1991) Non-deterministic exponential time has two-prover interactive protocols *Comput. Complexity*, **1**, 3–40.

Bárány I. and Fiala T. (1982) Többgépes ütemezési problémák közel optimális megoldása *Szigma-Mat.-Közgazdasági Folyóirat*, **15**, 177–191.

Bondy J. A. and Murty U. S. R. (1976) *Graph Theory with Applications* Elsevier, New York, and MacMillan, London.

Christofides N. (1976) Worst case analysis of a new heuristic for the travelling salesman problem. Technical Report 338, Graduate School of Industrial Administration, Carnegie-Mellon University, Pittsburgh.

Cook S. A. (1971) The complexity of theorem-proving procedures. In *Proceedings of the 3rd Annual ACM Symposium on Theory of Computing*, pp 151–158.

Feige U., Goldwasser S., Lovász L., Safra S. and Szegedy M. (1991) Approximating clique is almost NP-complete. In *Proceedings of the 32nd IEEE Symposium on Foundations of Computer Science*, pp 2–12.

Garey M. R. and Johnson D. S. (1978) 'Strong' NP-completeness results: motivation, examples and implications *J. Assoc. Comput. Mach.*, **25**, 499–508.

Gonzalez T., Ibarra O. H. and Sahni S. (1977) Bounds for LPT schedules on uniform processors *SIAM J. Comput.*, **6**, 155–166.

Gonzalez T. and Sahni S. (1976) Open shop scheduling to minimize finish time *J. Assoc. Comput. Mach.*, **23**, 665–679.

Graham R. L. (1966) Bounds for certain multiprocessing anomalies *Bell System Tech. J.*, **45**, 1563–1581.

Graham R. L. (1969) Bounds on multiprocessing timing anomalies *SIAM J. Appl. Math.*, **17**, 416–429.

Hochbaum D. S. and Shmoys D. B. (1986) A unified approach to approximation algorithms for bottleneck problems *J. Assoc. Comput. Mach.*, **33**, 533–550.

Hochbaum D. S. and Shmoys D. B. (1988) A polynomial approximation scheme for scheduling on uniform processors: using the dual approximation approach *SIAM J. Comput.*, **17**, 539–551.

Hoogeveen J. A., Lenstra J. K. and Veltman B. (1994) Three, four, five, six, or the complexity of scheduling with communication delays *Oper. Res. Letters*, **16**, 129–137.

Johnson D. S. (1974) Fast algorithms for bin-packing *J. Comput. System Sci.*, **8**, 272–314.

Johnson D. S. and Papadimitriou C. H. (1985) Performance guarantees for heuristics. In Lawler E. L., Lenstra J. K., Rinnooy Kan A. H. G. and Shmoys D. B., editors, *The Traveling Salesman Problem: a Guided Tour of Combinatorial Optimization*, pp 145–180 Wiley, Chichester.

Karmarkar N. and Karp R. M. (1982) An efficient approximation scheme for the one-dimensional bin-packing problem. In *Proceedings of the 23rd Annual IEEE Symposium on Foundations of Computer Science*, pp 312–320.

Karp R. M. (1972) Reducibility among combinatorial problems. In Miller R. E. and Thatcher J. W., editors, *Complexity of Computer Computations*, pp 85–103 Plenum Press, New York.

Lenstra J. K. and Rinnooy Kan A. H. G. (1978) The complexity of scheduling under precedence constraints *Oper. Res.*, **26**, 22–35.

Lenstra J. K., Shmoys D. B. and Tardos É. (1990) Approximation algorithm for scheduling unrelated parallel machines *Math. Programming*, **46**, 259–271.

Picouleau C. (1991) Two new NP-complete scheduling problems with communication delays and unlimited number of processors. Technical Report RP91/24, MASI, Institut Blaise Pascal, Université Paris VI, Paris.

Rayward-Smith V. J. (1987) UET scheduling with unit interprocessor communication delays *Discrete Appl. Math.*, **18**, 55–71.

Röck H. and Schmidt G. (1983) Machine aggregation heuristics in shop scheduling *Methods Oper. Res.*, **45**, 303–314.

Sahni S. and Gonzalez T. (1976) P-complete approximation problems. *J. Assoc. Comput. Mach.*, **23**, 555–565.

Schmidt J. P., Siegel A. and Srinivasan A. (1993) Chernoff–Hoeffding bounds for applications with limited independence. In *Proceedings of the 4th Annual ACM–SIAM Symposium on Discrete Algorithms*, pp 331–340.

Shmoys D. B., Stein C. and Wein J. (1994) Improved approximation algorithms for shop scheduling problems *SIAM J. Comput.*, **23**, 617–632.

Williamson D. P., Hall L. A., Hoogeveen J. A., Hurkens C. A. J., Lenstra J. K., and Shmoys D. B. (1994) Short shop schedules. Submitted for publication.

Yue M. (1991) A simple proof of the inequality $FFD(L) \leq \frac{11}{9}OPT(L) +1$, $\forall\ L$, for the *FFD* bin-packing algorithm *Acta Mathematicae Applicatae Sinica*, **7**, 321–331.

---

*Department of Mathematics and Computing Science, Technische Universiteit Eindhoven, Postbus 513, 5600 MB Eindhoven, The Netherlands*

*School of OR & IE, E & TC Building, Cornell University, Ithaca, NY 14853, USA*

CHAPTER 2

# Recent Asymptotic Results in the Probabilistic Analysis of Schedule Makespans

E. G. Coffman, Jr. and Ward Whitt
*AT&T Bell Laboratories*

**Abstract**

Makespan scheduling problems are in the mainstream of operations research, industrial engineering, and computer science. A basic multiprocessor version requires that $n$ tasks be scheduled on $m$ identical processors so as to minimize the makespan, i.e. the latest task finishing time. In the standard probbility model considered here, the task durations are i.i.d. random variables with a distribution $F$, and the objective is to estimate the distribution of the makespan as a function of $m$, $n$, and $F$. This paper surveys probabilistic results for the multiprocessor scheduling problem and an important variant known as the permutation flow-shop problem. Several of the results are new; the others have appeared in the last few years.

Because of the difficulty of exact analysis, the results take the form of limits as $n \to \infty$ or as both $m \to \infty$ and $n \to \infty$ with $m < n$. Some highlights of the survey are: a new asymptotic analysis of the on-line greedy scheduling policy, the resolution of a longstanding open problem in the analysis of off-line policies, new applications of central limit theorems to makespan scheduling, and limit theorems giving the asymptotic behavior under the greedy and optimal policies for the flow-shop problem. Open problems and modeling issues are also discussed.

## 2.1 INTRODUCTION

An integer $m \geqslant 2$ together with positive task running times $T_1, \ldots, T_n$ defines an instance of the *multiprocessor scheduling problem*: Schedule $T_1, \ldots, T_n$ on $m$ identical processors $P_1, \ldots, P_m$ so as to minimize the latest task finishing time or *makespan*; i.e., partition the set $\{T_1, \ldots, T_n\}$ into subsets $P_1, \ldots, P_m$ so as to minimize the maximum subset sum

$$L_{m,n} \equiv \max_{1 \leq j \leq m} \sum_{\{i: T_i \in P_j\}} T_i.$$

To avoid trivialities, we assume that $n \geq m$ unless stated otherwise. The problem finds application in operations research as a model of scheduling parallel machines in industrial job shops. It has also had a prominent role in computer science, where the term multiprocessor originates. Along with a number of other fundamental $\mathcal{NP}$-complete problems, it has served as a theoretical testbed for the development of new ideas in the design and analysis of algorithms (see Garey and Johnson 1979).

Because of the problem's complexity, several heuristic policies have been studied. Our interest here is in simple but effective techniques (as illustrated in the next paragraph) rather than elaborate heuristic search techniques. The combinatorial worst-case analysis of such policies dates back over 25 years (see Graham (1966) and, for a general treatment, Blazewicz *et al.* (1993)). More recently, the competitive analysis of on-line algorithms has been applied to the problem (for recent results and references to others, see Phillips and Westbrook 1993). In this setting, the problem has also been called *load balancing*, a term that suggests broader applications. For example, in computer storage allocation it may be necessary to distribute $n$ files among $m$ identical storage units so as to minimize the maximum of the total file sizes.

For the purposes of defining multiprocessor scheduling policies, it is convenient to assume that the tasks are presented in the form of a list $(T_1, \ldots, T_n)$. The on-line *greedy* policy is arguably the simplest (and fastest) heuristic for finding approximate solutions to the multiprocessor scheduling problem. This policy uses no advance information on the number or durations of tasks. The policy begins by assigning the first $m$ tasks $T_1, \ldots, T_m$ to the $m$ processors $P_1, \ldots, P_m$; the processors start running these tasks at time 0, while the remaining tasks wait. Thereafter, whenever a processor finishes its current task, the next waiting task, if any, is assigned to the idle processor. In queueing terminology, the system operates as an $m$-server queue with a first-come first-served service discipline; $n$ customers arrive to an empty system at time 0, and the latest of their departure times is the makespan. The rule for resolving ties among processors is immaterial, so we leave it unspecified. The off-line greedy policy operates just as the on-line version, except that the list $(T_1, \ldots, T_n)$ is first sorted into decreasing order. The off-line version is also called the *largest-processing time* (LPT) policy, a term we use hereinafter; the term 'greedy' by itself refers to the on-line policy.

Understandably, the greedy and LPT policies were among the first policies studied when the probabilistic analysis of scheduling algorithms began some 15 years ago. In the standard probability model considered here the task durations $T_i$ are independent and identically distributed (i.i.d.) with distribution $F(t) = P(T_i \leq t)$. The problem is to find the distribution of the makespan $L_{m,n}$ as a function of the number $m$ of processors, the number $n$ of tasks and the distribution $F$. The general aim is to bring out typical behavior rather than the worst-case behavior, which can be highly unlikely. With explicit formulas in mind, probabilistic analysis is usually quite difficult, so research has often turned to large-$n$ asymptotics.

This chapter surveys new probabilistic results, concentrating on those of the past few years; earlier research is covered in Coffman and Lueker (1991). We do not claim that our survey is exhaustive; rather, our goal is to illustrate current directions, mathematical approaches and open problems in a field that is quite active.

Section 1.2 covers the greedy policy, presenting new results of the authors in collaboration with L. Flatto, A. Weiss and P. E. Wright. The analysis here is self-contained, but Coffman *et al.* (1993) study theoretical questions in more depth. Section 2.3 discusses off-line policies, concentrating on the differencing methods of Karmarkar and Karp (1982). Yakir (1993) recently solved an intriguing open problem set by Karmarkar and Karp's original analysis. The principal new insight in Yakir's approach is described.

Central limit theorems are natural tools for asymptotic makespan analysis. Section 2.4 applies these tools in a policy-free set-up, i.e. limit theorems are proved that hold simultaneously for all scheduling policies. These results are new.

Research on the permutation flow-shop problem, a fundamental variant of makespan scheduling, is surveyed in Section 2.5. In this problem each task consists of $m$ *operations*, one for each processor, i.e. $T_i \equiv (T_{i1}, \ldots, T_{im})$, $1 \leqslant i \leqslant n$. A permutation of the task indices $(1, 2, \ldots, n)$ defines a schedule because the operations of each task must be performed on processors in the sequence $P_1, \ldots, P_m$, and because the $n$ operations must be performed in the same task order on every processor. Now, the $nm$ operation times $T_{ij}$ are regarded as i.i.d. with distribution $F$. Thus under the greedy policy the queueing system with analogous dynamics is a network with $m$ single-server queues in tandem, the first-come first-served service discipline and $n$ customers initially at the first queue ready to begin service. In this case all service times are i.i.d. Asymptotic behavior is described for the greedy policy when either or both of $m$ and $n$ are large, and for an optimal policy when $m = 2$ and $n$ is large. The results for the greedy policy are contained in Glynn and Whitt (1991), Greenberg *et al.* (1993) and Srinivasan (1993); the results for the optimal policy when $m = 2$ are due to Ramudhin *et al.* (1993). Section 2.6 caps off the chapter with a discussion of open problems and modeling issues.

This section concludes with matters of convention. Probabilistic results have varied widely in the classes of distributions $F$ allowed. *A convenient, common subset consists of those distributions supported on a finite interval, with a positive continuous density $f$. In what follows, $F$ has these properties, unless stated otherwise.* The mean and variance of $F$ are denoted by $\tau$ and $\sigma^2$. Our uniform treatment simplifies the presentation of the basic ideas; but while such distributions are adequate as models of most practical situations, many of the results hold for broader classes of distributions. Details on these technical matters can be found in the references.

Because of the form of the results in Sections 2.2–2.4, policies will also be assessed in terms of the *error* $\alpha_{m,n} = L_{m,n} - n\tau/m$, where $n\tau/m$ is an obvious lower bound on $E[L_{m,n}]$. An analogous normalization is introduced in Section 2.5 for the flow-shop problem. The notation $L_{m,n}$, $\alpha_{m,n}$ will be used generically; in any given instance the problem and policy being considered will be clear in context.

## 2.2  THE ON-LINE GREEDY POLICY

In the past decade several papers have been devoted to an asymptotic analysis of the greedy policy for general $m$ (see Boxma 1985; Bruno and Downey 1986; Coffman and Gilbert 1985; Han et al. 1992; Loulou 1984). None of this work has led to limiting behavior as precise as that found for $m = 2$ in an early result of Feller (1971, p. 208); for $m \geqslant 3$, the analysis has resorted to various bounding techniques. However, Feller's result can in fact be generalized, as shown below. (See Coffman et al. (1993) for extensions.)

Fix $m \geqslant 2$, and, for convenience, extend the greedy process to the infinite time horizon, i.e. construct a greedy schedule from the infinite sequence $T_1, T_2, \ldots$. Let $C_n$ denote the $n$th completion time; let $R_i^n$, $1 \leqslant i \leqslant m-1$, denote the residual times of those tasks still running at time $C_n$, ordered by increasing processor index; and let $R_{(1)}^n \leqslant \cdots \leqslant R_{(m-1)}^n$ denote the order statistics of the $R_i^n$. Without loss of generality, assume that $C_n \neq C_{n+1}$ for all $n \geqslant 1$. At time $C_{n-m+1}$, tasks $T_1, \ldots, T_n$ have all started, $n - m + 1$ of them have finished and $m - 1$ are still running with residual times $R_i^{n-m+1}$, $1 \leqslant i \leqslant m-1$. Then the tasks $T_1, \ldots, T_n$ have a latest finishing time

$$L_{m,n} = C_{n-m+1} + R_{(m-1)}^{n-m+1}, \tag{2.1}$$

and a sum of running times that can be expressed as

$$\sum_{i=1}^n T_i = mC_{n-m+1} + \sum_{i=1}^{m-1} R_{(i)}^{n-m+1} = mC_{n-m+1} + \sum_{i=1}^{m-1} R_i^{n-m+1}. \tag{2.2}$$

Combining (2.1) and (2.2) gives

$$L_{m,n} = \frac{1}{m}\left[\sum_{i=1}^n T_i + mR_{(m-1)}^{n-m+1} - \sum_{i=1}^{m-1} R_i^{n-m+1}\right]. \tag{2.3}$$

To proceed, we need information on the random variables $R_{(i)}^n$. A direct approach analyzes the Markov chain $\{(R_{(1)}^n, \ldots, R_{(m-1)}^n), n \leqslant 1\}$. Let $p_n$ denote the density at epoch $C_n$ of the Markov chain, and define

$$\mathbf{t}_0(y) = (t_1 + y, \ldots, t_{m-1} + y),$$
$$\mathbf{t}_i(y) = (y, t_1 + y, \ldots, t_{i-1} + y, t_{i+1} + y, \ldots, t_{m-1} + y), \quad 1 \leqslant i \leqslant m-1,$$

with $\mathbf{t} \equiv \mathbf{t}_0(0)$. Let $t_0 = 0$. A straightforward analysis then shows that, with $p_0$ given,

$$p_{n+1}(\mathbf{t}) = \sum_{i=0}^{m-1} \int_0^\infty p_n(\mathbf{t}_i(y)) f(t_i + y) dy, \quad n \geqslant 0. \tag{2.4}$$

An application of the theory of Harris-recurrent Markov chains (see Asmussen 1987, pp. 150–158) proves convergence of the distributions to a proper limit independent of the starting state. In terms of random variables, we can write

$$(R_{(1)}^n, \ldots, R_{(m-1)}^n) \Rightarrow (R_{(1)}^*, \ldots, R_{(m-1)}^*) \quad \text{as } n \to \infty, \tag{2.5}$$

where $\Rightarrow$ denotes convergence in distribution. Indeed, there is convergence of the probability measures in total variation as $n \to \infty$.

To obtain an explicit formula for the limiting distribution, i.e. the distribution of $(R_{(1)}^*, \ldots, R_{(m-1)}^*)$ in (2.5), the Markov chain analysis now requires that we solve the stationary version of the rather awkward recurrence in (2.3). A key observation allows us to side-step this difficulty by applying the theory of stationary point processes, in an argument that makes no direct use of properties already established by the Markov chain approach. The observation is that the sequence $\{C_n\}$ generated by the greedy rule is equal stochastically to the superposition of $m$ i.i.d. ordinary renewal processes, i.e. each is defined independently by $F$ and each starts with a point at 0. The time-stationary version of each renewal process is the familiar equilibrium renewal process, in which the distance to the first point has the equilibrium residual-life distribution $G$ with density $g(t) = [1 - F(t)]/\tau$. Each original renewal process is the Palm (or synchronous) version of its time-stationary version.

Now consider the superposition of $m$ i.i.d. copies of this time-stationary renewal process. This is a time-stationary point process with the distance to the first point from each component stream having distribution $G$. Since we want to look at the superposition process at completion times, we are interested in the Palm (synchronous) version of this stationary point process. Section 5.1 of Baccelli and Brémaud (1987) characterizes this Palm version in terms of the Palm and stationary versions of the component processes. However, from this superposition process alone, we cannot extract the stationary distribution of $(R_{(1)}^n, \ldots, R_{(m-1)}^n)$ directly. To do this, we mark the points of each component stream with the index of the processor on which it occurs, and then apply the corresponding superposition result for stationary marked point processes in Section 1.3.5 of Franken et al. (1982). This result shows that each of the $m - 1$ streams is equally likely to produce the current point (i.e. the $m$ possible marks of the current point are equally likely), and that the residual time to the next point in each of the remaining $m - 1$ streams has the distribution $G$. It follows that the stationary version of $(R_{(1)}^n, \ldots, R_{(m-1)}^n)$ at completion times coincides with the order statistics of $m - 1$ i.i.d. random variables with distribution $G$. This implies that a stationary solution to (2.4) is

$$p(\mathbf{t}) = (m-1)! \prod_{i=1}^{m-1} g(t_i), \tag{2.6}$$

which a substitution into (2.4) will verify.

As a consequence, we have the limit

$$L_{m,n} - \frac{1}{m}\sum_{i=1}^{n} T_i \Rightarrow \alpha_m \equiv R^*_{(m-1)} - \frac{1}{m}\sum_{i=1}^{m-1} R^*_i \quad \text{as } n \to \infty, \tag{2.7}$$

where $R^*_1, \ldots, R^*_{m-1}$ are i.i.d. random variables with distribution $G$.

We now consider expected values. Note that the residual-life distribution $G$ has mean $(\sigma^2 + \tau^2)/2\tau = \frac{1}{2}\tau(v^2 + 1)$, where $v \equiv \sigma/\tau$ is the coefficient of variation of $F$ (recall that $\tau$ and $\sigma$ are the mean and standard deviation of $F$). Uniform integrability follows from our assumptions on $F$, so, from (2.7), we obtain

$$E[\alpha_{m,n}] \to E[\alpha_m] = E[R^*_{(m-1)}] - \frac{1}{m}\sum_{i=1}^{m-1} E[R^*_i] \quad \text{as } n \to \infty \tag{2.8}$$

or, equivalently,

$$E[\alpha_{m,n}] = \int_0^\infty [1 - G^{m-1}(x)]dx$$
$$- \left(\frac{m-1}{m}\right)\tau\left(\frac{v^2+1}{2}\right) + o(1) \quad \text{as } n \to \infty. \tag{2.9}$$

An important special case is the uniform distribution, $F(t) = t$, with $g(t) = 2(1-t)$, $0 \leq t \leq 1$. From (2.9), we obtain

$$E[\alpha_m] = \int_0^1 [1 - (2t - t^2)^{m-1}]dt - \frac{(m-1)}{3m}. \tag{2.10}$$

A direct calculation gives

$$E[\alpha_{m,n}] = \frac{2m+1}{3m} - \frac{\Gamma(\frac{3}{2})\Gamma(m)}{\Gamma(m+\frac{1}{2})} + o(1) \quad \text{as } n \to \infty. \tag{2.11}$$

Coffman et al. (1993) also consider rates of convergence. For our distributions $F$, it is easily verified that, from any point of the (compact) state space to any other such point, the $r$-step transition density is strictly positive for at least one $r \leq 2(m-1)$. Then Doeblin's condition holds and convergence to the stationary distribution is geometrically fast (see e.g. Meyn and Tweedie 1993, Section 16.2). Similarly, it can be shown that the $o(1)$ term in (2.9) can be sharpened to $O(\rho^n)$ for some $\rho$, $0 < \rho < 1$.

A simplified analysis applies to the exponential distribution $F(t) = 1 - e^{-t/\tau}$, $t \geq 0$, which falls outside our standard class of distributions. In this case the $C_n$, $n \geq m$, are the epochs of a Poisson process at rate $m/\tau$, so that for all $i$ and $n \geq m$, the $R^n_i$ and thus $R^*_i$ are $m-1$ i.i.d. random variables with the distribution $G = F$. Then (2.3) gives

$$E[\alpha_{m,n}] = \tau[H_m - 1] \quad \text{for all } m \text{ and } n \geq m, \tag{2.12}$$

where $H_m = \sum_{j=1}^m 1/j$ (see also Coffman and Gilbert 1985).

The exponential case with $m = 2$ was examined by Coffman and Wright (1992) under more general assumptions, namely an initial delay $x$ (release time) on one of the processors and a random number $N$ of tasks having either a geometric or a Poisson distribution with mean $n$. Explicit, though complicated, expressions for the moments $E[L_{2,n}^k(x)]$ were studied by computing a variety of asymptotics as $n \to \infty$ and $x \to \infty$ at different rates.

Because of the increased use of massively parallel computers, it is natural to consider asymptotics as $m \to \infty$. From expressions like (2.11) and (2.12), large-$m$ asymptotics for the mean of the time-stationary random variable $\alpha_m$ can be obtained directly. For example, (2.11) and the asymptotics of the gamma function give

$$E[\alpha_m] = \frac{2}{3} - \frac{\sqrt{\pi}}{2\sqrt{m}} + O\left(\frac{1}{m}\right) \quad \text{as } m \to \infty \tag{2.13}$$

when $F$ is the uniform distribution on $[0, 1]$; similarly, when $F$ is the exponential distribution, (2.12) and the asymptotics of $H_m$ give

$$E[\alpha_m] = \tau(\ln m - 1 - \gamma) + o(1) \quad \text{as } m \to \infty, \tag{2.14}$$

where $\gamma = (0.5772\ldots)$ is Euler's constant.

More generally, we can obtain asymptotic properties of $E[\alpha_m]$ from (2.9). When $F$ has support $[0, b]$, (2.7) and the strong law of large numbers imply that

$$\alpha_m \to \alpha \equiv b - \tfrac{1}{2}\tau(v^2 + 1) \quad \text{w.p.1} \quad \text{as } m \to \infty. \tag{2.15}$$

From (2.15), we can see how $F$ influences the asymptotic error $\alpha$. For a given bound $b$, $\alpha$ *decreases* in $\tau$ and $v^2$. For given $b$ and $\tau$, the lowest value of $\alpha$ is $\tfrac{1}{2}b$, which is approached by the two-point distribution with mass $\tau/b$ on $b$ and mass $(b - \tau)/b$ on 0 (see e.g., Whitt 1984, p. 120).

It is interesting that for this extremal two-point distribution the greedy policy is optimal for all $m$ and $n$; i.e. *there is a distribution with finite positive variance for which greedy gives the minimum expected error*. The optimality of the greedy policy in this case is easy to see because the makespan is the same as for a random number of tasks, each with a constant running time $b$. In this case all work-conserving policies (in which no processor is idle when there is a task that has not started) are optimal. This two-point distribution is not in our class of distributions, but it is approached by such distributions.

The term $R^*_{(m-1)}$ in $\alpha_m$ obviously becomes even more important if $F$ does *not* have finite support. The asymptotic behavior of $R^*_{m-1}$ is described by the classical extreme-value theory (see Leadbetter et al. 1983; Reiss 1989). This extreme value theory applies to the iterated limit as first $n \to \infty$ and then $m \to \infty$ provided (2.9) is still valid.

Since the superposition of $m$ i.i.d. renewal processes, appropriately scaled, converges to a Poisson process as $m \to \infty$ (see e.g. Çinlar 1972), one might expect that the general formula for $E[\alpha_m]$ in (2.8) and (2.9) would in some sense approach the formulas for the exponential distribution in (2.12) and (2.14), but this is *not* the case. For the question here, the superposition limit theorem does not apply. The superposition limit theorem implies that the distribution of $R^*_{(1)} \equiv R^{*m}_{(1)}$ is asymptotically exponential as $m$ gets large, but in (2.8) we focus on $R^*_{(m-1)}$ and $\sum_{i=1}^{m-1} R^*_{(i)}$.

An interesting open problem is the joint limiting behavior as $m \to \infty$ and $n \to \infty$. Above, we considered only the iterated limit in which first $n \to \infty$ and then $m \to \infty$. If $m = n$ then the extreme-value theory for i.i.d. random variables with distribution $F$ describes the makespan. It would be interesting to develop different asymptotics in intermediate cases.

## 2.3 OFF-LINE POLICIES

The off-line component of the LPT policy introduced in Section 2.1 is simply an initial sorting of the list $(T_1, \ldots, T_n)$ into decreasing order. Results for LPT as precise as those for the greedy policy in Section 2.2 have not been obtained. On the other hand, asymptotic behavior for fixed $m$ is rather well understood. For comparison with Section 2.2, we illustrate the main results by the following, taken from Rinnooy Kan and Frenk (1986) and Frenk and Rinnooy Kan (1987). For any fixed $m$, $\alpha_{m,n} \Rightarrow 0$ a.s., $n \to \infty$, so long as $F$ is strictly positive in a neighborhood of the origin. Moreover, if $F(t) = t^a$, $0 \leqslant t \leqslant 1$, for some $0 < a < \infty$, then the convergence rate is $O(\log \log n / n)^{1/a}$, and the moments are bounded by $E[\alpha^k_{m,n}] = O(n^{-k/a})$. Results in a similar vein were presented by Boxma (1985) and Coffman et al. (1984a,b).

Consider next the superior differencing methods of Karmarkar and Karp (1982) with the restriction to $m = 2$, for simplicity. We describe the largest-first differencing method (LDM); it is a particularly simple differencing method, and no other such method is known to have a better asymptotic performance. Other methods that have been successfully analyzed are either much more elaborate (Karmarkar and Karp 1982) or have worse performance (Lueker 1987); see also Coffman and Lueker (1991) for a discussion of these methods.

LDM starts by computing the absolute difference $d$ between the largest two tasks in the current list; it then replaces these two tasks by a single task of duration $d$, leaving a list with one fewer task. LDM iterates this procedure $n - 2$ more times until a single task remains; the duration of this task is $2\alpha_{2,n} = 2L_{2,n} - S_n$. Noting that the largest two tasks being differenced at each step are to be put on different processors, it is a simple exercise to work backward through the differencing sequence to determine a partition of $\{T_1, \ldots, T_n\}$ that gives the final difference $2\alpha_{2,n}$. An intriguing problem set by Karmarkar and Karp over 10 years ago was a proof that, under LDM with $F$ the uniform distribution on $[0, 1]$,

$$E[\alpha_{2,n}] = O(n^{-c \log n}) \tag{2.16}$$

for some constant $c > 0$. The recent work of Yakir (1993) provides an elegant solution to this problem. We give below Yakir's important new insight into the structure of LDM.

The new insight is based on Lueker's (1987) initial transformation of the problem. This transformation uses the well-known fact that, if

$$S_j = \sum_{i=1}^{j} X_i, \quad 1 \leq j \leq n+1$$

are the partial sums of $n+1$ i.i.d. exponentials $X_i$ with parameter 1 then the ratios $S_1/S_{n+1}, \ldots, S_n/S_{n+1}$ are independent of $S_{n+1}$ and equal in distribution to $T_{(1)}, \ldots, T_{(n)}$, i.e. to the order statistics of $n$ independent samples from the uniform distribution on $[0, 1]$. Let $\alpha_{2,n}, \hat{\alpha}_{2,n}$ denote the errors produced by LDM from the respective lists $(T_1, \ldots, T_n), (S_1, \ldots, S_n)$, and let $\stackrel{d}{=}$ denote equality in distribution. Then one obtains $\hat{\alpha}_{2,n} \stackrel{d}{=} S_{n+1}\alpha_{2,n}$, with $S_{n+1}$ and $\alpha_{2,n}$ independent, so $E[S_{n+1}] = n+1$ gives $E[\alpha_{2,n}] = (n+1)^{-1}E[\hat{\alpha}_{2,n}]$. Thus, to prove (2.16), it is sufficient to prove that $E[\hat{\alpha}_{2,n}] = O(n^{-c \log n})$ for some $c > 0$.

Let $X_1^r, \ldots, X_r^r$ denote the spacings between the tasks $S_j^r = \sum_{i=1}^{j} X_i^r$, $1 \leq j \leq r$, just after the $(n-r)$th iteration of LDM, $1 \leq r \leq n-1$; the initial spacings $X_i^n = X_i$ are i.i.d. parameter-1 exponentials. The $(n-r)$th iteration of LDM inserts the difference $X_r^r = S_r^r - S_{r-1}^r$ into the sequence $S_1^r, \ldots, S_{r-2}^r$ to form the sequence $S_1^{r-1}, \ldots, S_{r-1}^{r-1}$. The key result, easily proved by induction on $r$, is that, if the $X_i^r$, $1 \leq i \leq r$, are independent exponentials with parameters $\lambda_i^r$ then the following hold.

(i) Given the event $\{S_{i-1}^r \leq X_r^r < S_i^r\}$, $i = 1, \ldots, r-2$, the spacings in the list $(S_1^{r-1}, \ldots, S_{r-1}^{r-1})$, i.e.

$$\{X_k^r\}_{1 \leq k \leq i-1}, \quad X_r^r - S_{i-1}^r, \quad S_i^r - X_r^r, \quad \{X_k^r\}_{i+1 \leq k \leq r-2},$$

are (conditionally) independent and exponential with parameters $\{\lambda_k^r\}_{1 \leq k \leq i}$, $\{\lambda_k^r + \lambda_r^r\}_{i+1 \leq k \leq r-2}$. Given the remaining possibility $\{X_r^r \geq S_{r-2}^r\}$, the $X_i^r$, $1 \leq i \leq r-2$, and $X_r^r - S_{r-2}^r$ are independent exponentials with parameters $\{\lambda_k^r + \lambda_r^r\}_{1 \leq k \leq r-2}$ and $\lambda_r^r$.

(ii) The event probabilities are

$$\Pr\{S_{i-1}^r \leq X_r^r < S_i^r\} = \frac{\lambda_r^r}{\lambda_i^r + \lambda_r^r} \prod_{j=1}^{i-1} \frac{\lambda_j^r}{\lambda_j^r + \lambda_r^r}, \quad 1 \leq i \leq r-2, \quad (1.17a)$$

$$\Pr\{X_r^r \geq S_{r-2}^r\} = \prod_{j=1}^{r-2} \frac{\lambda_j^r}{\lambda_j^r + \lambda_r^r}. \quad (1.17b)$$

It follows immediately that the sequence $(\lambda_1^n, \ldots, \lambda_n^n), (\lambda_1^{n-1}, \ldots, \lambda_{n-1}^{n-1}), \ldots, (\lambda_1^1)$ has the Markov property with transition probabilities given by (1.17). By (i) and the initial state $\lambda_1^n = \cdots = \lambda_n^n = 1$, we note that $\lambda_1^n \leq \lambda_1^{n-1} \leq \cdots \leq \lambda_1^1$, and for each $r = 1, \ldots, n$, we have $\lambda_1^r \geq \lambda_2^r \geq \cdots \geq \lambda_r^r$. The proof of (2.16) now reduces to an estimate of the growth rate of $\lambda_1^r$. Yakir's (1993) bounds establish that, to within a constant factor, $\lambda_1^r$ grows as $n^{c\log(n-r)}$ for some $c > 0$, as desired.

Recent simulations by J. L. Bentley (1993 personal communication) suggest that the value of $c$ in (2.16) is close to $\frac{1}{2}$, with 2 being the base of the logarithm. Johnson et al. (1991) ran experiments comparing LDM with simulated-annealing approaches to the makespan minimization problem. The clear superiority of LDM in this comparison contributed to their conclusion that the local optimization framework of simulated annealing was not well suited to makespan minimization problems.

With $m = 2$ and $F$ the uniform distribution on $[0, 1]$, a rough summary of the expected-error results is that, for the greedy, LPT, LDM and an optimal policy, we have respectively

$$E[\alpha_{2,n}] = O(1), \quad O(n^{-1}), \quad O(n^{-c\log n}), \quad O(\rho^n)$$

for some $c > 0, 0 < \rho < 1$. The $O(\rho^n)$ result for optimal scheduling is not discussed here (see Coffman and Lueker 1991, Section 4.3), and in fact remains a conjecture. The strongest results of this type are those of Karmarkar et al. (1986), who showed that the median of the final difference, $2\alpha_{2,n}$, is bounded by a constant times $n/2^n$.

## 2.4 POLICY-FREE ERROR ASYMPTOTICS

From (2.7) it is clear that the relative size of the error $\alpha_{m,n}$ compared with the makespan $L_{m,n}$ itself is asymptotically negligible as $n \to \infty$ for the greedy policy. For large $n$, obviously the dominant part of $L_{n,m}$ for any policy is the normalized sum of all the processing times. In this section we establish a stronger result. We show that *for any policy the limiting behavior of the error* $\alpha_{m,n}$ *as* $n \to \infty$ *is independent of the policy*. In particular, the expected error $E[\alpha_{m,n}]$ for a given policy is asymptotically negligible compared with the standard deviation of the makespan (which is the same as the standard deviation of the error). In a probabilistic setting, what we can gain from a good policy is asymptotically negligible as $n \to \infty$ compared with our degree of uncertainty about the makespan.

Central limit theorems (CLTs) and functional central limit theorems (FCLTs) provide asymptotics that exhibit this property for general distributions and a policy-free set-up, i.e. a model yielding results simultaneously valid for all policies. The policies to be considered in the illustrations below are those in the class of *list scheduling* (LS) policies. Such a policy begins by computing a permutation $\pi_n = (\pi(1), \ldots, \pi(n))$ of the integers $1, \ldots, n$, and then schedules the ordered list $(T_{\pi(1)}, \ldots, T_{\pi(n)})$ by the greedy rule. For any given sequence $T_1, T_2, \ldots$, an LS policy defines a sequence of permutations $\{\pi_n, n \geq 1\}$. Note that the policies of earlier sections are all LS policies.

Let $S_n$ and $M_n$ be the sum and maximum of $T_1, \ldots, T_n$, and note that both quantities are invariant under permutations of $T_1, \ldots, T_n$. Let the number $m_n$ of processors be a nondecreasing function of $n$, and denote the makespan and error under permutation $\pi_n$ by $L_n^{\pi_n}$ and $\alpha_{m,n}^{\pi_n}$. From (2.3), we obtain the basic inequality

$$|S_n - m_n L_n^{\pi_n}| \leq m_n M_n \quad \text{for all } \pi_n, \tag{2.17}$$

from which we see that the limiting behavior of $L_n^{\pi_n}$ and $\alpha_{m,n}^{\pi_n}$ is determined by the asymptotics of $(S_n, M_n)$. Typically, when a CLT holds for $S_n$, $M_n$ is asymptotically negligible compared with $S_n$. (See Resnick (1986, Section 4.5) for further discussion of the asymptotic behavior of $(S_n, M_n)$.) In our case, we have the CLT

$$n^{-1/2}(S_n - n\tau) \Rightarrow N(0, \sigma^2) \quad \text{as } n \to \infty, \tag{2.18}$$

where $\Rightarrow$ denotes convergence in distribution and $N(a,b)$ denotes a normally distributed random variable with mean $a$ and variance $b$. It then follows from (2.17) and Theorem 4.1 of Billingsley (1968) that if

$$m_n n^{-1/2} M_n \Rightarrow 0 \quad \text{as } n \to \infty$$

then, for any sequence of permutations $\{\pi_n, n \geq 1\}$,

$$n^{-1/2} m_n \alpha_n^{\pi_n} \Rightarrow N(0, \sigma^2) \quad \text{as } n \to \infty. \tag{2.19}$$

For example, suppose $T$ is exponentially distributed with mean $\tau$. Since $M_n / \ln n \Rightarrow \tau$, $n \to \infty$ (see e.g. Leadbetter 1983), (2.19) holds if $m_n = o(n^{1/2} / \ln n)$, $n \to \infty$.

For a fixed number $m_n = m$, $n \geq 1$, of processors, no explicit assumption about $M_n$ needs to be made. This can be seen in the general setting of the following FCLT for $S_n$. In terms of the usual diffusion-limit scalings, define the normalized processes

$$\mathbf{S}_n \equiv \mathbf{S}_n(t) = \frac{S_{\lfloor nt \rfloor} - \tau nt}{n^{1/2}}, \quad t \geq 0$$

$$\boldsymbol{\alpha}_n^{\pi_n} \equiv \boldsymbol{\alpha}_n^{\pi_n}(t) = \frac{L_{\lfloor nt \rfloor}^{\pi_n} - (\tau/m)nt}{n^{1/2}}, \quad t \geq 0.$$

If $\mathbf{B}$ denotes standard (zero drift, unit diffusion) Brownian motion then we have

$$\mathbf{S}_n \Rightarrow \sigma \mathbf{B} \quad \text{as } n \to \infty, \tag{2.20}$$

where $\Rightarrow$ denotes weak convergence in the Skorohod space $D \equiv D([0,1], \mathbb{R})$ (for technical details see Ethier and Kurtz 1986). By the continuous mapping theorem with the maximum jump functional, we deduce from (2.20) that $n^{-1/2} M_n \Rightarrow 0$ as $n \to \infty$. Hence, for any sequence of permutations $\{\pi_n, n \geq 1\}$,

$$\boldsymbol{\alpha}_{m,n}^{\pi_n} \Rightarrow \frac{\sigma}{m} \mathbf{B} \quad \text{in } D \quad \text{as } n \to \infty. \tag{2.21}$$

This gives the approximation

$$L_{m,n}^{\pi_n} \approx \frac{n\tau}{m} + n^{1/2} N(0, \sigma^2/m^2), \tag{2.22}$$

in which $\pi_n$ does not appear, since the effect of the permutation $\pi_n$ is of order $M_n$, which is asymptotically negligible compared with $n^{1/2}$.

We remark that the setting for the above limit laws can be broadened considerably, covering interesting cases where the independence assumption or the identical-distribution assumption does not hold.

## 2.5 FLOW SHOPS

In a variant of multiprocessor scheduling, called *permutation flow shop scheduling*, the processors are connected in tandem and tasks consist of ordered sets of operations $(T_{i1}, \ldots, T_{im})$, $1 \leq i \leq n$, to be done in sequence on $P_1, \ldots, P_m$. A schedule is determined by a permutation $\pi_n = (\pi(1), \ldots, \pi(n))$; on processor $j$ for each $j$ the operation $T_{\pi(i),j}$ must precede the operation $T_{\pi(i+1),j}$, $1 \leq i \leq n-1$. Thus, we have a tandem queueing system with $m$ single-server queues and $n$ arrivals, all available at time 0. The makespan $L_{m,n}^{\pi_n}$ is the finishing time of $T_{\pi(n),m}$. In the problem considered here all $mn$ operations are i.i.d. random variables; for simplicity the distribution $F$ and its properties will be carried over to the flow shop problem, but it will refer to operation times rather than entire task times.

The combinatorial problem of selecting an optimal permutation $\pi_n$ is $\mathcal{NP}$-complete for $m \geq 3$, as shown by Garey *et al.* (1976), but Johnson (1954) proved that the following simple rule is optimal for $m = 2$: if $\min(T_{i1}, T_{j2}) < \min(T_{i2}, T_{j1})$ then schedule $T_i$ before $T_j$. We return to an analysis of Johnson's rule after discussing the greedy rule for general $m$.

As before, the greedy rule sequences tasks in the order given, i.e. with $\pi_n = (1, \ldots, n)$. The makespan is again denoted by $L_{m,n}$; the error is now defined to be $\alpha_{m,n} \equiv L_{m,n} - (n + m - 1)\tau$, where $(n + m - 1)\tau$ is a trivial lower bound to the expected makespan. Glynn and Whitt (1991), motivated by earlier work of Srinivasan (1993), studied in depth the asymptotic behavior of $L_{m,n}$ as $m$, $n$ or both tend to infinity. An immediate dual is obtained for each limit by the easily proved symmetry

$$\{L_{i,j} : 1 \leq i \leq m, 1 \leq j \leq n\} \stackrel{d}{=} \{L_{j,i} : 1 \leq j \leq n, 1 \leq i \leq m\}. \quad (2.23)$$

Theorem 2 of Iglehart and Whitt (1970) gives an FCLT for each $m$,

$$n^{-1/2}\alpha_{m,n} \Rightarrow \sigma\widehat{\alpha}_m \quad \text{as } n \to \infty, \quad (2.24)$$

where $\widehat{\alpha}_m$ is a functional of $m$-dimensional Brownian motion and convergence is in the appropriate Skorohod space. By applying the subadditive ergodic theorem (Liggett 1985, p. 277), Glynn and Whitt established that

$$m^{-1/2}\widehat{\alpha}_m \Rightarrow \gamma \quad \text{as } m \to \infty, \quad (2.25)$$

where $\gamma$ is a positive constant. Analysis has so far yielded little information about $\widehat{\alpha}_m$ or $\gamma$. This prompted a simulation study by Greenberg *et al.* (1993) which provided further insights, e.g. the simulations suggest that $\gamma = 2$.

By means of a strong approximation theorem, Glynn and Whitt proved a result more general than the above iterated limit: if for any $\epsilon$, $0 < \epsilon < 1$, $m_n = n^{1-\epsilon}$, then

$$(nm_n)^{-1/2}\alpha_{m_n,n} \Rightarrow \gamma \quad \text{as } n \to \infty, \tag{2.26}$$

with $\gamma$ as in (2.25).

If $m, n \to \infty$ at comparable rates then hydrodynamic limits for $L_{m,n}$ emerge. By applying the results in Section 4.2 of Srinivasan (1993), Glynn and Whitt showed that if the $T_i$ are exponentially distributed with mean 1 then

$$n^{-1}L_{\lfloor xn \rfloor, n} \to (1 + \sqrt{x})^2 \quad \text{w.p.1} \quad \text{as } n \to \infty \tag{2.27}$$

for any $x > 0$. Glynn and Whitt also extend this result to any distribution with an exponential tail; they get a deterministic function $\gamma(x)$ as the limit, which, according to simulations, depends on the distribution. They verify that $\gamma(x)$ is strictly increasing and concave, and provide upper and lower bounds.

We return now to an analysis of optimal scheduling, and discuss results recently given by Ramudhin et al. (1993) for the case $m = 2$ with $F$ uniform on a finite interval. Ramudhin et al. begin by introducing the following more easily analyzed, stochastically symmetric version of the optimal policy. Partition $\{T_1, \ldots, T_n\}$ into the sets $\mathcal{T}_1$ and $\mathcal{T}_2$ of tasks with shorter operations on $P_1$ and $P_2$ respectively, i.e. $\mathcal{T}_1 = \{T_i : T_{i1} \leq T_{i2}\}$ and $\mathcal{T}_2 = \{T_i : T_{i2} < T_{i1}\}$. An optimal policy first schedules the tasks in $\mathcal{T}_1$ in increasing order of the $T_{i1}$ and then schedules the tasks of $\mathcal{T}_2$ in decreasing order of the $T_{i2}$. An asymptotic analysis of this schedule shows that

$$L_{2,n} \stackrel{d}{=} T_{11} + \sum_{i=1}^n T_{i2} + I, \quad \text{if } \sum_{i=1}^n T_{i1} < \sum_{i=1}^n T_{i2}$$

$$L_{2,n} \stackrel{d}{=} \sum_{i=1}^n T_{i1} + T_{n2} + I, \quad \text{otherwise,}$$

where $I \Rightarrow 0$ a.s. as $n \to \infty$. The asymptotic makespan then becomes

$$\max\left\{\sum_{i=1}^n T_{i1} + T_{n2}, T_{11} + \sum_{i=1}^n T_{i2}\right\},$$

in the sense of the limit law

$$\alpha_{2,n} \Rightarrow \sigma \max\{N_1, N_2\} \quad \text{a.s., } n \to \infty, \tag{2.28}$$

where $N_1, N_2$ are i.i.d. standard normal random variables. This yields the estimate

$$E[\alpha_2, n] = \sigma\sqrt{n/\pi} + o(\sqrt{n}) \quad \text{as } n \to \infty \tag{2.29}$$

which differs only in the multiplicative constant from that obtainable from (2.24).

Ramudhin et al. present several other results on queue lengths, workloads, and waiting times under the optimal policy and a simpler, near-optimal policy. A nontrivial lower bound on expected optimal makespans for $m \geq 3$ remains an open problem. In particular, it would be interesting to see whether $E[\alpha_{m,n}]$ grows at least as fast as $\sqrt{mn}$ under an optimal policy. The upper bound provided by the greedy policy would then show that this growth rate is exact within a constant factor.

## 2.6 FINAL REMARKS

When viewed against the much broader and more varied background of combinatorial makespan scheduling problems, probabilistic analysis appears to be in its infancy. A few of the many variants, scarcely touched at present, are task precedence constraints, dedicated processors, set-up times, interprocessor transfer times, variable profiles, and preemptive sequencing policies (see Lawler et al. 1992). Modeling issues are often an initial hurdle in problems with additional structure. To obtain a tractable model, the uniform or exponential is often the distribution of choice. A case in point is the recent work of Dell'Olmo et al. (1993) on dedicated three-processor systems. In this variant of multiprocessor scheduling, each task specifies a nonempty subset of the processors that it requires throughout its running time. In a uniform model of the processors required by tasks, Dell'Olmo et al. show that, for $m = 3$ and for all $n$ sufficiently large, optimal schedules can be computed in linear time for over 95% of the instances; this property of an instance is checkable in advance.

However, uniform assumptions do not always lead to interesting structures, as the following classical model with precedence constraints illustrates. Add to the problem instance of Sections 2.2–2.4 a random irreflexive partial order $\prec$ on $\{T_1, \ldots, T_n\}$ representing precedence relations ($T_i \prec T_j$ means that $T_j$ cannot begin until $T_i$ is finished). To define the term 'random', a natural first assumption would be that $\prec$ is chosen uniformly at random among all partial orders on $\{T_1, \ldots, T_n\}$. However, a typical such partial order has height 3 (see Kleitman and Rothschild 1975), and hence yields a simplistic model for many applications of makespan scheduling. A more promising model is that studied by Winkler (1985), in which random orders are constructed from the intersection of $k \geqslant 2$ random linear orders, or, equivalently, the random orders induced by the ordinary product order on $n$ points chosen independently and uniformly at random from the unit hypercube $[0, 1]^k$. In the induced order $(x_1, \ldots, x_k) \prec (y_1, \ldots, y_k)$ if and only if $x_i \leqslant y_i$ for all $i = 1, \ldots, k$, with strict inequality holding for at least one $i$. For large $n$, typical such random orders have the following properties:

(i) there are no isolated tasks;
(ii) there are $(\ln n)^{k-1}/(k-1)!$ minimal and (by symmetry) maximal tasks,
(iii) the height is $c_k n^{1/k}$ for some $c_k$, $0 < c_k < e$,
(iv) the width is approximately $c'_k n^{(k-1)/k}$ for some constant $c'_k$; in the balanced case, with $k = 2$, $c_k = c'_k = 2$.

An interesting sample problem might be to determine asymptotic expected makespans under a greedy policy with random partial orders as above and all $T_i = 1$. In this case a greedy policy could be highest-level-first (the next task to be scheduled is one that dominates a longest chain), with some rule for resolving ties amongst highest-level tasks.

Many other open problems exist in the same settings as discussed here, but with different performance metrics, such as the sum, possibly weighted, of task finishing times, and tardiness measures; again see Lawler et al. (1992). One fundamental variant that has received a great deal of attention is the following dual of the multiprocessor scheduling problem: for a fixed makespan (deadline) that exceeds the longest task running time, determine the least $m$ such that $\{T_1, \ldots, T_n\}$ can be scheduled on $P_1, \ldots, P_m$ with all tasks finishing by the deadline. This is the one-dimensional bin packing problem; the monograph by Coffman and Lueker (1991) covers much of the probabilistic analysis of this problem.

## REFERENCES

Asmussen S. (1987) *Applied Probability and Queues* Wiley, New York.
Baccelli F. and Brémaud P. (1987) *Palm Probabilities and Stationary Queueing Systems* Springer, New York.
Billingsley P. (1968) *Convergence of Probability Measures* Wiley, New York.
Blazewicz J., Ecker K., Schmidt G. and Weglarz J. (1993) *Scheduling in Computer and Manufacturing Systems* Springer, Berlin.
Boxma O. J. (1985) A probabilistic analysis of multiprocessor list scheduling: the Erlang case *Stochastic Models*, **1**, 209-220.
Bruno L. L. and Downey P. J. (1986) Probabilistic bounds on the performance of list scheduling *SIAM J. Comput.*, **15**, 409-417.
Çinlar E. (1972) Superposition of point processes. In Lewis P. A. W., editor, *Stochastic Point Processes: Statistical Analysis, Theory and Applications*, pp 549-606 Wiley, New York.
Coffman E. G., Jr. and Gilbert E. N. (1985) On the expected relative performance of list scheduling *Oper. Res.*, **33**, 548-561.
Coffman E. G., Jr. and Lueker G. S. (1991) *Probabilistic Analysis of Packing and Related Partitioning Problems* Wiley, New York.
Coffman E. G. Jr. and Wright P. E. (1992) Load balancing on two identical facilities. Technical Memorandum, AT&T Bell Laboratories, Murray Hill, NJ.
Coffman E. G. Jr., Flatto L. and Lueker G. S. (1984a) Expected makespans for largest-first multiprocessor scheduling. In *Proceedings of 10th International Symposium on Models of Computer System Performance*, pp 491-506, North-Holland, Amsterdam.
Coffman E. G. Jr., Frederickson G. N. and Lueker G. S. (1984b) A note on expected makespan for largest-first sequences of independent tasks on two processors *Math. Oper. Res.*, **9**, 260-266.
Coffman E. G., Jr., Flatto L. and Whitt W. (1993) Limit laws for multiprocessor scheduling, AT&T Bell Laboratories, Murray Hill, NJ 07974 (to be published) (1994).
Dell'Olmo P, Speranza M. G. and Tuza Zs. (1993) Polynomial instances and approximation results for the scheduling problem on three dedicated processors. Technical Report IASI-CNR, Rome.
Ethier S. N. and Kurtz T. K. (1986) *Markov Processes: Characterization and Convergence* Wiley, New York.
Feller W. (1971) *An Introduction to Probability Theory and Its Applications*, Vol. II Wiley, New York.
Franken P., König D., Arndt U. and Schmidt V. (1982) *Queues and Point Processes* Wiley, New York.

Frenk J. B. G. and Rinnooy Kan, A. H. G. (1987) The asymptotic optimality of the LPT rule *Math. Oper. Res.*, **12**, 241–254.

Garey M. R. and Johnson D. S. (1979) *Computers and Intractability: A Guide to the Theory of NP-Completeness* Freeman, New York.

Garey M. R., Johnson D. S. and Sethi R. (1976) The complexity of flow shop and job shop scheduling *Math. Oper. Res.*, **1**, 117–129.

Graham R. L. (1966) Bounds for certain multiprocessing anomalies *Bell. Syst. Tech. J.*, **45**, 1563–1581.

Greenberg A. G., Schlunk O. and Whitt, W. (1993) Using distributed-event parallel simulation to study departures from many queues in series *Prob. Eng. Inf. Sci.*, **7**, 159–186.

Glynn P. W. and Whitt W. (1991) Departures from many queues in series *Ann. Appl. Prob.*, **1**, 546–572.

Han S., Hong D. and Leung J. Y.-T. (1992) On the asymptotic optimality of heuristic multiprocessor scheduling algorithms. Technical Report, University of Nebraska, Computer Science Department, Lincoln.

Iglehart D. L. and Whitt W. (1970) Multiple channel queues in heavy traffic. II, Sequences, networks, and batches *Adv. Appl. Prob.*, **2**, 355–369.

Johnson D. S., Aragon C. R., McGeoch L. A. and Schevon C. (1991) Optimization by simulated annealing: an experimental evaluation; Part II, Graph coloring and number partitioning *Oper. Res.*, **39**, 378–406.

Johnson S. M. (1954) Optimal two and three stage production schedules with set-up times included. *Nav. Res. Log. Quart.*, **1**, 61–68.

Karmarkar N. and Karp R. M. (1982) The differencing method of set partitioning. Technical Report UCB/CSD 82/113, Computer Science Division (EECS), University of California, Berkeley.

Karmarkar N., Karp R. M., Lueker G. S. and Odlyzko A. M. (1986) Probabilistic analysis of optimum partitioning *J. Appl. Prob.*, **23**, 626–645.

Kleitman D. J. and Rothschild B. L. (1975) Asymptotic enumeration of partial orders on a finite set *Trans. Am. Math. Soc.*, **205**, 205–210.

Liggett T. M. (1985) *Interacting Particle Systems* Springer, New York.

Leadbetter M. R., Lindgren G. and Rootzén H. (1983) *Extremes and Related Properties of Random Sequences and Processes* Springer, New York.

Lawler E. L., Lenstra J. K., Rinnooy Kan A. H. G. and Shmoys D. B. (1992) Sequencing and scheduling: algorithms and complexity. Report BS-R8909, Centre for Mathematics and Computer Science, 1009AB Amsterdam. (To appear in *Handbook of Operations Research and Management Science* North-Holland, Amsterdam).

Loulou R. (1984) Tight bounds and probabilstic analysis: two heuristics for parallel processor scheduling *Math. Oper. Res.*, **9**, 142–150.

Lueker G. S. (1987) A note on the average-case behavior of a simple differencing method for partitioning *Oper. Res. Lett.*, **6**, 285–287.

Meyn S. P. and Tweedie R. L. (1993) *Markov Chains and Stochastic Stability* Springer, London.

Phillips S. and Westbrook J. (1993) Online load balancing and network flow. In *Proceedings of 25th Annual ACM Symposium in the Theory of Computing*, pp 402–411, ACM Press, New York.

Ramudhin A., Bartholdi J. J. III, Calvin J. M., Vande Vate J. H. and Weiss G. (1993) A probabilistic analysis of 2-machine flowshops. Technical Report, School of Industrial and System Engineering, Georgia Institute of Technology, Atlanta.

Reiss R.-D. (1989) *Approximate Distributions of Order Statistics* Springer, New York.

Resnick S. I. (1986) Point processes, regular variation and weak convergence *Adv. Appl. Prob.*, **18**, 66–138.

Rinnooy Kan, A. H. G. and Frenk J. B. G. (1986) On the rate of convergence to optimality of the LPT rule *Discrete Appl. Math.*, **14**, 187–198.

Srinivasan R. (1993) Queues in series via interacting particle systems *Math. Oper. Res.*, **18**, 39–50.

Whitt W. (1984) On approximations for queues, I: Extremal distributions *AT&T Bell Lab. Tech. J.*, **63**, 115–138.

Winkler P. (1985) Random orders *Order*, **4**, 317–331.

Yakir B. (1993) The differencing algorithm LDM for partitioning: a proof of Karp's conjecture. Technical Report 93/03, Department of Biostatistics, University of Rochester, New York.

*AT&T Bell Laboratories, 600 Mountain Avenue, Murray Hill, NJ 07974, USA*

CHAPTER 3

# A Tutorial in Stochastic Scheduling

**Gideon Weiss**
*Georgia Tech, Atlanta, USA; Haifa University, Israel*

**Abstract**

We survey some recent progress in stochastic scheduling. We concentrate on results that can be derived without detailed assumptions on the processing time distributions. Consequently we present mainly heuristics and approximations. We show how such heuristics can be analyzed. We discuss some sophisticated methods for deriving priority rules. We explore the complexities that arise in scheduling the simultaneous operation of several servers, in parallel, in series and in a network. The emphasis is on how the theory of stochastic scheduling can provide insights and proper perspectives for the practitioner.

## 3.1 INTRODUCTION

### 3.1.1 Role of scheduling

The scheduling of jobs and the control of their flow through a production process is essential to modern manufacturing. Increasing complexity of these processes, rising capital outlay and greater automation will make it even more crucial in the foreseeable future.

Three distinct approaches to these problems are deterministic scheduling theory, queueing networks and software engineering.

*Deterministic scheduling*

Here problems are formulated and solved using the methods of combinatorial optimization, for introductions and surveys see Conway *et al.* (1967), Baker (1974) and Lawler *et al.* (1989); for recent significant work see Adams *et al.* (1988). Most problems are $\mathcal{NP}$-hard, and require heuristic methods for solution. Even so, problems are often too big to be solved satisfactorily, especially since the static formulation necessitates frequent resolving.

---

*Scheduling Theory and its Applications* Edited by P. Chrétienne, E. G. Coffman, Jr., J. K. Lenstra and Z. Liu
© 1995 John Wiley & Sons Ltd

*Queueing networks*

Here scheduling is modeled as a dynamic and stochastic process (see Kleinrock 1976; Kelly 1979; Harrison 1988). Recent work includes heuristics for scheduling and control of networks in heavy traffic (see Wein 1990a,b,1992; Laws and Louth 1990; Kelly and Laws 1993; Laws 1995—Chapter 16 of this volume). These methods are very promising though they have currently only been applied to relatively simple models. Some theoretical problems that arise in these models are currently being studied (see Dai and Wang 1993; Kumar 1993).

*Software engineering*

These encompass all software-based methods that are used to solve scheduling problems, including proprietary programs such as OPT (1989) (see also Goldratt and Cox 1986), FACTOR (see Pritsker Corporation 1989), Leitstand (1991) (see also Kanet and Adelsberger 1991), generic hierarchical approaches such as those of Gershwin (1989), Leachman (1987), Roundy *et al.* (1988), and plant-specific programs such as those of Adler *et al.* (1988, 1993) and Fraiman *et al.* (1989). These include such features as consideration of data flow, knowledge-based decision rules, and user interaction, which are ignored by the other approaches. The theoretical challenge here is to incorporate enough modeling and algorithmic sophistication.

### 3.1.2 Stochastic scheduling

This attempts to bridge some of the gaps between the three approaches, and in doing so provide some useful insights. Work on stochastic scheduling was spurred on by the development of deterministic scheduling theory in the 1960s, and in the 1970s there was a great amount of work done to extend the deterministic models to the stochastic case, in particular to models that assume that processing times are unknown random variables drawn from an exponential distribution (Bruno and Hofri 1975; Cunningham and Dutta 1973; Pinedo 1981,1983; Pinedo and Ross 1982; Pinedo and Weiss 1979,1984,1985; Weiss 1984; Weiss and Pinedo 1980). Some other important developments go back further to the 1950s, when some rules on optimal scheduling of queueing systems were discovered (Cox and Smith 1961; Cobham 1954; Schrage 1968; Schrage and Miller 1966). Later work on optimal scheduling and control of queueing systems includes that of Serfozo (1981), Hajek (1983,1984,1985), Stidham (1985) and Stidham and Weber (1989). The theory of comparison of random variables, developed by Stoyan (1983), Shaked and Shanthikumar (1987) and Shanthikumar and Yao (1991), has been used to obtain greatly simplified proofs of many earlier results (see Righter 1992).

Even though the field of stochastic scheduling is relatively self-contained and small, the number of papers on various aspects of it is measured in the hundreds, covering a wide variety of problems and models. We do not attempt to give an exhaustive list of references or a comprehensive survey of all the topics here. Rather than doing so, we try to illustrate some specific points, which from our subjective point of view we feel are of particular interest to the general scheduling community. We want to stress the following points.

*How important is scheduling?*

A simple calculation in Section 3.2 shows that optimal scheduling can improve the schedule by an arbitrarily large factor. The calculation is performed by a probabilistic analysis. We also assess the cost of imprecise information, thus tying together deterministic scheduling, with full information and stochastic scheduling, with imprecisely known processing times.

*General processing time distributions*

One of our premises is to present results that do not depend on specific probabilistic assumptions. Thus we wish to look at general distributions of service times, rather than assume that processing times are drawn, for example, from an exponential distribution. This means that the results we present have a wider range of applicability—but on the other hand fewer exact optimality results can be obtained.

*Heuristics*

Many deterministic scheduling problems are computationally $\mathcal{NP}$-hard. Without specific distributional assumptions, stochastic scheduling problems are no easier. This means that an optimal solution is very difficult to find. Furthermore, an optimal solution may have few features that will be attractive to the scheduler on intuitive grounds, or that will lend themselves to further analysis. Also, an optimal solution may depend on a great deal of information about the system, including values of a large number of parameters of the processing time distributions, which in practice are rarely known with any accuracy. The degree to which an optimal solution depends on such detailed but inaccurate data is hard to assess. Consequently, it can be argued that a simple heuristic policy, based on some simple plausible rules, making use of the values of only a small number of clearly relevant parameters, and using these values in a way that allows sensitivity analysis, is preferable to an algorithm that would produce an optimal solution.

*Assessment of heuristics*

An important part of studying heuristic rules is assessing their performance. Two approaches exist—worst case analysis, and average case (probabilistic) analysis. With deterministic scheduling problems, it is typical that the worst case analysis presents the heuristic as much worse than it actually is, while a probabilistic analysis, which appears closer to reality, suffers from unjustifiable assumptions on the population of problems, assumptions that do not form part of the deterministic model. Happily, in stochastic scheduling a probabilistic framework is part of the model, and a worst case analysis of heuristics is therefore much closer to reality. We believe that this is a very important reason for looking at stochastic scheduling problems. This is demonstrated in Sections 3.4 and 3.7. Probabilistic average-case analysis of scheduling is the subject of a recent book by Loffman and Lueker (1991); see also Coffman and Whitt (1995)—Chapter 2 of this volume.

*Approximate and turnpike optimality*

The performance of a heuristic relative to the optimal policy can be assessed in two ways: by comparing the (expected) values of the objective functions, and by comparing the (expected) decisions taken by the two policies. If the (expected) values of the objective functions are close, we say the heuristic is approximately optimal. If the majority of the (expected number of the) decisions taken by the heuristic are in fact optimal, we say it has turnpike optimality. This latter type of optimality is of great practical importance—in particular when problems are resolved frequently to adjust for new data and unforeseen changes. We illustrate both types of optimality for some stochastic scheduling problems in Section 3.4.

*Heuristics based on priority rules*

Very simple priority rules are optimal for some of the basic scheduling models. Examples include SPT (shortest processing time first) and Smith's rule for minimizing flowtime and weighted flowtime, and Jackson's EDD (earliest due date) rule for minimizing maximal tardiness, in scheduling a single machine, and Johnson's rule for minimizing makespan for a two-machine flow shop. These rules are so simple and well known that they appear to be almost trivial. Yet some of these rules are surprisingly robust: they can be used as heuristics for more general models, and they can be generalized to stochastic models. In fact, some very sophisticated priority rules, namely Gittins index priorities, can be thought of as generalizations of such simple rules. The Gittins index can be described as a ratio of expected reward per expected unit time, maximized over all possible stopping times, which is a vastly generalized version of Smith's rule in which the priority index is the cost rate of a job divided by its processing time. We take up this point, because we believe that it is fruitful to trace the origin of the Gittins index to simple, well-understood rules. We discuss such rules in Sections 3.3, 3.4, 3.6 and 3.7. See also Kaspi *et al.* (1995)—Chapter 17 of this volume.

*Simultaneous service*

It is well known that some single-server deterministic scheduling problems are $\mathcal{NP}$-hard. Nevertheless, many single-server deterministic scheduling problems are easy. Furthermore, the stochastic versions of these problems are also often easily solved by generalizing the solution of the deterministic problems. In fact, problems of scheduling and of controlling some very complex service stations that have only a single-server in them are often tractable. We illustrate this point amply in Sections 3.2, 3.3, and 3.6; the Gittins index and the Klimov model are striking examples. This tractability is lost the moment that there are several servers that operate simultaneously. We try in this chapter to gain some insights into the nature of this increase in complexity. The increase in complexity is moderate and quite well understood for parallel servers, as shown in Section 3.4, where we discuss a case for which the single-server optimal policy provides an excellent heuristic for parallel servers. Substantially more complex is the situation of machines in series, flow shops and tandem queues—only some partial results are available (see Section 3.5). Much more complex still is the case of general job shops and general queueing networks. We give some brief pointers to possible heuristics for those, in Sections 3.6–3.9.

## 3.2 SINGLE-MACHINE NONPREEMPTIVE SCHEDULING

We consider jobs which undergo a single processing stage. Job $j$ arrives at time $A_j$, waits for a duration $W_j$, then starts its processing at time $S_j$ for a processing duration $X_j$, and is completed and departs at time $C_j$. The flowtime of the system is often defined as the average of $C_j - A_j = W_j + X_j$, the sojourn time of job $j$ in the system. The flowtime is one important measure of a system's performance. By Little's law, for a given throughput, it is equivalent to the average WIP (work in process) or inventory in the system. It may sometimes be possible to control the arrival times and the processing times of jobs—however, we assume here that this is not the case, and the only control we have is in the order in which jobs are processed. Minimization of flowtime is therefore equivalent to minimizing the average or the sum of the starting times, the waiting times, or the completion times; we shall consider the average waiting time in this section.

A single processor is available to process all the jobs, and jobs are processed with no preemption. We shall look at two cases: scheduling a batch of $n$ jobs that are all available at time 0, and scheduling an infinite stream of jobs with Poisson arrivals. In both cases the flowtime is minimized by the SPT (shortest processing time first) priority rule. In a stochastic version of this problem, if $X_j$ are only known stochastically, SEPT (shortest expected processing time first) minimizes the expected flowtime.

While the rule is similar whether the processing times are known precisely or approximately, the expected flowtime is not the same. In this section we give some approximation formulas for the effect of lack of information on the flowtime. The results go back to Conway *et al.* (1967), Cobham (1954) and Schrage and Miller (1968), but they have been significantly sharpened recently by Matloff (1988).

### 3.2.1 Batch jobs

We shall assume the following framework: the processing times of the $n$ jobs, $X_1, \ldots, X_n$, are drawn randomly from a population with distribution $F$, with mean $m_1$ and standard deviation $\sigma$. This assumption can often be made, even in deterministic scheduling, and in practice one usually has a fairly good idea about $F$, which summarizes what jobs the system handles. The scheduler has some information about each job, from which he can predict the job's processing time as $Y_j = E(X_j \,|\,information)$. We assume that $Y_1, \ldots, Y_n$ are again a random sample, from a distribution $G$ with smaller variance than $F$ (this assumption means that we have the same kind of random information on all the jobs). Using this information, the scheduler sequences the jobs optimally from smallest to largest $Y_j$. Taking expectation of the average waiting time, first conditionally on $Y_1, \ldots, Y_n$, and then over the variables $Y_1, \ldots, Y_n$, one gets the following approximation formula:

$$E(\text{Average wait}) \approx \frac{1}{2}(n-1)m_1\left(1 - rd\frac{\sigma}{m_1}\right),$$

where $\sigma/m_1$ is the coefficient of variability of $F$, $d$ is a shape parameter of the distribution, defined by

$$m_1 - d\sigma = m_{1,2} = E(\min(X_1, X_2))$$

that is $m_{1,2}$ is the expected value of the smaller of two independent observations from $F$. Typically, $d \approx \frac{1}{2}$ (for the exponential distribution $d = \frac{1}{2}$, for the normal distribution $d = 0.5642$ and for the uniform distribution $d = 0.5774$). $r$ is the correlation coefficient between the processing time $X_j$ and the predicted processing time $Y_j$.

The two special cases of $r = 1$ and of $r = 0$ correspond to exact knowledge and to no knowledge at all about $X_1, \ldots, X_n$; In those two cases the formula is exact. For $0 < r < 1$ the formula is only exact under linear model assumptions—one case of a linear model is when processing times are normally distributed. If one thinks of $r$ as a measure of the information available about the processing times then one can say that the flowtime decreases approximately linearly as the amount of information increases.

To examine the formula further, we note that $0 \leqslant r \leqslant 1$, $d \approx \frac{1}{2}$, and (for positive random variables) usually $\sigma/m_1 \leqslant 1$. Hence the ratio of average waiting time per job without or with information is typically $\leqslant 2$; the value 2 is obtained for example with processing times drawn from an exponential distribution, when full information and no information at all are compared. We note that 2 is not an upper bound, and examples of arbitrarily high ratios can be constructed by using extremely overdispersed $F$.

### 3.2.2 Stream of jobs

An infinite stream of jobs are arriving at times $0 \leqslant A_1 \leqslant A_2 \leqslant \cdots$. We assume that $A_1, A_2, \ldots$ form a Poisson process with rate $\lambda$. This assumption means that arrival times are completely unpredictable, and no information about previous arrival times can prepare us for the next arrival—all we know is that it will occur in an expected time of $1/\lambda$, or that it will occur within the next short time interval $\Delta$ with a probability $\lambda\Delta$. Because the arrival stream is so unpredictable and yet so uniform, it can be ignored when scheduling those jobs that are present in the system.

The arriving jobs require processing times $X_1, X_2, \ldots$ which we assume are independent identically drawn from the processing time distribution $F$, with mean $m_1$. Another parameter of $F$ is relevant here, namely $B = \sup\{x : F(x) < 1\}$; in words, $B$ is the longest processing time that jobs drawn from $F$ are ever likely to have.

We assume that the arrival rate $\lambda$ is smaller than the processing rate $\mu = 1/m_1$, so that the traffic intensity $\rho = \lambda/\mu$ is less than 1, and therefore the process is stable. We are interested in the average waiting time per job (as surrogate to flowtime). We note that in a more deterministic world, if arrivals are equally spaced and all service times are equal, there is no waiting at all so long as $\rho \leqslant 1$. Therefore, unlike the batch case, the waiting here is purely the result of stochastic variability. We again present an approximation formula for the effect of partial information on the average waiting time. This formula is based on heavy traffic arguments and holds as $\rho \to 1$, when the average number of jobs queueing for the processor and the average waiting time are very large; exact formulas for any $0 < \rho < 1$ were derived in Schrage and Miller (1968), but these are less easy to interpret.

Full information in this case means that each job's processing time becomes known upon the job's arrival. For a Poisson arrival stream the optimal policy is to start the shortest waiting job whenever the processor becomes available, that is, the SPT rule. With partial information, predicted processing time replace the unknown actual processing time, and SEPT is optimal. With no information, all policies lead to the same average waiting time, though policies may differ on other criteria. For instance, FIFO (first in first out) will give a smaller waiting time variance than LIFO (last in first out).

The heavy traffic formula is

$$\frac{E(\text{waiting time} \mid SPT)}{E(\text{waiting time} \mid FIFO)} = \frac{m_1}{B}$$

An intuitive argument to derive this formula is as follows. By work conservation (that is, no unnecessary idling of the processor), the amount of work in the system at any time $t$ is independent of the policy. Let $S$ be the long time temporal average amount of work in the system and let $L$ denote the long time temporal average number of jobs in the system. If no information is available on the job processing times then the amount of work $S$ represents an average of $L(FIFO) = S/m_1$ jobs in the queue. If, however, information on processing times is available and long jobs have lower priority than shorter jobs then in heavy traffic almost all the jobs queueing up will be $B$ long, so the workload $S$ represents only $L(SPT) = S/B$ waiting jobs. Applying Little's formula which says that $L = \lambda W$ where $W$ is the average waiting time of a job, yields the above formula. Note that this crude analysis does not yield the value of partial information.

The important thing to note here is that the ratio can be much higher than in the batch case, where $\leqslant 2$ was typical; indeed, if job processing times are not essentially bounded, so that $B = \infty$, the ratio will be arbitrarily large as $\rho \to 1$. Of course one pays for the gain in average flowtime by extremely long waiting times of long jobs.

### 3.2.3 Polling systems

Somewhere between the batch and the stream models are polling models. Here a single server is serving several streams, and visits those in a cyclic order. At each visit the server serves all the customers in the queue at his arrival (gated service) or all the customers until the queue is empty (exhaustive service). Hence batches of jobs are allowed to accumulate, while the server is making his cycle of the other streams. For further discussion of polling models see Takagi (1986) and Browne and Yechiali (1989).

## 3.3 SINGLE-MACHINE PREEMPTIVE SCHEDULING

We consider the batch scheduling problem of Section 3.2.1, but we now allow preemption of jobs. In the deterministic case, when all processing times are known in advance, it never pays to use preemptions—if a job needs to be preempted, it should not have been started. In the stochastic case, when the processing time of a job becomes known only as it is being processed, we can make use of this dynamic information, and improve the flowtime by using preemptions. It turns out that there is a priority index that we can calculate for each job and that changes dynamically as the job is processed. We describe this index and the resulting optimal priority policy next.

This priority index is an example of a Gittins dynamic allocation index, which is defined for so called bandit processes, and solves the multiarmed bandit problem. We discuss the Gittins index in Section 3.3.2, and return to it in later sections. For the scheduling problem see Sevcik (1974), Harrison (1975) and Meilijson and Weiss (1975), and for bandit problems and the Gittins index see Gittins (1979,1989), Whittle (1980) and Weber (1992). See also Kaspi et al. (1995)—Chapter 17 of this volume.

### 3.3.1 Priority index and optimal preemption of jobs

Consider a job whose processing time $X$ is drawn from a distribution $F$. Assume the job has been processed for a duration $x$ and has not yet been completed; call $x$ the age of the job. The information obtained by watching the job for a duration $x$ and knowing that it is not yet complete is summarized by the conditional remaining processing time distribution $F(\cdot \mid X > x)$. We may now decide to give this job an additional period of processing of up to $y$ time units if required—of course the job may finish before using up all the time $y$. Define the pre-index

$$v(x,y) = \frac{\int_x^{x+y} dF(t)}{\int_x^{x+y}(1-F(t))dt} = \frac{P(completion\ in\ (x, x+y))}{E(time\ used\ in\ (x, x+y))}$$

and the index for the job at age $x$

$$v(x) = \max_{y>0} v(x,y)$$

In the batch scheduling problem, assume there are $n$ jobs, with processing time distributions $F_1, \ldots, F_n$, and current ages $x_1, \ldots, x_n$. Let $v_1(x_1), \ldots, v_n(x_n)$ denote the current values of their indices. Then

**Theorem 3.1 (Meilijson and Weiss 1975)** *To minimize expected flowtime it is optimal to always schedule the job with the highest current index.*

### 3.3.2 Bandit processes and the Gittins index

Consider $n$ processes that at time $t$ ($t = 0, 1, \ldots$) are in states $X_1(t), \ldots, X_n(t)$. At any time $t$, exactly one of the processes, say $i(t) = i$, is made active; all other processes are passive. The active process undergoes a state transition according to Markov transition probabilities

$$P_i(x, y) = P(X_i(t+1) = y \mid X_i(t) = x)$$

while the passive states are frozen: $X_j(t+1) = X_j(t)$, $j \neq i$. A reward $R(t) = R_i(X_i(t))$ is earned by the active process; the passive processes earn nothing.

This system of $n$ processes is called by Gittins a system of alternative bandit processes. The problem is to decide at each time point which process to activate, so as to maximize $\sum_t \beta^t R(t)$, where $\beta \leq 1$ is a discount factor. Gittins has defined his index for each of the processes as

$$v_i(x) = \max_{\tau > 0} \frac{E(\sum_{t=0}^{\tau-1} \beta^t R_i(X_i(t)) \mid X_i(0) = x)}{E(\sum_{t=0}^{\tau-1} \beta^t \mid X_i(0) = x)}$$

where the maximization is taken over all possible stopping times (note that a stopping time $\tau$ is as a rule random—it can depend on where the process $X_i(t)$ will go; for example, it may be the first time that the process enters a certain set of states). In words, the index expresses the maximal reward per unit time (both discounted) that can be achieved by the process, starting at state $x$, and choosing the best amount of time to stay active. Gittins has shown the following.

**Theorem 3.2 (Gittins 1974, 1979)** *The optimal policy is at any time to activate the bandit process with the highest index.*

The importance of Gittins' result is that the indices are calculated for each process separately, and do not depend on the other $n - 1$ processes. The $n$-dimensional problem of what to do in state $X_1(t), \ldots, X_n(t)$ is uncoupled into the solution of $n$ one-dimensional problems. Gittins' result solved the multiarmed bandit problem posed by Bellman (1956). As we can see, it applies directly to Section 3.3.1, where the index of a job is an example of a Gittins index (with discount factor 1). It also solves Klimov's problem of the control of the M/G/1 queue and its extensions, as discussed in Section 3.6. Whittle (1980) has given an equivalent formulation of the Gittins index, in terms of a "retirement reward". For process $i$ consider the problem of choosing at each point in time either to be active and collect the usual reward, or to retire and collect a retirement reward $v$. This is a Markov decision problem with optimality equation

$$V(x, v) = \max\left(R_i(x) + \beta \sum_y P_i(x, y) V(y, v), \ v\right)$$

Whittle shows that if the value of $v$ is allowed to vary in this problem, and one starts when process $i$ is in state $x$, then the value $v$ for which both actions are optimal (one is indifferent between continuing and retiring) is the Gittins index $v_i(x)$. This alternative formulation is the starting point for restless bandits discussed in Section 3.6.

## 3.4 MACHINES IN PARALLEL

In this section we look at the simplest situation when there is simultaneous processing of more than one job: the operation of $M + 1$ identical machines which are working in parallel. We consider a batch of $n$ jobs, each of which can be done on any of the machines. In addition to flowtime, which we express as $\sum C_j$, we shall consider minimization of two other objective functions: weighted flowtime $\sum a_j C_j$, where $a_j$, the weight of job $j$, can be thought of as holding cost per unit time, and makespan $C_{\max} = \max (C_j)$, the time needed from the start of the first job to the completion of the last job. These problems become hard on parallel machines, even in the deterministic case.

We believe that the following discussion shows that analysis of the stochastic problems provides a deeper understanding of parallel operation, and a more realistic evaluation of heuristics, than can be obtained from the study of the deterministic problems alone.

### 3.4.1 Deterministic scheduling of parallel machines

Flowtime on parallel machines is minimized by SPT as on a single machine. Minimization of makespan (which for a single machine is independent of the schedule) on parallel machines is an $\mathcal{NP}$-hard combinatorial optimization problem. However, there is little doubt that it is an extremely easy problem to solve satisfactorily: if $n$ is small, it can be solved by enumeration; for moderate $n$ polynomial approximation schemes exist; for large $n$ LPT is an excellent heuristic; in fact, for large $n$ every non idling schedule will be almost optimal in practice, since the only optimization involved is how to make the $M + 1$ last jobs finish together, and the effect of $M + 1$ jobs is usually negligible for large $n$. See Karmarkar and Karp (1982), Karmarkar et al. (1986), Lueker (1987) and Frenk and Rinnooy Kan (1987) for details. Minimization of weighted flowtime on a single machine is achieved by Smith's rule: start the jobs in decreasing order of $a_j/X_j$. The problem for $M + 1$ parallel machines is however $\mathcal{NP}$-hard. Smith's rule is a natural heuristic for this problem. Intuitively, what it does is to reduce the holding cost rate of the remaining jobs fastest, and this holds for one machine as well as for $M + 1$ machines. However, on $M + 1$ machines, towards the end of the schedule, machines fall idle as the last $M + 1$ jobs are completed, and processing becomes inefficient. This end effect is the source of the difficulty in minimizing $\sum a_j C_j$, just as it is for the makespan problem. Worst case analysis shows that the SR (Smith's rule) heuristic can be 1.20 times worse than the optimal solution (see Kawaguchi and Kyan 1986).

### 3.4.2 Exact results for minimizing flowtime on parallel machines

The most general exact optimality result for parallel machines is the following.

**Theorem 3.3 (Weber et al. 1986)** *If the processing times of jobs are stochastically comparable random variables then SEPT minimizes the expected flowtime.*

We say $X$ and $Y$ are stochastically comparable if $X \leqslant_{ST} Y$ or $Y \leqslant_{ST} X$, where we say $X$ is stochastically smaller than $Y$, $X \leqslant_{ST} Y$, if $P(X > t) \leqslant P(Y > t)$ for all $t$.

This theorem subsumes several earlier results discovered for specific distributions (see Bruno et al. 1981; Weiss and Pinedo 1980; Weber 1982). No general optimality results exist for the stochastic problem of minimizing weighted flowtime. If Smith's rule agrees with SEPT then it is optimal for exponentially distributed processing times (see Weiss and Pinedo 1980; Kämpke 1987).

### 3.4.3 Additional optimality results for stochastic scheduling of parallel machines

LEPT (longest expected processing time first) minimizes the expected makespan if the processing times are exponentially distributed, and for a few other special cases (see Bruno et al. 1981; Van der Heyden 1981; Pinedo and Weiss 1979; Weiss and Pinedo 1980; Weber 1982). It seems remarkable that while the deterministic problem is $\mathcal{NP}$-hard, the same problem with exponentially distributed processing times is solved by the simple LEPT priority rule—the explanation to this lies in considering the preemptive problem. Here both the deterministic and the exponential problems are easily solved. In the deterministic case the optimal solution requires preemptions, to achieve a perfect partition of the jobs and hence a better makespan than is possible without preemption. In the exponential case the optimal solution to the preemptive problem is LEPT, which does not utilize the option of preemption, and is therefore also the solution to the nonpreemptive problem. Limited as these results for makespan are, they seem to be fairly close to exhaustive—we doubt whether there is much more to be discovered here. Several papers have been written on scheduling parallel machines that are working at different speeds. Here the problem is to decide whether to put a job on an available machine, or wait for a faster machine to become free; Kleinrock (1976), Agrawala et al. (1984), Coffman et al. (1987), Lin and Kumar (1984), Mirchandani and Xu (1989), Xu et al. (1990) and Righter (1988) consider various versions of this problem. An interesting feature in this model is that the individually optimal policy for each customer turns out to be the socially optimal policy for the whole problem. Presented in terms of the individual customers, there is a threshold for the queue length in front of each customer, which if exceeded will make the customer accept a slower processor. Scheduling with treelike precedence constraints was considered by Chandy and Reynolds (1977), Bruno (1985), Pinedo and Weiss (1985) and Frostig (1988). For two machines, intree precedence constraints, and i.i.d. processing times that are exponentially distributed, or that are drawn from an ILR (increasing likelihood ratio) distribution, the rule of highest level first is optimal, in analogy with the deterministic result of Hu (1961). Out-tree constraints are discussed by Coffman and Liu (1992).

### 3.4.4 Analysis of Smith's rule heuristic for minimization of weighted flowtime on parallel machines

For the remainder of this section we present an analysis of the performance of Smith's rule in minimizing the weighted flowtime of a batch of stochastic jobs. The conclusion is that, under very realistic assumptions, Smith's rule is very close to optimal, which we believe is a more realistic view of its performance than that given by the deterministic worst case analysis. Our conclusion therefore is that it is almost optimal to use the same priority order for scheduling jobs on parallel machines as one would use for a single machine. The actual expected flowtime for $M + 1$ parallel machines appears to be asymptotically equivalent to that of a single machine with their combined capacity. This is an example of resource pooling, which in this case is achieved by the simplest policy. The following formulas in Sections 3.4.4.1 and 3.4.4.2 (Weiss 1990,1992), for expected flowtime and expected weighted flowtime, are of interest in themselves. They are also needed to analyze the performance of Smith's rule, which is given in Sections 3.4.4.3 and 3.4.4.4.

*3.4.4.1 Expected flowtime for i.i.d. jobs*

Consider an infinite supply of jobs with processing times $X_1, X_2, \ldots$, i.i.d. drawn from a distribution $F$; let $m_1$, $\sigma^2$, and $m_3$ denote the mean, variance and third moment of $F$. A set of $M + 1$ machines become available to process these jobs at times $U_{00} \leqslant U_{10} \leqslant \ldots \leqslant U_{M0}$, assume $\sum_{i=0}^{M} U_{i0} = 0$. The stream of jobs is processed on the $M + 1$ machines in parallel, and so the completion times of jobs on the different machines form $M + 1$ independent renewal processes. Let $U_{0n} \leqslant U_{1n} \leqslant \ldots \leqslant U_{Mn}$ denote the times at which the $M + 1$ machines complete the first $n$ jobs, let $D_{in} = U_{in} - U_{0n}, i = 1, \ldots, M$, and let

$$S_n^2 = \frac{1}{M} \sum_{i=0}^{M} U_{in}^2 - \frac{1}{M(M+1)} \left( \sum_{i=0}^{M} U_{in} \right)^2$$

be the sample variance of the times at which the $M + 1$ machines complete these $n$ jobs.

**Lemma 3.1 (Weiss 1990)** *Consider $D_{1n}, \ldots, D_{Mn}$, the remaining processing times of the last $M$ jobs among jobs $1, \ldots, n$. As $n \to \infty$, these converge in distribution to an ordered sample from the equilibrium distribution of $F$ with probability density function $f_e(t) = [1 - F(t)]/m_1$ (one needs to impose the condition that $F$ is nonarithmetic for this theorem; $F$ is arithmetic if $F$ is concentrated on points $\delta_0 + k\delta$, $k = 0, \pm 1, \pm 2, \ldots$. However, a slightly more complicated version of the theorem holds when $F$ is arithmetic).*

The expected flowtime of $n$ i.i.d. jobs on $M+1$ machines follows from this lemma:

$$E\left(\sum_{j=1}^{n} C_j\right) = \frac{n(n+1)}{2(M+1)}m_1 + \frac{nM}{2(M+1)}m_1\left(1 - \frac{\sigma^2}{m_1^2}\right) - \frac{M}{2}\left(\frac{S_0^2}{m_1} - \frac{E(S_\infty^2)}{m_1}\right)$$
$$+ \frac{M}{2}\left(\frac{E(S_n^2)}{m_1} - \frac{E(S_\infty^2)}{m_1}\right)$$

Each of the four terms in this formula has a meaning: the first is equal to the expected flowtime on a single machine with $(M+1)$-fold speed—that is, a single machine that has the combined capacity of the $M+1$ parallel machines; the second accounts for a constant delay per job that is the result of parallel processing—this delay equals the difference between the first moments of $F$ and of $f_e$, we call it a synchronization delay; the third and fourth terms are contributions of starting conditions and ending conditions. We have added and subtracted from these terms the constant term defined by

$$\frac{M}{2}\frac{E(S_\infty^2)}{m_1} = \lim_{n\to\infty}\frac{M}{2}\frac{E(S_n^2)}{m_1} = \frac{M}{M+1}\frac{m_3}{3m_1^2} + \frac{M-1}{M+1}\frac{m_1}{4}\left(1 - \frac{\sigma^2}{m_1^2}\right)^2.$$

The four parts of the formula have orders of magnitude $O(n^2)$, $O(n)$, $O(1)$ and $o(1)$. We list some special cases of this formula:

(1) constant processing times $X_j = 1$. We assume that starting times of machines are staggered, with intervals $1/(M+1)$:

$$E\left(\sum_{j=1}^{n} C_j\right) = \frac{n(n+M+1)}{2(M+1)}.$$

(2) For exponential processing times, mean 1, all machines start at 0:

$$E\left(\sum_{j=1}^{n} C_j\right) = \frac{n(n+1)}{2(M+1)} + \frac{M}{2}.$$

(3) For uniformly distributed processing times, on $(0, 2)$, all machines start at 0.

$$E\left(\sum_{j=1}^{n} C_j\right) = \frac{n(n+\frac{2}{3}M+1)}{2(M+1)} + \frac{M(M+2)}{9(M+1)}.$$

(4) Decreasing hazard rate jobs—these are jobs that, as long as they are not completed, the longer one works on them, the longer their remaining processing time becomes. This is perhaps typical of R&D activities and not of manufacturing. Such jobs have $\sigma/m_1 > 1$, and so the synchronization delay is negative—in other words, it is best to process them in parallel and not one after the other on a single machine

### 3.4.4.2 Expected weighted flowtime

The formula (3.1) exhibits a large degree of insensitivity to the assumption of i.i.d. processing times. The formula remains unchanged if one dispenses with the assumption that the jobs are identically distributed, and only assumes they have equal first and second moments. If the means are equal and variances differ, the only change in the formula is to replace $\sigma^2$ by $\sum_{i=1}^{n} \sigma_j^2/n$. More generally, for any independent processing time distributions, we can get an expression of the expected weighted flowtime when the weights are proportional to the expected processing times.

**Theorem 3.4 (Weiss 1988)** *Let the processing times of the $n$ jobs be independent random variables $X_j$ with means $E(X_j) = 1/\mu_j$, and variances $\sigma_j^2$, and assume that the flowtime (completion time) of job $j$ has weight $1/\mu_j$, for $j = 1, \ldots, n$. Then*

$$E\left(\sum_{j=1}^{n} \frac{C_j}{\mu_j}\right) = \frac{1}{2(M+1)} \left(\sum_{j=1}^{n} \frac{1}{\mu_j}\right)^2 + \frac{1}{2}\sum_{j=1}^{n} \frac{1}{\mu_j^2} - \frac{M}{2(M+1)} \sum_{j=1}^{n} \sigma_j^2$$
$$- \frac{M}{2} S_0^2 + \frac{M}{2} E(S_n^2).$$

Two things are remarkable about this formula: except for the last term, it depends on the distributions of the processing times only through their means and variances; and, more surprisingly, except for the last term, it is completely independent of the schedule. The essence of the following analysis of Smith's rule is the way in which this formula pushes all the complications to the "end effects" term.

### 3.4.4.3 Approximate optimality

We now present a bound on the difference between expected weighted flowtime under Smith's rule, and under any other schedule. We consider $n$ jobs, with expected processing times $1/\mu_j$ and with weights $a_j$, $j = 1, \ldots, n$. Define

$$\overline{D}^2 = \max_{1 \leq j \leq n} \sup_{s>0} E((X_j - s)^2 \mid X_j > s)$$

In words, $\overline{D}^2$ measures how large the square of the remaining processing time of an unfinished job is expected to be, at the most.

**Theorem 3.5 (Weiss 1990)** *Let SR denote the use of Smith's rule, and let $\Pi$ be an arbitrary nonpreemptive scheduling strategy. Then*

$$E\left(\sum_{j=1}^{n} a_j C_j \mid SR\right) - E\left(\sum_{j=1}^{n} a_j C_j \mid \Pi\right) \leq \frac{M^2}{2(M+1)} \max_{1 \leq j \leq n} (a_j \mu_j) \overline{D}^2. \qquad (3.2)$$

This theorem holds for a wider class of strategies $\Pi$ than are usually considered. One can include strategies which allow the use of inserted idle time. One can also include strategies for which the actual processing time of each job becomes known at the instant at which the job starts being processed, and this knowledge can be used by the strategy. The bound given by the formula (3.2) is particularly useful when all $a_j\mu_j$ and $\overline{D}^2$ are bounded by a constant, independent of the number of jobs $n$. There are many practical examples when this is indeed the case; for instance,

- when the processing times are drawn from a fixed number of given distributions, $F_1, \ldots, F_K$ independent of $n$;
- when the distributions have hazard rates bounded from below by a constant $\Lambda$;
- when the distributions are all NBU (new better than used), and have bounded second moments, independent of $n$.

Note that the assumption of a finite $\overline{D}^2$ independent of $n$ imposes a uniform bound on the distributions of the processing times, but not on the processing times themselves, which may be unbounded. In fact, in the instances listed above one would, as a rule, still have that as $n \to \infty$ the processing times of the longest jobs will also $\to \infty$. Under this assumption, the worst case ratio is

$$\frac{E(\sum_{j=1}^n a_j C_j \mid SR)}{E(\sum_{j=1}^n a_j C_j \mid \Pi)} = 1 + O\left(\frac{1}{n^2}\right)$$

which is very different from the deterministic worst case ratio of 1.20. We believe that in practice the $1 + O(1/n^2)$ will be more realistic than the 1.20 ratio.

### 3.4.4.4 Turnpike optimality

In the previous section we saw that the expected value of the objective function under Smith's rule is close to optimal. We now show that more than that is true— most of the time the optimal action is to follow Smith's rule. We call this the "turnpike optimality" property of Smith's rule in analogy with turnpike optimality in optimal control problems and in discounted dynamic programming. Turnpike optimality has not received much attention in scheduling theory; we believe it is of great practical importance. In the context of the previous section, define

$$\delta^{(M)} = \min_{j_0 \ldots j_M} \inf_{s_0 \ldots s_M} E(\min(X_{j_0} - s_0, \ldots, X_{j_M} - s_M) \mid X_{j_0} > s_0, \ldots, X_{j_M} > s_M)$$

In words, $\delta^{(M)}$ measures how small the expected value of the time interval between two non-simultaneous decision moments can be. If the processing times have discrete probability distributions then $\delta^{(M)}$ is simply the unit of the discrete distributions. Let also

$$\mu_{\max} = \max(\mu_j)$$
$$(a\mu)_{\max} = \max(a_j\mu_j)$$
$$(\Delta a\mu)_{\min} = \min\{|a_j\mu_j - a_k\mu_k| \ : \ a_j\mu_j \neq a_k\mu_k\}$$

**Theorem 3.6 (Weiss 1992)** *Let $\Pi_0$ be a nonpreemptive, work-conserving, nonpredictive and nonrandomizing strategy. Let $L$ be the number of times that $\Pi_0$ starts a job not according to SR. If*

$$E\left(\sum_{j=1}^{n} a_j C_j \mid SR\right) - E\left(\sum_{j=1}^{n} a_j C_j \mid \Pi_0\right) \geq 0,$$

that is, if $\Pi_0$ is better than $SR$ then

$$E(L \mid \Pi_0) \leq \frac{M^2}{2}(a\mu)_{\max}\mu_{\max}\overline{D}^2/(\Delta a\mu)_{\min}\delta^{(M)} \tag{3.3}$$

Again, under reasonable assumptions, this bound does not grow with $n$, and so as $n \to \infty$,

$$\frac{E(L \mid \Pi_0)}{n} \to 0.$$

In words, the policy of using Smith's rule will schedule the optimal job at the majority of the decision moments, the fraction of optimal decisions that are not according to Smith's rule is vanishingly small. We believe that these non SR optimal decisions will occur only towards the end of the schedule.

**Conjecture** *Let $n_0$ be the first decision at which the optimal policy deviates from Smith's rule. Then under appropriate assumptions, $n - n_0$ or at least $E(n - n_0)$ is bounded, independent of $n$.*

### 3.4.4.5 Preemptive scheduling of stochastic jobs on parallel machines

Recently Weiss (1995) has shown that the results obtained for nonpreemptive scheduling of parallel machines in Section 3.4.4 extend to the preemptive case. In the preemptive case the optimal policy for a single machine is given by a Gittins priority policy as described in Section 3.3.1. For parallel machines this rule is not optimal but provides a heuristic. This heuristic is nearly optimal: essentially the same bounds as in equations (3.2) and (3.3) hold for the Gittins index heuristic, compared with any other preemptive policy.

## 3.5 MACHINES IN SERIES

Following the discussion of machines in parallel, we now turn to machines in series, sometimes called machines in tandem in the queueing terminology, or flow shops. In a flow shop of $M$ machines the machines are ordered $1, \ldots, M$, and each job needs to go through the machines in that order. We shall consider mainly permutation flow shops, in which there is no overtaking, so that the order of the jobs is the same on each of the machines. Permutation schedules are used for their simplicity, even though for some stochastic problems, as is also the case for deterministic problems, they are not optimal. As a general remark, one should note that even though each job sees the machines as a series, the $M$ machines of the flow shop are actually working simultaneously. The flow shop is much more complicated than machines in parallel, and we are much further from understanding it. Some of the research on deterministic flow shop scheduling is surveyed by Monma and Rinnooy Kan (1983); see also Barany (1981). Optimization and scheduling of tandem queues is considered by, among others, Tembe and Wolff (1974), Yamazaki and Sakasegawa (1975), Muth (1979,1984), Pinedo (1982), Wie and Pinedo (1986), Pinedo and Wie (1986), Greenberg and Wolff (1988), Wein (1988) and Harrison and Wein (1989). We make no attempt to discuss all this work here or to give an exhaustive survey. Our discussion is of two topics: probabilistic analysis of the two-machine flow shop, and choosing the order of machines in tandem queues. Further discussion of probabilistic analysis is presented by Coffman and Whitt (1995)—Chapter 2 of this volume. The topic of optimal design, control, and stochastic scheduling of flow shops or tandem queues is the subject of Weber (1995)—Chapter 15 of this volume.

### 3.5.1 Two-machine flow shop

In a batch of $n$ jobs, job $j$ requires $A_j$ processing on machine 1, and $B_j$ processing on machine 2. We assume that $A_1, \ldots, A_n$ are drawn i.i.d. from a distribution $F_A$, with mean $m_A$, and $B_1, \ldots, B_n$ are drawn independent of $A_1, \ldots, A_n$ and i.i.d. from a distribution $F_B$, with mean $m_B$. These distributional assumptions enable us to do a probabilistic analysis of the system. We assume, however, that the actual values of $A_1, \ldots, A_n$ and $B_1, \ldots, B_n$ are known to the scheduler, so we are analyzing a deterministic problem. Johnson's rule says: divide the jobs into two sets—those with $A_j \leqslant B_j$ are scheduled first, according to SPT order of the $A_j$; those with $A_j > B_j$ are scheduled next, according to LPT order of the $B_j$. Johnson's rule minimizes the makespan (the time until the last completion) of all the jobs. It achieves that by causing the least possible idling on machine 2. The problems of minimizing makespan on more than two machines, and of minimizing flowtime on two machines or more (and indeed almost all other scheduling problems involving flow shops) are $\mathcal{NP}$-hard. Perhaps because it is the only rule with some positive distinction, Johnson's rule has received a great deal of attention in literature, and is quite familiar to practitioners. Yet the following analysis (Bartholdi et al. 1995) shows that there is very little merit to its use. Consider first the case $m_A < m_B$. Then it is enough to do just a few jobs according to Johnson's rule and proceed arbitrarily thereafter, and with very high probability there will be no idling at all on machine 2, and the policy will achieve the makespan of Johnson's rule. In fact, for moderately large $n$, the makespan of a random order of the jobs is $nm_B + O(1)$, and Johnson's rule is superfluous. The case $m_B < m_A$ is the same by reversibility of flow shops—under fairly general conditions flow shops have the following reversibility property: running the jobs in reversed order, through the machines ordered from last to first, achieves the same makespan (or throughput) as the original (see Yamazaki and Sakasegawa 1975; Muth 1979). In the case $m_A = m_B = m$ (the balanced case), we assume for simplicity equal standard deviations $\sigma_A = \sigma_B = \sigma$. For moderately large

$n$ the expected makespan is bounded below by $nm + 0.56\sqrt{n}\sigma + O(1)$, which is the expectation (for large n) of $\max(\sum A_j, \sum B_j)$. Johnson's rule achieves this bound up to an $O(1)$ term, by virtue of having almost no idle time either on machine $A$ or on machine $B$ during the entire schedule. However, to achieve this, Johnson's rule typically builds up a queue of approximately $n/4$ jobs between the two machines, which persists for more than half the makespan. Some simple heuristics can achieve the same makespan (up to order $O(1)$), with a queue that is of order of magnitude $\sqrt{n}$. A schedule that starts the jobs in a random order (and therefore needs no information and can be applied online) gives a makespan of $nm + 1.13\sqrt{n}\sigma + O(1)$, which is not much worse than Johnson's rule.

### 3.5.2 Choosing the order of machines

In an $M$-machine permutation flow shop, job $j$ requires $M$ stages of processing, with processing times $X_{ij}$, $i = 1, \ldots, M$. Given a batch of $n$ jobs, one could seek the best order of jobs and the best order of the processing stages, among all the permutations of the jobs as well as all the permutations of the machines. However, there seems to be no quick way, short of enumerating the $n!\,M!$ possibilities, of finding optima, and not much else can be said on this general problem. However, in practice one often needs to pick the order of the machines, and hence of the processing stages of all the jobs, prior to obtaining the values of the job processing times. In that case the information about the jobs may be the distribution of the processing times—which is assumed to be an i.i.d. sample of the random processing time vector $X_1, \ldots, X_M$. The problem is to pick the best permutation of the components of this vector as the optimal order of the machines, on the assumption that jobs will be served in the order in which they arrive (or, equivalently, jobs will be scheduled on the machines with no additional information about their individual processing times). To take into account the future scheduling of batches of jobs, at the stage of designing the order of the machines, is surely intractable and we ignore it. The problem then is how to order a set of tandem machines, for the service of a stream of jobs with i.i.d. processing time vectors $X_1, \ldots, X_M$. By the duality of jobs and machines, this problem is equivalent to choice of the optimal order of jobs, when job $j$ needs processing $X_{1j}, \ldots, X_{Mj}$, which are an i.i.d. sample of a random $X_j$; in other words one has i.i.d. machines instead of i.i.d. jobs. In the following three subsections we discuss some aspects of the problem of choosing machine order. We discuss the case of unlimited waiting room between the machines first, and consider the case of no waiting room and blocking of machines next. We consider two criteria: the throughput, which is the average number of jobs that depart the system per unit time (similar to the makespan criterion), and the waiting time of the jobs in the system (equivalent to the flowtime).

#### 3.5.2.1 Interchangeability of tandem queues

We consider first a series of $M$ machines with infinite buffer space between the machines. Let $\mu_i = 1/E(X_i)$ be the rate of service of machine $i$, and denote by $\lambda$ the arrival rate (the number of job arrivals per unit time, over a long period). The throughput of the flow shop is governed by the slowest (the smallest) of these rates. In particular, if $\lambda < \mu_i$ for all $i$ then the system is "stable", with throughput $\lambda$. If $\lambda \geq \min_{1 \leq i \leq M} \mu_i$, in particular if $\lambda = \infty$ (which corresponds to an unlimited supply of jobs in front of the first machine), then the throughput is $\min_{1 \leq i \leq M} \mu_i$ irrespective of the order of the machines. However, the average waiting time of the jobs (or the average number of jobs queueing in the system, which by Little's formula is proportional to the waiting time), may be very different for different orders of the machines. In some special cases it turns out that the order of the machines is irrelevant. Those cases include

- *deterministic jobs*, where processing on machine $i$ lasts a deterministic amount of time $x_i$ for every job (Friedman 1965).
- *exponential jobs*, where processing on machine $i$ is exponentially distributed, independent of the other machines, with rate $\mu_i$ (Weber 1979; Lehtonen 1986; see also Chao et al. 1989).
- *0–1 single activity jobs*, where each job requires 1 unit of processing time on one machine, this being machine $i$ with probability $p_i$, and 0 processing time on all the other machines (Weber and Weiss 1994).

Interestingly enough, the last two results turn out to be special cases of a recent result of Weber (1992b). Assume that jobs numbered $1, 2, \ldots$, require total processing times $c_1, c_2, \ldots$. The actual processing times of job $j$ on machines $1, \ldots, M$, denoted by $X_{1j}, \ldots, _{Mj}$, follow a distribution

$$P(X_{1j}, \ldots, X_{Mj} \mid c_j) = \frac{p_1^{X_{1j}} \cdots p_M^{X_{Mj}}}{\sum p_1^{y_1} \cdots p_M^{y_M} \mid y_1 + \cdots + y_M = c_j}$$

that is, the processing times on the $M$ machines are distributed like $M$ independent geometric random variables, with parameters $p_1, \ldots, p_M$, conditional on the total processing time of the job on all the machines adding up to $c_j$. Let $a_1 \leq a_2 \leq \ldots$ and $d_1 \leq d_2 \leq \ldots$ denote the arrival and departure times of the jobs.

**Theorem 3.7 (Weber 1992b)** *The joint distributions of arrivals and departures,*

$$P(a_1, \ldots, a_n, d_1, \ldots, d_n \mid c_1, \ldots, c_n)$$

*are independent of the order of the machines.*

*3.5.2.2 Ordering machines to reduce waiting time*

As a rule, the average waiting time of jobs in a tandem queueing system depends on the order of machines. Unfortunately, there are no exact expressions for the average waiting times under general service time assumptions, which one can use to choose the best order. In practice, one can compare various orders through simulation studies—but these provide little insight. Tembe and Wolff (1974), Greenberg and

Wolff (1988), Whitt (1985) and Wein (1988) present approximate analyses of waiting times in tandem queues under various assumptions, and discuss some rules of thumb and heuristics for ordering the machines. One of the heuristics proposed by Wein (1988) is described next. Arrivals occur at i.i.d. intervals with rate $\lambda$, and services at the $i$th machine are i.i.d. with rate $\mu_i$, and with traffic intensity $\rho_i = \lambda/\mu_i < 1$. The distributions of the interarrival times and the service times at station $i$ have squared coefficients of variations $c_0^2$, and $c_i^2$ respectively. It is assumed that the stations operate under balanced heavy load, as expressed by the existence of a large integer $n$ for which $\max_{1 \leqslant i \leqslant M} \sqrt{n}(1 - \rho_i) < 1$. Then the heuristic for ordering the machines is to choose the order that minimizes

$$\sum \frac{c_i^2}{\mu_k - \lambda} \delta_{ik}$$

where $\delta_{ik} = 1$ if machine $i$ directly precedes machine $k$, $0 \leqslant i \leqslant M$, $1 \leqslant k \leqslant M$, and $\delta_{ik} = 0$ otherwise. Minimizing this sum over all permutations is a travelling salesman problem, which is easily solved for a reasonable number of machines.

### 3.5.2.3 Optimal order of machines with blocking

We consider $M$ machines in series with an infinite supply of jobs in front of the first machine, and no waiting room between the machines. The jobs are assumed to require i.i.d. processing times. Job $j$ requires processing for durations $X_{1j}, \ldots, X_{Mj}$ on the $M$ machines, which are assumed to be independent random variables with distributions $F_1, \ldots, F_M$. Jobs are fed into the system whenever the first machine is free. They move through the $M$ machines and out. Because there is no waiting room, there is no queueing, but jobs may be blocked in one machine, after processing is complete, because the next machine is not free. This form of blocking is called manufacturing blocking. The problem is to choose the right order of machines—or, in other words, the right permutation of $F_1, \ldots, F_M$. This problem has been studied by Yamazaki et al. (1989), with the objective of maximizing the throughput of the system. The problem is very hard—all that can be proved is that under suitable conditions one should have slower machines in the first and last position than in the second and penultimate positions respectively (see Huang and Weiss 1990; Shanthikumar et al. 1991). It is clear that this problem does not possess a "nice" solution beyond the rule of thumb that "slow machines should be kept as separate as possible". A special model of machines in series, where each job requires unit service at only one of the machines is studied by Weber and Weiss (1994), who have named it a "cafeteria" process. The optimal loads of the machines are shown to be of "bowl shape".

## 3.6 QUEUEING NETWORKS WITH A SINGLE SERVER

In Section 3.2.2 we considered jobs that arrive in a Poisson stream at a single-server queue, and are processed with no preemptions; scheduling waiting jobs in this system according to SEPT minimizes the expected flowtime (in the form of the average waiting or sojourn time). If jobs incur different holding costs then scheduling them according to Smith's rule, or, as it is called in the queueing literature, the "$c\mu$ rule", minimizes the expected weighted flowtime. For these two problems job $j$ has $E(X_j)$ and $a_j E(X_j)$ as priority index, and the optimal policy is a priority policy—at any decision time, schedule the job with the highest priority. In Section 3.3.1 we considered scheduling a batch of jobs, when preemptions were allowed: we found that for job $j$ of age $x_j$ we could calculate a priority index $v_j(x_j)$, and the priority policy based on this priority index minimizes the expected flowtime for the batch. It turns out that the same priority policy minimizes expected flowtime also for a Poisson stream of arriving jobs, and that the weighted flowtime problem has $a_j v_j(x_j)$ as the appropriate index. In this section we generalize these models further. We start with Klimov's model on control of the M/G/1 queue, and we then present a further generalization to branching bandit processes. One view of this general version is to regard it as a queueing network, with a single server that is moving around the nodes of the network. All these problems are solved optimally by a priority policy. The priority index for this policy is a Gittins type priority index, and the optimality results are generalizations of Gittins' original results.

### 3.6.1 Klimov's model for the control of the M/G/1 queue

Jobs of several types, denoted by $k$, $k = 1, \ldots, K$, arrive at a single-server station, in independent Poisson arrival streams; type $k$ have arrival rate $\lambda_k$, $\lambda_k \geqslant 0$. A job of type $k$ requires processing time of duration $X_k$, with mean $m_k$; all processing times are assumed to be independent. Jobs cannot be preempted while in process. On completing its processing, a job of type $k$ may rejoin the queue, with probability $P_{kk}$, or rejoin the queue as a job of a different type, say type $j$, with probability $P_{kj}$, or it may leave the system, with probability $1 - \sum_j P_{kj}$. A holding cost of $a_k$ per unit time is charged to any type $k$ job in the system. The problem is how to schedule jobs for processing, so as to minimize the long run average holding costs (this is equivalent to minimizing weighted flowtime). Klimov (1974) solved this problem, using queueing theory techniques; Sevcik (1974), Harrison (1975), Meilijson and Weiss (1977), Tcha and Pliska (1977) and Nash and Gittins (1977), and more recently Varaiya et al. (1985) and Weiss (1988), give various proofs and generalizations, using Markov decision processes. It is easy to see that the problems of Sections 3.2.2 and 3.3.1 are special cases. In the problem of Section 3.2.2 every customer leaves the system when his service is complete. For the preemptive scheduling problem of Section 3.3.1, assume job $j$ requires a processing time $X_j$ that is discrete with $P(X_j = x) = P_j(x)$, and with $P(X_j > x) = Q_j(x)$, $x = 0, 1, \ldots$. Denote by $(j, x_j)$, job $j$ at age $x_j$, and define $(j, x_j)$ as a type. The service time of this type is 1, and at the end of this service the job leaves the system (is completed) with probability $P_j(x_j)/Q_j(x_j - 1)$ and turns into type $(j, x_j + 1)$ with probability $Q_j(x_j)/Q_j(x_j - 1)$. In Klimov's problem, whenever a service is completed, the server is faced with a state given by $n_1, \ldots, n_K$ where $n_k$ is the number of type-$k$ customers in the system, and he needs to decide which of these to serve. At first glance, this appears to be a formidable problem because of the large $K$-dimensional state space. In fact, the optimal policy is a priority policy, where each type has a priority index and at any state the optimal policy is to start serving one of the customers with the highest priority index among those in the queue. The priority index of a type $k$ job is

$$v(k) = \max_{\tau > 0} \frac{E(\text{cost reduction} \mid \text{started as type } k)}{E(\tau \mid \text{started as type } k)},$$

where one works on a single job, starting as a type-$k$ job with cost rate $a_k$ and proceeding through a possible sequence of type changes until a stopping time $\tau$, at which point if the job left the system then the cost reduction is $a_k$, and if the job is now a type $j$ then the cost reduction is $a_k - a_j$. This is clearly a Gittins index. This definition appears unsuitable for calculation, since there is no indication of how to maximize over all $\tau > 0$. However, the following recursive procedure to calculate the indices of types $1, \ldots, K$ is straightforward.

- *Step 1*: determine the highest priority type. For each type $k$ calculate the expected cost rate reached at the end of $X_k$, call it $d_1(k)$, and calculate the ratio $[a_k - d_1(k)]/m_k$. Rename the type with maximal ratio type 1; it is the highest priority type, and the ratio is its index.
- *Step r* determine the $r$th priority type. Assume types $1, \ldots, r-1$ have been determined as top-priority types. For type $k$, $k \geqslant r$, calculate the expected time it takes for a type-$k$ customer until it leaves the system or until it changes into a customer of one of the types $r, r+1, \ldots, K$; call this expected time $h_r(k)$. Calculate the expected cost rate reached at that time, call it $d_r(k)$. Calculate the ratio $[a_k - d_r(k)]/h_r(k)$. Rename the type with maximal ratio type $r$; it is the $r$th priority type, and the ratio is its index.

These calculations are easily performed on the partitioned vector of cost rates, vector of mean processing times, and transition probability matrix. It is interesting to note that the arrival rates play no role in the index.

### 3.6.2 Branching bandit processes

The following problem formulation generalizes Klimov's setup considerably and recasts it in the form of bandit processes. We have a collection of bandit processes, as in Section 3.3.2. For simplicity, we assume that these processes are i.i.d., and that they move on a finite state space, with states $1, \ldots, K$. The state of the whole system can then be described at any time by $n_1, \ldots, n_K$ the number of arms in each of the states. At a decision moment an arm is chosen and made active, while all the other arms are passive. Say the active arm is in state $k$; then its activation (processing or any other interpretation) will last for a random duration $X_k$ with distribution $F_k$, at the end of which a reward $R_k$ will be earned (fixed or random), and the arm will be replaced by a new set of "descendant" arms, a random vector of arms, $N_{k1}, \ldots, N_{kK}$, so that $N_{kj}$ is the number of descendant arms that are type $j$. The passive arms remain frozen and earn no rewards. A discount factor $\beta$ is used to discount the rewards. It is desired to find a policy that will maximize the discounted sum of rewards. This problem was studied by Weiss (1988), who showed that a priority order of the states exists, and can be calculated recursively, and the priority policy based on that order is optimal. The descendants mechanism of branching bandit processes enables the modeling of arrival processes more general than Klimov's model, such as batch Poisson arrivals, as well as control of arrivals by the scheduler. It also allows a job to split into several jobs. With a discount factor $< 1$, the priority order of jobs is no longer independent of the arrival rates. If the discount rate is 1, one can add or eliminate Poisson streams of arrivals without changing the priority order.

### 3.6.3 Queueing networks with a single server

A standard formulation of a queueing network scheduling problem, which is used to model the flow of jobs in a factory, is as follows. The network consists of a set of nodes, which may be grouped in subsets of nodes; each subset forms a service station. There are several types of jobs that arrive from outside. Each type has a specified entry node, and it then follows type-specific Markov routing through the nodes until it leaves the network. The sojourn time at each node is also a type-specific random variable. One can now take each type–node combination and define it as a class. A job of class $k$ will be processed (at a specific node) for a random duration $X_k$ with mean $m_k$ and variance $\sigma_k^2$, and on completion of this time it will become a class-$j$ job with a probability $P_{kj}$ and leave the system with a probability $1 - \sum_j P_{kj}$. Type- or class-specific holding costs are denoted by $a_k$. In this form the problem looks very similar to the Klimov model. Indeed, if we make the (totally unrealistic) assumption that there is only a single operator in the whole factory and this operator moves about from node to node and can chose at any time which class to serve, while everything else is frozen, then we are back exactly to Klimov's model. In that case we know that there exists a priority order of the classes, and the single operator will always serve a job from the class with the highest priority present in the system. To relate the single-operator network to the conventional network, we endow the single operator with a service speed that is the combined speed of all the servers at all the nodes (as we did in Section 3.4.3 when we compared $M + 1$ parallel machines with a single machine with $(M + 1)$-fold speed). Unfortunately, it is not at all clear whether this single-operator network and its Klimov-model solution is in any sense an approximation to the multiple-node multiserver queueing network. It differs from the conventional network on two counts: first, service capacities at different nodes or service stations cannot be utilized anywhere else in the system; second, at a single node when several jobs are present it is not usually possible to concentrate all the service capacity at this node (say five identical machines) on a single high-priority job. Thus, to schedule the usual network model, at each node we need to activate several jobs—as many as we have machines. Furthermore, while we allocate machines at one specific node, the changes in state at that node are also affected by similar activities at the other nodes. In the next section we consider restless bandits, where we activate more than one arm (analogous to several machines at a node), and where the states of passive arms are not frozen (perhaps related to activities at other nodes).

## 3.7 RESTLESS BANDITS

### 3.7.1 Formulation of restless bandits

Recall the formulation of bandit processes in Section 3.3.2. Consider again $n$ processes whose states at time $t$, where $t = 0, 1, \ldots$, are given by $X_1(t), \ldots, X_n(t)$. At any time $t$ exactly $m$ of these processes, say $i_1(t), i_2(t), \ldots, i_m(t)$ are made active; all remaining $n - m$ processes remain passive. Active process $i$ undergoes state transition according to active Markov transition probabilities

$$P_i^{(A)}(x,y) = P^{(A)}(X_i(t+1) = y \mid X_i(t) = x)$$

and collects a reward $R_i^{(A)}(X_i(t))$. At the same time, the passive process $j$, instead of being frozen undergoes a state transition, according to the passive transition probabilities,

$$P_j^{(P)}(x,y) = P^{(P)}(X_j(t+1) = y \mid X_j(t) = x)$$

and collects a passive reward $R_j^{(P)}(X_j(t))$. The total reward at time $t$ is the sum of the rewards of the $m$ active and $n - m$ passive processes. The problem is to operate the system so as to maximize the long-term average expected reward per unit time. We denote this reward by $R_{OPT}(m,n)$. In what follows we shall for simplicity assume (as we did in Section 2.6.2) that all the processes are identically distributed, and that they move on a finite set of states, $1, \ldots, K$. We can then describe the system state by the number of processes in each of the states $1, \ldots, K$ as $n_1(t), \ldots, n_K(t)$, or by the vector $z_n(t) = z_{1n}(t), \ldots, z_{Kn}(t)$, where $z_{kn}(t) = n_k(t)/n$ is the fraction of the $n$ jobs that are in state $k$ at time $t$. Whittle (1988) introduced this model, and named it restless bandits because passive processes are not frozen. We gave some motivation for this type of model in the last section; Whittle lists some other applications. Unfortunately, the fact that $m$ processes are activated simultaneously, and a fortiori the restlessness of the passive processes, make this problem intractable. It is no longer true that a priority policy is optimal, the solution has to be sought in the full multidimensional state space, and there are no indications that the solution might have any "nice" features.

### 3.7.2 Whittle's relaxed problem and index policy

If one attempts to maximize the average reward from a single process, one obtains the optimality equation

$$g + h(k) = \max\left\{ R^{(A)}(k) + \sum_j P^{(A)}(k,j)h(j), R^{(P)}(k) + \sum_j P^{(P)}(k,j)h(j) \right\}.$$

In this equation $g$ is the long term average reward per unit time from the process; $h(k)$ is the so-called potential function, which for state $k$ is the relative transient reward of starting initially in that state. The solution to this optimality equation will induce an optimal partition of the states $1, \ldots, K$ into states where it is optimal to be active and into states where the optimal policy is to be passive. Whittle now introduces a "subsidy" for being passive, denoted by $v$, with the interpretation that when the process is made passive, the reward is augmented at a rate of $v$ per unit time. Including the subsidy, the optimality equation is now:

$$g(v) + h(k) = \max \left\{ R^{(A)}(k) + \sum_j P^{(A)}(k,j)h(j), v + R^{(P)}(k) + \sum_j P^{(P)}(k,j)h(j) \right\}$$
(3.4)

where the optimal long-term average reward rate per unit time, $g(v)$, as well as the potential function and the optimal partition of the states into passive and active states, are now functions of the subsidy $v$. Returning to the $n$ processes, Whittle introduces a relaxed problem: instead of having exactly $m$ active processes at all times, allow any number $m(t)$ of active processes at time $t$, but require that the long-term average of $m(t)$ be $m$. To solve the relaxed problem, the subsidy $v$ is treated like a Lagrange multiplier for the relaxed constraint, and the optimal reward per unit time for the relaxed problem is

$$R_{\text{REL}}(m, n) = \inf_v (ng(v) - v(n - m))$$
(3.5)

Let $v^*$ be the value that minimizes (3.5). When (3.4) is solved with $v^*$, a partition of the states into active states, passive states, and one state that is randomized, passive with probability $\theta$, and active with probability $1 - \theta$, is obtained. The relaxed policy operates each arm independently of all others, and makes it active or passive according to this partition. The subsidies can also be used to calculate an index for each state. For state $k$ define the index $v(k)$ as the smallest value of $v$ such that in solving (3.4) with that value, in state $k$ one is indifferent between active and passive. Note the similarity with Whittle's definition of the Gittins index in Section 3.3.2, and the analogy between "retirement reward" there and "subsidy" here. If the set of states for which it is optimal to retire is monotonically increasing in $v$ then the problem is called indexable, and the indifference values used as the indexes are unique. The index policy is at any time $t$ to activate the $m$ processes with the highest indices. Denote by $R_{\text{IND}}(m, n)$ its average reward per unit time. Clearly,

$$R_{\text{IND}}(m, n) \leq R_{\text{OPT}}(m, n) \leq R_{\text{REL}}(m, n)$$

### 3.7.3 Whittle's conjecture on asymptotic optimality

In his paper Whittle conjectures that for indexable problems, as $n \to \infty$, the rewards of the three policies converge to the same reward per unit time per arm, that is,

$$\lim_{n \to \infty} \frac{R_{\text{IND}}(m, n)}{n} = \lim_{n \to \infty} \frac{R_{\text{OPT}}(m, n)}{n} = \lim_{n \to \infty} \frac{R_{\text{REL}}(m, n)}{n}$$

This conjecture was investigated by Weber and Weiss (1990,1991) for i.i.d. processes on a finite state space, with the following results.

**Theorem 3.8 (Weber and Weiss 1990)** *Let $\alpha = m/n$ and $r(\alpha) = R_{\text{REL}}(m,n)/n$ (note that this is independent of n). Then*

$$\lim_{n \to \infty} \frac{R_{\text{OPT}}(m,n)}{n} = r(a).$$

The following is a sufficient condition for the second half of the conjecture.

**Theorem 3.9 (Weber and Weiss 1990)** *If the problem is indexable, and if the fluid approximation of the system under the index policy is globally asymptotically stable, then*

$$\lim_{n \to \infty} \frac{R_{\text{IND}}(m,n)}{n} = r(a)$$

The proof of Theorem 3.9 is based on studying the fluid approximations of the system under the relaxed policy and under the index policy. Under the sufficient conditions in the theorem, these two fluid approximations converge to the same point. Some large-deviation theorems are then used to show that the stochastic system stays close to this asymptotic point as $n$ becomes large, under both policies. Weber and Weiss (1990, 1991) show that stability always holds if there are three or fewer states, but construct a counterexample to the conjecture for four states. However, counterexamples of this kind seem very rare, and the asymptotic gap that they exhibit between the index and the relaxed policies is extremely small. In conclusion, it seems that the index policy combined with the upper bound of the relaxed policy provide an almost perfect heuristic for the restless bandit problem.

## 3.8 FLUID APPROXIMATIONS

Fluid approximations to queues, queueing networks or stochastic systems replace the stochastic systems by a deterministic trajectory that describes the motion of the mean of the stochastic system. Thus they agree with the stochastic system on first moments only. Fluid approximations are discussed in Kleinrock (1976) and Newell (1982). A. Weiss and Mitra (1988) have used them to analyze communication systems, and have derived large-deviation results. Chen and Mandelbaum (1987) have investigated convergence of queueing networks to their fluid approximations. Chen and Yao (1989) have investigated optimal control of fluid networks. The proof of Whittle's conjecture in the last section is based on fluid approximations. The following scheme seems natural: approximate a network scheduling problem by a fluid control problem, solve the latter, and use the solution to obtain a heuristic for the original problem. We conjecture that there is an intimate connection between fluid approximations and restless bandits, and that by exploiting it one may be able to investigate the optimality of the above scheme.

## 3.9 BROWNIAN APPROXIMATIONS

Brownian or diffusion approximations replace a queueing network, with its jump transitions, by a continuous diffusion process. The approximation holds under heavy traffic conditions, and agrees with the system on the two first moments, and is thus much finer than the fluid approximation (the latter is likened to the law of large numbers, the former to the more informative central limit theorem). Diffusion approximations to heavy traffic queueing systems are developed in Iglehart and Whitt (1970), Whitt (1974), Reiman (1984), Harrison (1985,1988), Harrison and Reiman (1981), Harrison and Williams (1987), Harrison and Nguyen (1990) and Dai and Harrison (1990). Diffusion approximations can be used to evaluate the performance of queueing, and can thus serve as an important tool for scheduling and control. Recently, Wein (1987,1990a,b,1992), Harrison and Wein (1989) and Laws and Louth (1990) have used diffusion approximations in a heuristic scheme to obtain scheduling and control rules for queueing networks. In this scheme the queueing network is approximated by a diffusion approximation, and the scheduling problem for the original network is replaced by a Brownian control problem. The Brownian control problem is solved—analytically for some special problems and numerically for more general problems. Finally, the solution to the Brownian control problem, which is not in itself a feasible policy for the original problem, is reinterpreted to produce a heuristic for the original problem. The method has been applied so far to several problems, and has produced heuristics that are very original and attractive. The heuristics are often a combination of priority rules and threshold rules. This is a rapidly developing area, and we cannot do it justice in this survey at this point in time.

## ACKNOWLEDGMENTS

This work was supported by NSF Grants ECS 8712798 and DDM 8914863, and by the fund for the promotion of research at Technion. A preliminary version of this chapter appeared in the supplement to the proceedings of the NSF Design and Manufacturing Systems Grantees Conference, sponsored by the National Science Foundation Mechanical and Aerospace Engineering Department and the Center for Professional Development of Arizona State University, 8–12 January, 1990, Tempe, Arizona.

## REFERENCES

Adams J., Balas E. and Zawack D. (1988) The shifting bottleneck procedure for job shop scheduling *Management Sci.*, **34**, 391–401.

Adler L., Fraiman N. and Pinedo M. (1988) An expert system for scheduling in a liquid packaging plant. In *Proceedings of the Third International Conference on Expert Systems and the Leading Edge in Production and Operations Management.*

Adler L., Fraiman N. M., Kobacker E., Pinedo M. L., Plotnicoff, J. C. and Wu T.-P. (1993) BPSS: a scheduling system for the packaging industry. *Oper. Res.,* **41**, 641–648.

Agrawala A. K., Coffman E. G., Garey M. R. and Tripathi S. K. (1984) A stochastic optimization algorithm minimizing expected flowtime on uniform processors *IEEE Trans. Comput.,* **33**, 351–357.

Baker K. R. (1974) *Introduction to Sequencing and Scheduling* Wiley, New York.
Barany I. (1981) A vector sum theorem and its application to improving flow shop guarantees *Math. Oper. Res.* **6**, 445–452.
Bartholdi J. J., Calvin J., Ramudhin A., Vande-Vate J. H. and Weiss, G. (1995) A probabilistic analysis of 2-machine flowshops. *Oper. Res.*, to be published.
Bellman R. (1956) A problem in the sequential design of experiments *Sankhia*, A **16**, 221–229.
Browne S. and Yechiali U. (1989) Dynamic priority rules for cyclic type queues. *Adv. Appl. Prob.* **21**, 432–450.
Bruno J. (1985) On scheduling tasks with exponential service times and in-tree precedence constraints *Acta Informatika*, **22**, 139–148.
Bruno J. and Hofri M. (1975) On the scheduling of chains of jobs on one processor with limited preemptions *SIAM J. Comput.* **4**, 478–490.
Bruno J., Downey P. and Frederickson, G. N. (1981) Sequencing tasks with exponential service times to minimize the expected flow time or makespan *J. Assoc. Comput. Mach.*, **28**, 100–113.
Chao X., Pinedo M. and Sigman K. (1989) On the interchangeability and stochastic ordering of exponential queues in tandem with blocking *Prob. Engng Inf. Sci.*, **3**, 223–236.
Chandy K. M. and Reynolds P. F. (1977) Scheduling partially ordered tasks with probabilistic execution times *Oper. Syst. Rev.*, **9**, 169–177.
Chen H. and Mandelbaum A. (1987) Discrete flow networks: bottleneck analysis and fluid approximations *Math. Oper. Res.*, **16**, 408–446.
Chen H. and Yao D. D. (1993) Dynamic scheduling of a multiclass fluid network *Oper. Res.*, **41**, 1104–1115.
Cobham A. (1954) Priority assignment in waiting line problems *Oper. Res.*, **2**, 70–76.
Coffman E. G. (1976) *Computer and Job Shop Scheduling Theory* Wiley, New York.
Coffman E. G. and Whitt W. (1995) Recent asymptotic results in the probabilistic analysis of schedule makespans. Chapter 2 of this volume.
Coffman E. G., Flatto L., Garey M. and Weber R. R. (1987) Minimizing expected makespan on uniform processors systems *Adv. Appl. Prob.* **19**, 177–201.
Coffman E. G. and Liu Z. (1992) On the optimal stochastic scheduling of out-forests *Oper. Res.*, **40** (Suppl.), S67–S75.
Coffman E. G. and Lueker G. S. (1991) *Probabilistic Analysis of Packaging and Partitioning Algorithms* Wiley, New York.
Conway R. W., Maxwell W. L. and Miller L. W. (1967) *Theory of Scheduling* Addison Wesley, Reading, MA.
Cox D. R. and Smith W. L. (1961) *Queues* Chapman & Hall, London.
Cunningham A. A. and Dutta S. K. (1973) Scheduling jobs with exponentially distributed processing times on two machines of a flowshop *Nav. Res. Log. Quart.*, **16**, 69–81.
Dai J. G. and Harrison J. M. (1991) Steady state analysis of RBM in a rectangle: numerical methods and a queueing application *Ann. Appl. Prob.*, **1**, 16–35.
Dai J. G. and Wang Y. (1993) Nonexistence of Brownian models of certain multiclass queueing networks *Queueing Theory Syst. Applic.*, **13**, 41–46.
Frenk J. B. G. and Rinnooy Kan A. H. G. (1987) The asymptotic optimality of the LPT rule *Math. Oper. Res.*, **14**, 241–254.
Friedman H. D. (1965) Reduction methods for tandem queueing systems *Oper. Res.*, **13**, 121–131.
Frostig E. (1988) A stochastic scheduling problem with intree precedence constraints *Oper. Res.*, **36**, 937–943.
Gershwin S. B. (1989) A hierarchical framework for discrete event scheduling in manufacturing systems *Proc. IEEE*, **77**, 195–209.

Gittins J. C. (1979) Bandit processes and dynamic allocations indices (with discussion) *J. R. Statist. Soc.*, B **41**, 148–177.
Gittins J. C. (1989) *Bandit Processes and Dynamic Allocation Indices* Wiley, New York.
Goldratt E. M. and Cox J. (1986) *The Goal—Excellence in Manufacturing (OPT)* North River Press, Croton on Hudson, NY.
Greenberg B. S. and Wolff R. W. (1988) Optimal order of servers for tandem queues in light traffic *Management Sci.*, **34**, 500–508.
Hajek B. (1983) The proof of a folk theorem on queueing delay with applications to routing in networks *J. Assoc. Comput. Mach.*, **30**, 834–851.
Hajek B. (1984) Optimal control of two interacting service stations *IEEE Trans. Autom. Control*, **29**, 491–499.
Hajek B. (1985) Extremal splittings of point processes *Math. Oper. Res.*, **10**, 543–556.
Harrison J. M. (1975) Dynamic scheduling of a multiclass queue: discount optimality *Oper. Res.*, **23**, 270–282.
Harrison J. M. (1985) *Brownian Motion and Stochastic Flow Systems* Wiley, New York.
Harrison J. M. (1988) Brownian models for queueing networks with heterogeneous customer populations. In Fleming W. and Lions P. L. editors, *Stochastic Differential Systems, Stochastic Control Theory and Applications*, pp 147–186 Springer, New York.
Harrison J. M. and Nguyen V. (1990) The QNET method for two moment analysis of open queueing networks *Queueing Syst. Theory Applic.*, **6**, 1–32.
Harrison J. M. and Reiman M. I. (1981) Reflected Brownian motion on an orthant *Ann. Prob.*, **9**, 302–308.
Harrison J. M. and Wein L. M. (1989) Scheduling of queues: heavy traffic analysis of a simple open network *Queueing Syst. Theory Applic.*, **5**, 265–280.
Harrison J. M. and Williams R. J. (1987) Brownian models of open queueing networks with homogeneous customer population *Stochastics*, **22**, 77–115.
Hu T. C. (1961) Parallel sequencing and assembly line problems *Oper. Res.*, **9**, 841–848.
Huang C. C. and Weiss G. (1990) On the optimal order of $M$ machines in tandem *Oper. Res. Lett.*, **9**, 299–303.
Iglehart D. L. and Whitt W. (1970) Multiple channel queues in heavy traffic, I and II *Adv. Appl. Prob.*, **2**, 150–177 and 355–364.
Kämpke T. (1987) On the optimality of static priority policies in stochastic scheduling on parallel machines *J. Appl. Prob.*, **24**, 430–447.
Kanet J. J. and Adelsberger H. H. (1991) The Leitstand, a new tool for computer integrated manufacturing *Prod. Invent. Management J., APICS*, **31**, 43–48.
Karmarkar N. and Karp R. (1982) The differencing method of set partitioning. Report UCB/CSD 82/113, Computer Science Division, EECS, University of California, Berkeley.
Karmarkar N., Karp R., Luecker G. S. and Odlyzko A. M. (1986) Probabilistic analysis of optimum partitioning *J. Appl. Prob.*, **23**, 626–645.
Kaspi H., Mandelbaum A. and Vanderbei R. J. (1995) Bandit processes: control, analysis and characterization. Chapter 17 of this volume.
Kawaguchi T. and Kyan S. (1986) Worst case bound of an LRF schedule for the mean weighted flowtime problem *SIAM J. Comput.*, **15**, 1119–1129.
Kelly F. P. (1979) *Reversibility and Stochastic Networks* Wiley, New York.
Kelly F. P. and Laws C. N. (1993) Dynamic routing in open queueing networks: Brownian models, cut constraints and resource pooling *Queueing Theory Syst. Applic.*, **13**, 47–86.
Kleinrock L. (1976) *Queueing Systems*, Vol. 2: *Computer Applications* Wiley, New York.
Klimov G. P. (1974) Time sharing service systems, *Prob. Theory Applic.*, **19** 532–551.
Kumar P. R. (1993) Re-entrant lines *Queueing Theory Syst. Applic.*, **13**, 87–110.
Lawler E. L., Lenstra J. K., Rinnooy Kan A. H. G. and Shmoys D. B. (1989) Sequencing and scheduling algorithms and complexity. Matematisch Centrum, Center for Mathematics and Computer Science Report BS-R8909, Amsterdam, The Netherlands.

Laws C. N. and Louth G. M. (1990) Dynamic scheduling of a four station queueing network *Prob. Engng Inf. Sci.*, **4**, 131–156.

Laws C. N. (1995) Dynamic routing and sequencing in open queueing networks. Chapter 16 of this volume.

Leitstand (1991) *AHP—Leitstand Users Manual* AH Havermann and Partner gmbh, Munich.

Leachman R. C. (1987) Preliminary design and development of a corporate level production planning system for the semiconductor industry. Technical Report ORC 86-11, Department of Industrial Engineering and Operations Research, University of California, Berkeley.

Lehtonen T. (1986) On the ordering of tandem queues with exponential servers *J. Appl. Prob.*, **23**, 115–129.

Lin W. and Kumar P. R. (1984) Optimal control of a queueing system with two heterogeneous servers *IEEE Trans. Autom. Control*, **29**, 696–703.

Lueker G. S. (1987) A note on the average case behavior of a simple differencing method for partitioning *Oper. Res. Lett.*, **6**, 285–287.

Matloff N. (1988) On the value of predictive information in a scheduling problem *Performance Eval.*, **10**, 309–315.

Meilijson I. and Weiss G. (1975) A simple solution to a time sharing problem. Technical Report, Department of Statistics, Tel Aviv University.

Meilijson I. and Weiss G. (1977) Multiple feedback at a single server station *Stochastic Proc. Applic.*, **5**, 195–205.

Mirchandani P. B. and Xu S. H. (1989) Optimal dispatching of multi-priority jobs to two heterogeneous workstations *J. Flexible Manuf. Syst.*, **2**, 25–42.

Monma C. L. and Rinnooy Kan A. H. G. (1983) A concise survey of efficiently solvable special cases of the permutation flow shop problem *RAIRO Rech. OpÇr.*, **17**, 105–119.

Muth E. J. (1979) The reversibility property of production lines *Management Sci.*, **25**, 152–158.

Muth E. J. (1984) Stochastic processes and their network representations associated with a production line queueing model *Eur. J. Oper. Res.*, **15**, 63–83.

Nash P. and Gittins J. (1977) A Hamiltonian approach to optimal stochastic resource allocation *Adv. Appl. Prob.*, **9**, 55–68.

Newell G. F. (1982) *Applications of Queueing Theory* Chapman & Hall, London.

OPT (1989) A case for MRPII, OPT, or JIT (manufacturing resource planning, optimized production technology, just in time) *Ind. Comput.*, P8(1).

Pinedo M. L. (1981) A note on the two machine job shop with exponential processing times *Nav. Res. Log. Quart.*, **28**, 693–696.

Pinedo M. L. (1982) Minimizing the expected makespan in stochastic flowshops *Oper. Res.*, **30**, 148–162.

Pinedo M. L. (1983) Stochastic scheduling with release dates and due dates *Oper. Res.*, **31**, 559–572.

Pinedo M. L. and Ross S. L. (1982) Minimizing expected makespan in stochastic open shops *Adv. Appl. Prob.*, **19**, 898–911.

Pinedo M. L. and Weiss G. (1979) Scheduling of stochastic tasks on two parallel processors *Nav. Res. Log. Quart.*, **26**, 527–535.

Pinedo M. L. and Weiss G. (1984) Scheduling jobs with exponentially distributed processing times on two machines with resource constraints *Management Sci.*, **30**, 883–889.

Pinedo M. L. and Weiss G. (1985) Scheduling jobs with exponentially distributed processing times and intree precedence constraints *Oper. Res.*, **33**, 1381–1388.

Pinedo M. L. and Wie S.-H. (1986) Inequalities for stochatic flow shops and job shops *Appl. Stochastic Models Data Anal.*, **2**, 61–69.

Pritsker Corporation (1989) sl FACTOR implementation guide, version 4.0. West Lafayette, IN.
Reiman M. I. (1984) Open queueing networks in heavy traffic *Math. Oper. Res.*, **9**, 441–458.
Righter R. (1988) Job scheduling to minimize expected weighted flowtime on uniform processors *Syst. Control Lett.*, **10**, 211–216.
Righter R. (1994) Applications of stochastic ordering ideas to scheduling. In Shaked M. and Shanthikumar J. G., editors, *Stochastic Orders and Their Applications*, Academic, New York.
Roundy R. O., Maxwell W. L., Herer Y. T., Tayur S. R. and Getzler A. (1988) A price directed approach to real time scheduling of production operations. Technical Report 823, School of Operations Research and Industrial Engineering, Cornell University.
Schrage L. (1968) A proof of the optimality of the shortest remaining processing time discipline *Oper. Res.*, **16**, 687–690.
Schrage L. and Miller L. W. (1966) The queue M/G/1 with the shortest remaining processing time discipline *Oper. Res.*, **14**, 670–685.
Serfozo R. (1981) Optimal control of random walks, birth and death processes, and queues *Adv. Appl. Prob.*, **13**, 61–83.
Sevcik K. C. (1974) Scheduling for minimum total loss using service time distributions *J. Assoc. Comput. Mach.*, **21**, 66–75.
Shaked M. and Shanthikumar J. G. (1987) Multivariate hazard rates and stochastic ordering. *Adv. Appl. Prob.*, **19**, 123–137.
Shanthikumar J. G. and Yao D. (1991) Bivariate characterization of some stochastic order relations *Adv. Appl. Prob.*, **23**, 642–659.
Shanthikumar J. G., Yamazaki G. and Sakasegawa H. (1991) Characterization of optimal order of servers in a tandem queues with blocking *Oper. Res. Lett.*, **10**, 17–22.
Stidham S. (1985) Optimal control of admission to a queueing system *IEEE Trans. Autom. Control*, **30**, 705–713.
Stidham S. and Weber R. R. (1989) Monotonic and insensitive optimal policies for the control of queues with undiscounted costs *Oper. Res.*, **37**, 611–625.
Stoyan D. (1983) *Comparison Methods for Queues and Other Stochastic Models* Wiley, New York.
Takagi H. (1986) *Analysis of Polling Systems* MIT Press, Cambridge, MA.
Tcha D. W. and Pliska S. R. (1977) Optimal control of single server queueing networks and multi class M/G/1 queues with feedback *Oper. Res.*, **25**, 248–258.
Tembe S. V. and Wolff R. W. (1974) The optimal order of service in tandem queues *Oper. Res.*, **24**, 824–832.
Van der Heyden L. (1981) Scheduling jobs with exponential processing and arrival times on identical processors so as to minimize the expected makespan *Math. Oper. Res.*, **6**, 305–312.
Varaiya P. P., Walrand J. and Buyukkoc C. (1985) Extensions of the multiarmed bandit problem: the discounted case *IEEE Trans. Autom. Control*, **30**, 426–439.
Weber R. R. (1979) The interchangeability of tandem $|M|$ 1 queues in series *J. Appl. Prob.*, **16**, 690–695.
Weber R. R. (1982) Scheduling jobs with stochastic processing requirements on parallel machines to minimize makespan or flowtime *J. Appl. Prob.*, **19**, 167–182.
Weber R. R. (1992a) On the optimality of the Gittins index *Ann. Appl. Prob.*, **2**, 1024–1033.
Weber R. R. (1992b) The interchangeability of tandem queues with heterogeneous customers and dependent service times *Adv. Appl. Prob.*, **24**, 727–737.
Weber R. R. (1995) Scheduling and interchangeability in tandem queues. Chapter 15 of this volume.
Weber R. R. and Weiss G. (1990) On an index policy for restless bandits *J. Appl. Prob.*, **27**, 637–648.

Weber R. R. and Weiss G. (1991) Addendum to on an index policy for restless bandits *Adv. Appl. Prob.*, **23**, 429–430.

Weber R. R. and Weiss G. (1994) The cafeteria process—tandem queues with 0–1 dependent service times and the bowl shape phenomena *Oper. Res.*, **42**, 895–912.

Weber R. R., Varaiya P. and Walrand J. (1986) Scheduling jobs with stochastically ordered processing times on parallel machines to minimize expected flowtime *J. Appl. Prob.*, **23**, 841–847.

Wein L. M. (1987) Asymptotically optimal scheduling of a two station multiclass queueing network. PhD Dissertation, Department of Operations Research, Stanford University.

Wein L. M. (1988) Ordering tandem queues in heavy traffic. Technical Report, Sloan School of Management, MIT.

Wein L. M. (1990a) Optimal control of a two station Brownian network *Math. Oper. Res.*, **15**, 215–242.

Wein L. M. (1990b) Scheduling networks of queues, heavy traffic analysis of a two station network with controllable inputs *Oper. Res.*, **38**, 1065–1078.

Wein L. M. (1992) Scheduling networks of queues: heavy traffic analysis of a multistation network with controllable inputs, *Oper. Res.*, **40**, S312–S334.

Weiss A. and Mitra D. (1988) A transient analysis of a data network with a processor sharing switch. Bell Laboratories Technical Report.

Weiss G. (1984) Scheduling spares with exponential lifetimes in a two component parallel system *Nav. Res. Log. Quart.*, **31**, 431–446.

Weiss G. (1988) Branching bandit processes *Prob. Engng Inf. Sci.*, **2** 269–278.

Weiss G. (1990) Approximation results in parallel machines stochastic scheduling *Ann. Oper. Res.*, **26**, 195–242.

Weiss G. (1992) Turnpike optimality of Smith's rule in parallel machines stochastic scheduling *Math. Oper. Res.*, **17**, 255–270.

Weiss G. (1995) On almost optimal priority rules for preemptive scheduling of stochastic jobs on parallel machines *Adv. Appl. Prob.*, to appear.

Weiss G. and Pinedo M. L. (1980) Scheduling tasks with exponential service times on non identical machines to minimize various cost functions *J. Appl. Prob.* **17**, 187–202.

Whitt W. (1974) Heavy traffic theorems for queues, a survey. In Clark, A. B., editor, *Mathematical Methods in Queueing Theory*, Springer, Berlin.

Whitt W. (1985) The best order for queues in series *Management Sci.*, **31**, 475–487.

Whittle P. (1980) Multi armed bandits and the Gittins index *J. R. Statist. Soc.*, B **42**, 143–149.

Whittle P. (1988) Restless bandits: activity allocation is a changing world *J. Appl. Prob. (Special Volume)*, **25**A, 287–298.

Wie S. H. and Pinedo M. L. (1986) On minimization of the expected makespan and flowtime in stochastic flowshops with blocking, *Math. Oper. Res.* **11**, 336–342.

Xu S. H., Mirchandani P. B., Srikantakumar P. S. and Weber R. R. (1990) Stochastic dispatching of multi priority jobs to heterogeneous processors *J. Appl. Prob.*, **27**, 852–862.

Yamazaki G. and Sakasegawa H. (1975) Properties of duality in tandem queueing systems *Ann. Inst. Statist. Math.*, **27**, 201–212.

Yamazaki G., Sakasegawa H. and Shanthikumar J. G. (1989) On optimal arrangement of stations in tandem queueing systems with blocking. Technical Report, University of California, Berkeley.

---

*School of ISyE, Georgia Tech, Atlanta, GA 30332, USA*

*Department of Statistics, University of Haifa, Haifa 31905, Israel*

CHAPTER 4

# Scheduling with Communication Delays: A Survey

P. Chrétienne and C. Picouleau
*Institut Blaise Pascal, Laboratoire LITP, Université Pierre et Marie Curie*

**Abstract**

The study of scheduling problems with interprocessor communication delays is a new research domain of growing interest resulting from the development of distributed memory computers. Finding good solutions to these problems is a necessary condition to get good performance from such computers. We present the underlying scheduling problems and give an overview of their complexity and solutions. These problems are classified according to three main characteristics: task duplication, an unlimited number of processors and communication times. For each class, the present knowledge of the borderline between easy and difficult problems is first described, specific properties of the optimal schedules are given, and the most significant exact or approximated solution algorithms are presented.

## 4.1 INTRODUCTION

The study of scheduling problems with interprocessor communication delays is a new research area interest in which is rapidly increasing with the development of distributed memory computers. Indeed, to get the best performance from such computers, one needs to search for the best compromise between parallelism and communication delays. For a given task graph to be executed, this problem is a scheduling problem where, in addition to the constraints of a classical scheduling problem, the communication times between dependent tasks assigned to distinct processors must be taken into account.

This chapter presents the main results obtained so far in this field. Task duplication, an unlimited number of processors, and communication times are the main characteristics chosen to classify these problems. For each class, we first

---

*Scheduling Theory and its Applications* Edited by P. Chrétienne, E. G. Coffman, Jr., J. K. Lenstra and Z. Liu
© 1995 John Wiley & Sons Ltd

describe the present knowledge of the borderline between easy and difficult problems. We give specific properties of the optimal schedules, and finally we present the most significant exact or approximation algorithms.

## 4.2 PROBLEM DEFINITION AND NOTATION

An instance of a scheduling problem is specified by a set $J = \{J_1, \ldots, J_n\}$ of $n$ nonpreemptive tasks, a set $U$ of $q$ precedence constraints $(J_i, J_j)$ such that $G = (J, U)$ is an acyclic directed graph (called the precedence graph), the processing times $p_j$, $J_j \in J$, and the communication times $c_{jk}$, $(J_j, J_k) \in U$. A set $M = \{M_1, \ldots, M_m\}$ of $m$ identical processors is available to process the tasks.

A *copy* of $J_i$ is a triple $(i, \pi, t)$, where $J_i$ is the task from which the copy is issued, $\pi$ is the processor assigned to the copy and $t$ is the time at which the copy is scheduled.

An arc $(J_j, J_k)$ means that *any* copy of $J_k$ must wait for data from a copy of $J_j$ (that copy being called its *supplier*) before being processed. The duration of this data transfer is $c_{jk}$ if the two copies are not assigned the same processor and zero otherwise.

A schedule $S$ of the problem is a *finite* set of *copies* that must satisfy the following conditions:

(1) for each task $J_j$, there is at least one copy $(j, \pi, t)$;
(2) at any time, each processor executes at most one copy;
(3) if $(J_j, J_k) \in U$ then each copy $(k, \pi, t)$ has a supplier $(j, \pi', t')$ such that $t \geq t' + p_j$ if $\pi = \pi'$ and $t \geq t' + p_j + c_{jk}$ otherwise.

So, in a schedule, at least one copy of each task has to be processed; each copy of a given task has to be supplied by one copy of each of its immediate predecessors in $G$; if one copy and its supplier are not assigned the same processor, the communication delay must be taken into account. Figure 4.1 illustrates how task duplication may yield a smaller makespan.

The objective is to determine a schedule whose makespan, defined as the largest completion time of a copy, is minimum. Following the notation of Veltman *et al.* (1990), this problem is denoted by $P|prec, p_j, c_{jk}, dup|C_{\max}$.

When duplication is not allowed, each task must be processed only once. So a schedule is entirely defined by assigning to each task $J_j$ a starting time $t_j$ and a processor $\pi_j$ such that the processor constraint is satisfied and for any $(J_j, J_k) \in U$, $t_k \geq t_j + p_j$ if $\pi_j = \pi_k$ and $t_k \geq t_j + p_j + c_{jk}$ otherwise. This problem is denoted by $P|prec, p_j, c_{jk}|C_{\max}$.

When an unlimited number of processors is assumed (i.e. $m \geq n$), the first field of the problem notation is denoted by $P\infty$. Note that in this case we only need to consider connected precedence graphs, since solving the restriction of the scheduling problem to each connected component and processing each component on disjoint processor subsets obviously yield an optimal schedule.

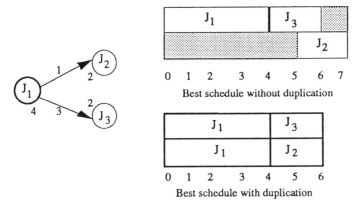

**Figure 4.1** Task duplication.

The set of the immediate predecessors (respectively successors) of $J_j$ in $G$ is denoted by $IN(j)$ (respectively $OUT(j)$). The largest and smallest processing times are denoted respectively by $p_{max}$ and $p_{min}$, and the notation $c_{max}$ and $c_{min}$ is used in the same way for communication times. The SCT (small communication time) assumption means that $c_{max}/p_{min} \leq 1$. The $r$-SCT ($0 < r \leq 1$) assumption is the special case of SCT when $c_{max}/p_{min} \leq r$.

## 4.3 DUPLICATION AND AN UNLIMITED NUMBER OF PROCESSORS

### 4.3.1 The SCT $P\infty|prec, p_j, c_{jk}, dup|C_{max}$ problem

Some important application programs (e.g. those issued from numerical analysis problems) involve tasks with a large granularity, i.e. tasks whose execution times are greater than communication times. This section presents a polynomial algorithm developed by Colin and Chrétienne (1991) providing an optimal schedule in the special case of $P\infty|prec, p_j, c_{jk}, dup|C_{max}$ when the SCT assumption is satisfied. The complete proof may be found in that paper. An instance $P_0$ of this problem is shown in Figure 4.2 and will serve to illustrate how the algorithm works.

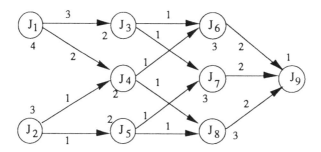

**Figure 4.2** Problem $P_0$.

| $J_i$ | 1 | 2 | 3 | 4 | 5 | 6 | 7 | 8 | 9 |
|---|---|---|---|---|---|---|---|---|---|
| $b_i$ | 0 | 0 | 4 | 4 | 3 | 7 | 6 | 6 | 11 |

**Figure 4.3** The release times of $P_0$.

In its first step, the algorithm computes a release time $b_i, J_i \in J$, of the execution time of the copies of $J_i$. The second step determines a *critical* spanning outforest of the precedence graph to build up a schedule in which any copy of a task is scheduled at its release time.

The first step uses a topological order of the tasks to compute the release times $b_i$ as follows :

$$b_i = \begin{cases} 0 & \text{if IN}(i) = \emptyset \\ b_s + p_s & \text{if IN}(i) = \{s\} \\ \max\{b_s + p_s, \max_{J_k \in IN(i) - \{s\}}\{b_k + p_k + c_{ki}\}\} & \text{otherwise} \end{cases} \quad (4.1)$$

where $J_s$ satisfies $b_s + p_s + c_{si} = \max_{J_k \in IN(i)} \{b_k + p_k + c_{ki}\}$. Note that the value of $b_i$ does not depend on which task $J_s$ satisfies the preceding equality.

The release times for problem $P_0$ are shown in Figure 4.3.

An arc $(J_i, J_j)$ of the precedence graph is said to be *critical* (with respect to the release times $b_i$) if $b_i + p_i + c_{ij} > b_j$. Clearly, in any schedule whose copies are scheduled at their release times a copy and its supplier must be assigned the same processor if the corresponding arc is critical.

By removing all the noncritical arcs from $G$, we get the so-called *critical subgraph* of $G$. Figure 4.4 reports the critical subgraph for the scheduling problem $P_0$.

Not every task $J_i$ has a critical ingoing arc. But if $J_i$ has one, there is *only one $J_s$* that satisfies $b_s + p_s + c_{si} = \max_{J_k \in IN(i)} \{b_k + p_k + c_{ki}\}$ and this critical arc is the arc $(J_s, J_i)$. So the critical subgraph is a spanning outforest.

The optimal (and earliest) schedule is finally built by assigning one processor to each critical path (i.e. a path from a root to a leaf in the critical subgraph) and by processing all the corresponding copies at their release times. Note that at most $n$ processors are used by this schedule. The optimal schedule for problem $P_0$ is shown in Figure 4.5.

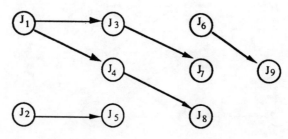

**Figure 4.4** The critical subgraph of $P_0$.

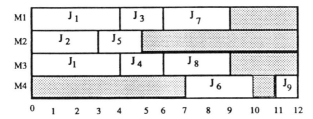

**Figure 4.5** The earliest schedule of $P_0$.

The algorithm provides an earliest schedule, but, as shown by Figure 4.5, it does not necessarily minimize the number of processors. The example shows that tasks $J_6$ and $J_9$ could just as well have been assigned to processor $M_2$. However, it has been shown by Picouleau (1992) that minimizing the number of processors is an $\mathcal{NP}$-hard problem when the minimum makespan must be guaranteed.

Note finally that the above algorithm also works if for any $J_j$ the largest communication time of an ingoing arc of $J_j$ is at most the smallest processing time of a task in $IN(j)$—a weaker assumption than SCT.

### 4.3.2 The $P\infty|prec, dup|C_{max}$ problem

We now consider the case when the communication times may be larger than the processing times. Papadimitriou and Yannakakis (1990) have shown that the special case $P\infty|prec, p_j = 1, c > 1, dup|C_{max}$ is $\mathcal{NP}$-hard and have proposed a sophisticated polynomial approximation algorithm with performance ratio two for the general case. The complexity proof as well as the approximation algorithm are worth being explained to get a good insight of the use of task duplication.

#### 4.3.2.1 The $P\infty|prec, p_j = 1, c > 1, dup|C_{max}$ problem

*Complexity*

Note first that the problem is in $\mathcal{NP}$ since it is easy to show that there is an optimal schedule with makespan at most $n$ that needs at most $n$ processors and at most $n^2$ copies. The polynomial reduction, which is illustrated by Figure 4.6, is from the problem CLIQUE (see Garey and Johnson 1979).

Let $(G, k)$ be an instance of CLIQUE, where $G = (V, E)$ is an undirected graph (with $p$ vertices and $q$ edges) and $k$ is a natural number $(q \geq \binom{k}{2})$. The corresponding instance of the scheduling problem is illustrated in Figure 4.6. With each vertex $v$ of $G$ is associated a path $P_v$ of $p^2$ tasks. With each edge $e = \{u, v\}$ of $G$ is associated an "edge-task", which is a direct successor of the last tasks of $P_u$ and $P_v$. Every edge-task is a direct predecessor of a terminal task $J_t$. The communication time $c$ is set to $(k-1)p^2 + \binom{k}{2}$. The question is to decide whether

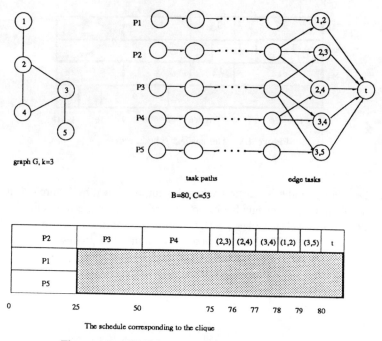

**Figure 4.6** CLIQUE $\prec P_\infty, p_j = 1, c > 1, dup_{\max}$.

the terminal task may be scheduled by time $B = kp^2 + q$ or less. The auxiliary value $\theta = kp^2 + \binom{k}{2}$ will be used in the proof.

Assume that $S$ is a schedule matching the deadline $B$. Let $\gamma_0$ be the copy (only one is needed) of the terminal task and let $\pi$ be the processor assigned to $\gamma_0$. The suppliers of $\gamma_0$ (one for each edge-task) are then separated into two disjoint subsets $T_1$ and $T_2$: those completed by time $\theta$ define $T_1$ while the others define $T_2$. From the definition of $c$ and $\theta$, each copy of $T_2$ is processed by $\pi$ and no more than $q - \binom{k}{2}$ copies are in $T_2$ since $\gamma_0$ must meet its deadline $B = \theta + q - \binom{k}{2}$. Consider now one copy $\gamma_1$ of $T_1$ and let $J_{(u,v)}$ be its corresponding edge-task. The tasks in $P_u$ and $P_v$ make $2p^2$ "path" ancestors of $\gamma_1$. One copy of each of these ancestors must be processed by the same processor as $\gamma_1$ since, by the definition of $T_1$, the copy $\gamma_1$ is completed by time $\theta$. This in turn implies that $\gamma_1$ is also processed by $\pi$. It comes finally that processor $\pi$ executes $\gamma_0$, all the copies in $T_2$ and each copy in $T_1$ together with all its "path" ancestors (Figure 4.6). Let $E$ be the set of endpoints of the edges associated with the copies of $T_1$ and denote by $e$ the number of vertices in $E$. Since $\gamma_0$ is scheduled by time $B$, we have $ep^2 + q \leq kp^2 + q$, which implies $e \leq k$. Since there are at least $\binom{k}{2}$ tasks in $T_1$, we have $e = k$. So the subgraph induced by $E$ is a clique of $G$.

Conversely, assume that the graph $G$ has a clique with $k$ vertices. The schedule shown in Figure 4.6 is feasible and meets the deadline $B$.

*Bounding the number of copies*

Using a dynamic programming approach, an exact algorithm has been developed by Jung et al. (1993) that provides an optimal schedule for the $P\infty|prec, p_j = 1, c > 0, dup|C_{\max}$ problem in $O(n^{c+1})$ time. Moreover, it is shown that an optimal schedule of an arbitrary task graph needs at least $O(\log n)$ additional copies, which answers the question on the minimum number of copies needed to guarantee optimality asked by Papadimitriou and Yannakakis (1990).

#### 4.3.2.2 An approximation algorithm

An approximation algorithm with performance ratio two has been developed in Papadimitriou and Yannakakis (1990) for $P\infty|prec, dup|C_{\max}$. In addition to the fact that approximation algorithms with performance guarantee are still rare for scheduling problems with communication delays, this algorithm is quite interesting by the way task duplication is exploited and release times are computed. We thus explain it in this section.

Like the exact algorithm solving the problem $P\infty|prec, c_{jk}, dup|C_{\max}$ under the SCT assumption (Section 4.3.1), the approximation algorithm first computes a *release time* $r_i$ for each copy of $J_i$. Then it builds up a schedule such that each task $J_i$ has at least one copy scheduled by $2r_i$. Note that a task will be said to be *scheduled by time t* if at least one of its copies is.

*Release times*

The argument yielding the release time computation is the following: if there is a schedule such that one copy $\gamma$ of $J_k$ is scheduled by time $t$ then, for each task in a subset of *t-critical ancestors* of $J_k$, one copy must be assigned the *same processor* as $\gamma$. Now if time $t$ is the smallest time such that $\gamma$ and the copies of its critical ancestors may be scheduled on a single processor while respecting their release times and precedence constraints then $t$ is a release time for the copies of $J_k$.

The algorithm uses a topological order of the tasks to determine release times. If $IN(k) = \emptyset$ then $r_k$ is set to 0. Otherwise, let $G_k$ be the subgraph of $G$ induced by the ancestors of $J_k$ and assume that, except for $J_k$, the release time of any ancestor has already been determined. An arc $(J_i, J_j)$ of $G_k$ is said to be *t-critical* (i.e. with respect to a given time $t$) if $r_i + p_i + c_{ij} > t$. A task $J_i$ of $G_k$ is said to be *t-critical* if there is at least one path from $J_i$ to $J_k$ in $G_k$ *all of whose arcs* are $t$-critical. Clearly, if a copy $\gamma$ of $J_k$ is scheduled by time $t$ then one copy of each $t$-critical task must be scheduled on the same processor as $\gamma$. As, for a given $t$, deciding whether such a schedule exists is the polynomial $1|prec, r_j|C_{\max}$ problem, the smallest feasible $t$ may be computed in polynomial time using binary search. The smallest value of $t$ defines $r_k$.

On the example shown in Figure 4.7, assume that we already know that $r_1 = r_2 = r_3 = 0$, $r_4 = 3$ and $r_5 = 4$. $G_6$ is the subgraph induced by $\{J_1, \ldots, J_6\}$. It is easy to see that $8 \leqslant r_6 \leqslant 11$. Choosing 9 as a candidate value for $r_6$ yields $J_5$ as the single critical task and a feasible one-machine schedule of $\{J_5, J_6\}$. Testing now

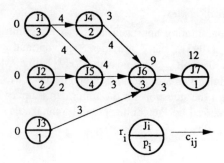

**Figure 4.7** The release times.

for $r_6 = 8$, we get the two critical tasks $J_4$ and $J_5$ and no feasible one-machine schedule of $\{J_4, J_5, J_6\}$. So $r_6$ is set to 9.

*The approximation algorithm*

Once each task has been assigned a release time, the algorithm computes a schedule whose copies are processed *not later than twice their release times*. Let $L$ be a topological list of the tasks and assume we are given a schedule $S'$ that meets the required property for the restriction of the problem to the tasks standing before $J_k$ in $L$.

To get a schedule $S$ with the required property for all the tasks in $L$ up to $J_k$, the algorithm uses $S'$ as a partial schedule and assigns a new (i.e. not busy in $S'$) processsor $\pi$ to a copy $\gamma$ of $J_k$ and to a new copy of each $r_k$-critical task. By the definition of $r_k$, we know that there is a one-machine schedule of task $J_k$ and of each $r_k$-critical task such that their precedence constraints are satisfied, task $J_k$ is scheduled at time $t \leqslant r_k$ and each $r_k$-critical task $J_i$ is scheduled at time $t_i \geqslant r_i$.

We transform $S'$ into $S$ by scheduling the copy $c$ of $J_k$ at time $t + r_k$ and each $r_k$-critical task $J_i$ at time $t_i + r_k$. To prove that $S$ is feasible, we have only to show that each copy executed by $\pi$ may be assigned suppliers. Consider the copy $d$ of the $r_k$-critical task $J_j$ (or the copy $c$ itself) and assume that $(J_i, J_j) \in G$. If $(J_i, J_j)$ is $r_k$-critical then so is $J_i$, and the supplier of $d$ is defined as the copy of $J_i$ executed by $\pi$. Assume now that $(J_i, J_j)$ is not $r_k$-critical; then the supplier of $d$ is defined as the copy of $J_i$ executed in $S'$ at $\theta \leqslant 2r_i$. As $(J_i, J_j)$ is not $r_k$-critical, we have $r_i + p_i + c_{ij} \leqslant r_k$. So $\theta + p_i + c_{ij}$ is at most $r_i + r_k$, which, in turn, is less than $r_j + r_k$, which, in turn, is less than $t_j + r_k$, the time at which $d$ is scheduled on $\pi$.

**Remark** It can be proved that, when applied to the problem $\bar{P}\infty|prec, p_j, c_{jk}, dup|C_{\max}$ under the SCT assumption, this approximation algorithm and the exact algorithm presented in Section 4.3.1 provide the same (optimal) schedule.

## 4.4 NO DUPLICATION AND AN UNLIMITED NUMBER OF PROCESSORS

The $P\infty|prec|C_{\max}$ problem is now considered. In Section 4.4.1 it is shown that very simple precedence graphs lead to $\mathcal{NP}$-complete problems. In Section 4.4.2 it is shown in contrast that the SCT $P\infty|prec, p_j, c_{jk}|C_{\max}$ is polynomial when $G$ is an in-tree, an out-tree, a series–parallel or a bipartite graph, but that it is $\mathcal{NP}$-complete for an arbitrary precedence graph. Three approximation algorithms with performance guarantees, applying to in-trees or out-trees, SCT instances and the general case respectively are described in Section 4.4.3.

### 4.4.1 $\mathcal{NP}$-hard problems

Let a *SEND–RECEIVE* graph be defined as shown in Figure 4.8, where a root task $J_0$ sends data to each of $n$ independent tasks $J_1, \ldots, J_n$, which in turn send data to a sink task $J_{n+1}$.

Using the reduction of PARTITION (see Garey and Johnson 1979) illustrated by Figure 4.8, the SEND–RECEIVE problem has been shown by Chrétienne (1992) to be $\mathcal{NP}$-complete. The proof is based on the dominance of the schedules whose structure is shown in Figure 4.8.

When the precedence graph is an out-tree, Chrétienne (1994) has shown that the problem is $\mathcal{NP}$-complete, even in the very specific case of a HARPOON problem.

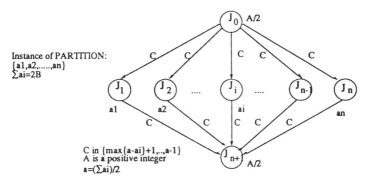

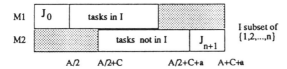

**Figure 4.8** PARTITION $\prec$ SEND–RECEIVE.

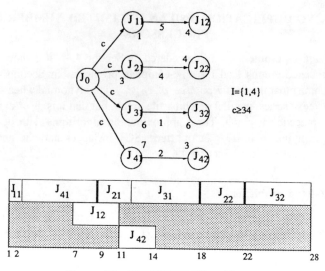

**Figure 4.9** The HARPOON problem.

The HARPOON problem consists of one root task $J_0$ sending data to the head tasks of $n$ independent arrows $(J_{i_1}, J_{i_2}), i \in \{1, \ldots, n\}$ (Figure 4.9). For simplicity, the head-task processing time, the tail-task processing time and the communication time of the arrow $i$ are denoted respectively by $a_i$, $b_i$ and $c_i$. It is further assumed that for any arrow $i$ $c_{0i_1} = c \geqslant \sum_{i \in \{1,\ldots,n\}} (a_i + b_i)$, so that no gain may result from processing the head tasks in parallel. We may also assume without loss of generality that $b_1 + c_1 \geqslant b_2 + c_2 \geqslant \cdots \geqslant b_n + c_n$.

The polynomial transformation is from the KNAPSACK problem (see Garey and Johnson 1979) and the basic feature of the proof is that there is an optimal schedule such that (Figure 4.9):

(a) the processor assigned to the root executes without delay the root, a subset $I$ of the head tasks in decreasing $b_i + c_i$ order, the head tasks of $\bar{I}$ (in any order) and finally the tail tasks of $\bar{I}$,
(b) the tail tasks of $I$ are processed as soon as possible by distinct processors.

Finding a subset $I$ that yields a makespan at most $B$ is shown to be polynomially equivalent to the solution of the following linear system, to which the KNAPSACK problem is, in turn, polynomially reduced.

$$x_i \in \{0, 1\} \quad i \in \{1, \ldots, n\},$$

$$\sum_{i=1}^{n} [a_i + (1 - x_i) b_i] \leqslant B - p_0,$$

$$\sum_{j=1}^{k} [a_j x_j + (b_k + c_k) x_k] \leqslant B - p_0 \quad k \in \{1, \ldots, n\}$$

Let us now consider problems with restricted processing and communication times. Using a reduction of 3-SAT, Picouleau (1995) has shown that $P\infty|prec, p_j = 1, c = 1|C_{\max}$ is $\mathcal{NP}$-complete. By a similar polynomial transformation also provided by Picouleau (1995), it is proved that the $r$-SCT special case of $P\infty|prec, p_j, c_{jk}|C_{\max}$ is $\mathcal{NP}$-complete.

Since the $P\infty|prec, p_j = 1, c = 1|C_{\max}$ problem is $\mathcal{NP}$-complete, it is quite important for the purpose of approximation to know the complexity of the decision problem of whether there exists a schedule with a fixed length $B$. Hoogeveen et al. (1992) have shown that the problem is polynomial for $B = 5$ and $\mathcal{NP}$-complete for $B = 6$ (by a reduction of 3-SAT). So, unless $P = NP$, there is no approximation algorithm with a worst case ratio smaller than $\frac{7}{6}$.

### 4.4.2 Polynomial problems

From the complexity results mentioned in the preceding section, it appears that, despite the fact that the number of processors is unlimited, most of the special cases are likely to be $\mathcal{NP}$-complete. We show in this section which restrictions are known to make the problem easy.

#### 4.4.2.1 The SEND problem

The SEND problem is defined in Figure 4.10, and it is further assumed without any loss of generality that

$$p_1 + c_{01} \geqslant p_2 + c_{02} \geqslant \cdots \geqslant p_n + c_{0n}.$$

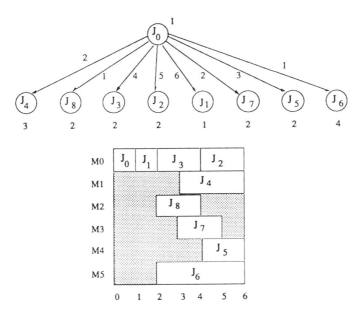

**Figure 4.10** The send problem.

Apart from the processor to which the root is assigned, any processor need not process more than one task, which, in turn, may be scheduled as soon as the message from the root has been transferred. Moreover, if $J_i$ is not assigned the same processor as the root then none of the tasks in $\{J_{i+1}, \ldots, J_n\}$ need be processed by that processor. By this simple argument, we know that in an optimal schedule the processor assigned to the root executes without any delay one of the $n$ task subsets $\{J_1, \ldots, J_i\}$, and that the remaining tasks are each processed by distinct processors. Thus the algorithm has only to find the task $J_i$ that minimizes $\max\{\sum_{k=1}^{i} p_k, p_{i+1} + c_{0,i+1}\}$. As this algorithm will be used in Section 4.4.3.1, it will be convenient to name it the **SD** *algorithm*.

Note that if the tasks of $J$ are further constrained by ready times, a similar polynomial algorithm can be used (see Picouleau 1992).

### 4.4.2.2 The SCT $P\infty|prec, c_{c_{jk}, p_j}|C_{\max}$ problem

This section shows that the SCT assumption leads to easy scheduling problems when the precedence graph is sufficiently structured. Two efficient dominance properties are quite useful to solve these problems.

**Proposition 4.1** *There is an optimal schedule that assigns independent tasks to distinct processors.*

**Proof** Let $S$ be a schedule such that one processor $\pi$ executes two independent tasks $J_i$ and $J_j$. Assume that $J_k$ is processed between $J_i$ and $J_j$ by $\pi$. Then either $\{J_i, J_k\}$ or $\{J_k, J_j\}$ is a pair of independent tasks. So we may assume $J_j$ is the next task executed by $\pi$ after task $J_i$. Let $A$ (respectively $B$) be the subset of the tasks processed by $\pi$ up to $J_i$ (respectively after $J_j$). By assigning to a free processor the tasks in $A$ and scheduling them at the same time as in $S$, we get a schedule whose makespan is the makespan of $S$ since SCT is satisfied and at least $p_i$ time units separate the completion of any task in $A - \{J_i\}$ from the starting time of any task in $B$. The transformation is then repeated until the condition is met. □

The second property concerns the directed bridges of $G$, where a bridge is an arc that is the only outgoing arc of its origin node and the only ingoing arc of its ending node.

**Proposition 4.2** *There is an optimal schedule that processes without any delay the two tasks of any directed bridge.*

**Proof** Let $S$ be a schedule such that the two tasks $J_i$ and $J_j$ of a directed bridge are assigned two distinct processors $\pi_i$ and $\pi_j$. Using a free processor to schedule at their scheduling times in $S$ the first tasks scheduled on $\pi_i$ (up to $J_i$) followed by the last tasks scheduled on $\pi_j$ (after task $J_j$), we get a new feasible schedule, since, from SCT, $p_i$ and $p_j$ are at least the largest communication time. It now only remains to make $J_j$ follow $J_i$ without delay. The transformation is then repeated until the condition is met. □

### 4.4.2.3 The SCT rooted tree

Chrétienne (1989) has developed an $O(n)$ algorithm solving the special case when $G$ is an in-tree or an out-tree. Let us consider an intree with root $J_r$ and denote by $A_i$ the subtree of $G$ whose root is the task $J_i$ of $IN(r)$. From Proposition 4.1, we know that there is an optimal schedule such that distinct subtrees are processed by disjoint processor subsets (Figure 4.11). The algorithm then proceeds in a recursive way as follows: assuming that an optimal partial schedule is known for each subtree $A_i$, it executes these partial schedules from time 0 on disjoint processor subsets and then processes the root as soon as possible on the processor already assigned to the root of the subtree $A_i$ such that a minimum makespan is obtained.

By reversing the arc orientation and reading the timing diagram from right to left, the preceding algorithm provides an optimal schedule for an out-tree.

### 4.4.2.4 The SCT series–parallel graph

Series–parallel graphs are recursively defined as follows: the atom graph is a single node, which is both the input and the output node of the atom graph; if $G_1, \ldots, G_q$

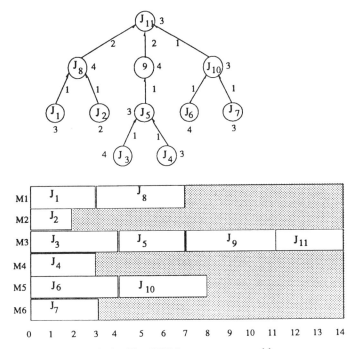

**Figure 4.11** The SCT $P\infty, p_j, c_{j,k}$,max problem.

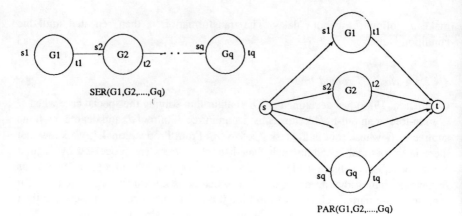

**Figure 4.12** Series–parallel graphs.

are series–parallel graphs, then the graphs $SER(G_1, \ldots, G_q)$ and $PAR(G_1, \ldots, G_q)$ shown in Figure 4.12 are series–parallel graphs.

Let $G = PAR(G_1, \ldots, G_q)$ be a series–parallel graph with $J_s$ and $J_t$ as input and output nodes, and let $C^*_1, \ldots, C^*_q$ denote the makespans of the optimal schedules of the series–parallel graphs $G_1, \ldots, G_q$. From Proposition 4.1, we know there is an optimal schedule of $G$ such that the subgraphs $G_1, \ldots, G_q$ are processed by disjoint processor subsets. If $J_s$ is assigned the same processor as $J_{s_i}$ (the input node of $G_i$) and $J_t$ is assigned the same processor as $J_{t_j}$ (the ouput node of $G_j$), then the associated makespan $M_{ij}$ is given by (Figure 4.13):

$$M_{ij} = p_s + \max_{k \in \{1,\ldots,q\}} \{c_{ss_k}(1 - \delta_{s_k s_i}) + C^*_k + c_{t_k t}(1 - \delta_{t_k t_j})\} + p_t$$

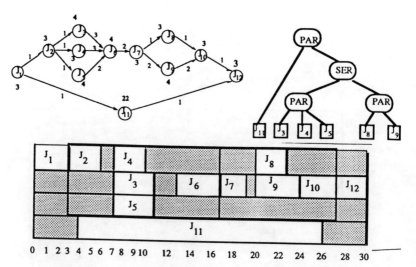

**Figure 4.13** An optimal schedule of a series–parallel graph.

where $\delta_{ij} = 1$ if $i = j$ and $\delta_{ij} = 0$ otherwise. The optimal schedule is thus obtained by computing the minimum $M_{ij}$ value.

Let $G = SER(G_1, \ldots, G_q)$ be a series–parallel graph with $J_s$ and $J_t$ as input and output nodes and let $C^*_1, \ldots, C^*_q$ denote the makespans of the optimal schedules of the series–parallel graphs $G_1, \ldots, G_q$. It is clear from the structure of $SER(G_1, \ldots, G_q)$ that there is no schedule of $G$ whose makespan is less than $\sum_{k \in \{1, \ldots, q\}} C^*_k$. The schedule we get by simply sticking together from left to right the optimal schedules of $G_1, \ldots, G_q$ is an optimal schedule of $G$ since its makespan is $\sum_{k \in \{1, \ldots, q\}} C^*_k$.

The algorithm follows directly from the two preceding properties. Once the parse tree of $G$ has been computed (Figure 4.13), which takes linear time (see Valdes et al. 1982) the algorithm proceeds bottom up in the tree using at each iteration either the *PAR* or *SER* schedule construction until the root of the tree has been reached.

### 4.4.2.5 *The SCT directed bipartite graph*

Let us now consider the case of a directed bipartite graph such as that in Figure 4.14, where $J = A \cup B$ ($A \cap B = \emptyset$), $U = \{u_1, \ldots, u_q\}$ ($U \subset A \times B$) and there is no isolated task. If $u = (J_a, J_b)$ is an edge, we denote by $\theta(u)$ the value $p_a + c_{ab} + p_b$ and we assume without loss of generality that

$$\theta(u_1) \geqslant \theta(u_2) \cdots \geqslant \theta(u_q).$$

With Proposition 4.1, we need only look for a schedule with smallest completion time among all the schedules $S$ in which each processor executes either a single task or the two end-tasks of one edge. With such a schedule $S$ we can associate a matching $U_S$ of those edges whose end-tasks are allocated to the same processor. $S$ has completion time $C(S)$ given by

$$C(S) = \max_{(a,b) \in U} \{p_a + c_{ab}(1 - \delta_{\pi_a, \pi_b}) + p_b\} = \max \{\max_{(a,b) \in U_S} \{p_a + p_b\}, \max_{u \notin U_S} \{\theta(u)\}\}$$

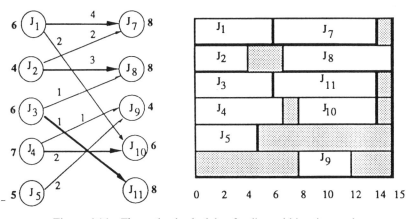

**Figure 4.14** The optimal schedule of a directed bipartite graph.

The matching of an optimal schedule should contain those edges with the largest $\theta(u)$ values. This is proved by the following proposition where $K$ is defined as the largest integer such that $U_K = \{u_1, \ldots, u_K\}$ is a matching of $G$.

**Proposition 4.3** *The schedule $S_K$ that assigns a distinct processor to each edge of $U_K$ and to each remaining task is optimal.*

**Proof** From the definition of the schedule $S_K$ we have

$$C(S_K) = \max \{ \max_{(a,b) \in U_K} \{p_a + p_b\}, \theta(u_{K+1}) \}$$

Assume that $S$ is a schedule such that the edge $u_k, k \in \{1, \ldots, K\}$, is not processed by a single processor. It follows that the makespan $C(S)$ satisfies $C(S) \geqslant \theta(u_k) \geqslant \theta(u_{K+1})$ and that for any edge $(J_a, J_b)$ in $U_K$ we have $C(S) \geqslant p_a + p_b$. So the schedule $S$ cannot be better than $S_K$. Suppose now that $S$ is a dominant schedule that does contain $U_K$ in its matching. From the definition of $K$, the matching of $S$ cannot contain the edge $u_{K+1}$. It follows that $C(S) \geqslant C(S_K)$. So $S_K$ is an optimal schedule □

The algorithm solving this special case follows directly from Proposition 4.3 (Figure 4.14). It first sorts the edges in decreasing $\theta(u)$ order and then computes the largest $K$ such that $U_K$ is a matching. Its complexity is $O(q \log q)$.

### 4.4.3 Approximation algorithms

Good approximation algorithms seem to be very difficult to design, since the compromise between parallelism and communication delays is not easy to handle. We present in this section three approximation algorithms: one for the general case, one for rooted trees and one for the $r$-SCT special case.

#### 4.4.3.1 Rooted trees

Using a reduction from X3C (see Garey and Johnson 1979), Jakoby and Reischuk (1992) have shown that the special case of $P\infty|tree, c, p_j = 1|C_{\max}$ is $\mathcal{NP}$-hard even when the in-degree of each node is at most two. Using a similar reduction, they also proved that for a binary tree, unit processing times and arbitrary communication times the problem is $\mathcal{NP}$-complete (see Jakoby and Reischuk 1992).

Let $G$ be an out-tree with depth $h$ ($h > 1$). The algorithm, developed by Picouleau (1992), finds a schedule whose makespan does not exceed the minimum makespan by more than $(h-1)c_{\max}$ time units. This algorithm, called **SDR**, relies on the exact algorithm **SD** solving the SEND problem (Section 44.4.2).

**SDR** works bottom up from the nodes with height one to the root. For any task $J_j$ with height one, **SDR** applies **SD** to the SEND problem whose root is $J_j$. Let $J_i$ be a task with height $h > 1$ and assume that **SDR** has already computed a schedule $\hat{S}_j$ for each subtree $A_j$, $J_j \in OUT(i)$. For any $J_j$ in $OUT(i)$, let us denote by $\hat{C}_j$ the makespan of $\hat{S}_j$ and by $C_j^*$ the minimum makespan of the problem associated with the subtree $A_j$.

The first step of **SDR** applies **SD** to an auxiliary SEND problem with $J_i$ as the sending task and $|OUT(i)|$ fictive receiving tasks. The processing time of the fictive task $J_j'$ is set to $\hat{C}_j$ and the communication time from $J_i$ to $J_j'$ remains equal to $c_{ij}$. The second step of **SDR** uses the optimal schedule of the auxiliary *SEND* problem to build the schedule $\hat{S}_i$ as shown in Figure 4.15. The following proposition proves the performance guarantee.

**Proposition 4.4** *For any subtree $A_i$ the makespan $\hat{C}_i$ of $S_i$ satisfies $\hat{C}_i - C_i^* \leqslant (h(A_i) - 1)c_{\max}$.*

**Proof** Let $h(A_i)$ be the height of the subtree $A_i$. If $h(A_i) = 1$ then $\hat{S}_i$ is an optimal schedule and $\hat{C}_i - C_i^* = 0$. Assume now that $h(A_i) > 1$, $C_k^* = \max_{J_j \in OUT(i)} \{C_j^*\}$ and $\hat{C}_l = \max_{J_j \in OUT(i)} \{\hat{C}_j\}$. For any $J_j$ in $OUT(i)$ induction yields $\hat{C}_j - C_j^* \leqslant [h(A_j) - 1]c_{\max}$. In an optimal schedule of $A_i$, $J_i$ is completed by the starting time of $A_j, j \in OUT(i)$, so we have $C_i^* \geqslant p_i + C_k^*$. From the definition of **SDR**, we get $\hat{C}_i \leqslant p_i + \hat{C}_l + c_{\max}$. It follows that

$$\hat{C}_i - C_i^* \leqslant c_{\max} + \hat{C}_l - C_k^* \leqslant C_l^* - C_k^* + h(A_l)c_{\max} \leqslant (h(A_i) - 1)c_{\max} \quad \Box$$

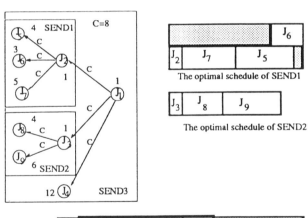

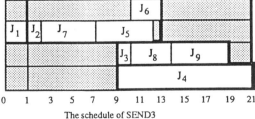

**Figure 4.15** The SDR approximation algorithm.

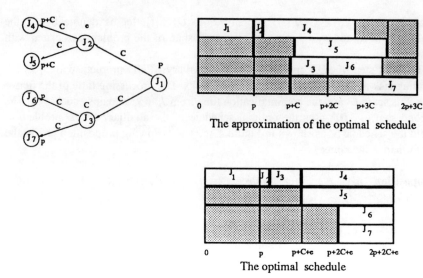

**Figure 4.16** The instance achieving the bound.

The bound $(h-1)c_{\max}$ is the best one for **SDR** since it is asymptotically achieved when the algorithm is applied to the instance reported in Figure 4.16 and the processing time $e$ of task $J_2$ tends to 0.

### 4.4.3.2 The $P\infty|prec|C_{\max}$ problem

The first approximation algorithm for the $P\infty|prec|C_{\max}$ problem was proposed by Sarkar (1989) and uses task clustering. Using the same basis, a more efficient algorithm, called the **DSC** algorithm, has been developed and analyzed by Gerasoulis and Yang (1992,1993). We present **DSC** in this section, since it provides a rather good schedule for a reasonably low complexity and finds the optimal schedule for some specific problems such as the SEND problem and the SCT $P\infty|tree, p_j, c_{jk}|C_{\max}$ problem.

Let us define a *cluster* as an ordered list of distinct tasks whose order is compatible with the precedence graph $G$. A partition into clusters is a set of task-disjoint clusters such that any task belongs to one cluster. Clearly, with a partition into clusters is associated

- a CPM-graph where each arc $(J_i, J_j) \in U$ is valued either by $p_i + c_{ij}$ if $J_i$ and $J_j$ are not in the same cluster or by $p_i$ otherwise;
- a CPM-schedule where clusters are assigned distinct processors.

A CPM-graph is an edge-valued graph with no positive-directed cycles and a CPM-schedule associates with each node the longest path to that node.

**DSC** builds up a sequence of at most $n$ partitions such that the makespans of the associated CPM-schedules are not increasing. At each step of the algorithm the partition contains two types of clusters: the *examined* clusters, whose content may only increase, and the *non-examined* clusters, which are singletons that may be removed later from the partition. By extension, each task of an examined cluster is said to be *inserted*. In the initial partition each cluster contains a single task, and is examined if the task it contains has no ancestor; otherwise it is nonexamined. Each foregoing partition will be such that

- all the ancestors of an inserted task are inserted;
- each unexamined cluster contains a single task.

Assume we are given a partition. The noninserted task $J_i$ is said to be *free* if all its ancestors are inserted, and is said to be *partially free* if at least one of its immediate predecessors is inserted and at least one is not. The priority of the free task $J_i$ is the sum $r_i + q_i$, where $r_i$ is the value of the longest path in the CPM-graph ending at node $J_i$ and $q_i$ is the value of the longest path in the CPM graph starting at node $J_i$. The priority of the partially free task $J_j$ is $s_j + q_j$, where $s_j$ is the value of the longest path in the CPM-graph ending at node $J_j$ whose last but one task is inserted.

The current step first determines the free task $J_{i^*}$ with highest priority (say $\alpha$) and the partially free task $J_{j^*}$ with highest priority (say $\beta$). If $\alpha \geq \beta$ then $J_{i^*}$ belongs to a critical path of the CPM-graph. For any (inserted) task $J_k$ in $IN(i^*)$, the algorithm tests if by making $J_{i^*}$ the *last task* of the cluster containing $J_k$, $J_{i^*}$ may be scheduled strictly sooner in the new partition. If none of these tests allows such a gain, $J_{i^*}$ remains alone in its cluster, which becomes examined. Otherwise $J_{i^*}$ is removed from its ancient cluster and added as the last task of its new cluster. If $\alpha < \beta$ then $J_{j^*}$ belongs to a critical path of the CPM-graph. Assume that the edge $(J_k, J_{j^*})$ is critical. If assigning tasks $J_k$ and task $J_{j^*}$ the same processor yields a smaller value of $s_{j^*}$ then the cluster containing $J_k$ remains unchanged until $J_{j^*}$ becomes free. Using the same procedure as before except that the cluster containing $J_k$ is no longer a candidate, the algorithm tries to add $J_{i^*}$ as the last task of an examined cluster.

In Figure 4.17 the algorithm is applied to a small example. The inserted tasks are marked by shaded circles containing the name of the task and its processing time. Each free (respectively partially free) task is marked by an "F" (respectively a "PF"), followed by its priority. For clarity, on each arc of the current CPM graph only the current communication time is inscribed. Finally, the arcs of the critical path yielding the clustering decision are shown as heavy lines.

**DSC** takes $O((n + q) \log n)$ time and satisfies $C_{\max}(DSC) \leq C^*_{\max}(1 + 1/g(G))$, where $g(G)$ is the *granularity* of $G$, which is defined as follows: for any task $J_j$, let $u_j$, $v_j$ and $g_j$ be defined by

$$u_j = \min_{J_i \in IN(j)} \{p_i\} / \max_{J_i \in IN(j)} \{c_{ij}\}$$

$$v_j = \min_{J_k \in OUT(j)} \{p_k\} / \max_{J_k \in OUT(j)} \{c_{jk}\}$$

$$g_j = \min\{u_j, v_j\}$$

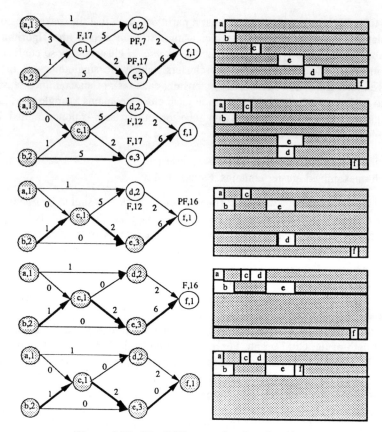

**Figure 4.17** The DSC approximation algorithm.

then $g(G) = \min_{j \in J}\{g_j\}$. Note that for the SCT $P\infty|prec, p_j, c_{jk}|C_{max}$ problem, **DSC** satisfies $C_{max}(DSC) \leqslant 2C^*_{max}$.

#### 4.4.3.3 The r-SCT $P\infty|prec, p_j, c_{jk}|C_{max}$ problem

We present in this section a simple approximation algorithm **IC**, developed by Picouleau (1992) for the r-SCT $P\infty|prec, p_j, c_{jk}|C_{max}$ problem. The guarantee provided by **IC** is $C_{max}(IC) \leqslant C^*_{max}(1 + r)$.

The basic idea of **IC** is to solve by the CPM method an auxiliary problem with the same precedence graph but where the processing time $p'_i$ of $J_i$ is set to $p_i(1 + r)$ and there are no communication delays. Since for any $(J_i, J_j) \in U$ we have $c_{ij} \leqslant rp_i$, the CPM-schedule of the auxiliary problem may be transformed into a feasible schedule $\hat{S}$ of the original problem by assigning each task a distinct processor and scheduling it at the same time as in the CPM-schedule of the auxiliary problem. If $\lambda$ is the longest path of $G$ when there are no communication delays, we have by construction

$C_{rmmax}(IC) \leq \lambda(1+r)$ and $C^*_{max} \geq \lambda$. Hence we have the expected performance guarantee.

## 4.5 NO DUPLICATION AND A LIMITED NUMBER OF PROCESSORS

In this section we first consider the complexity of problems with a restricted number $m > 1$ of processors. Most of these problems are $\mathcal{NP}$-complete, but some quite interesting questions remain open, especially for $m = 2$. In the second part of this section we present an interesting approximation algorithm for the UET-UCT problem, and we relate other approximation algorithms developed for more realistic cases.

### 4.5.1 Complexity results

Using a reduction of $P|prec, p_j = 1|C_{max}$, Rayward-Smith (1987) has shown that $P|prec, p_j = 1, c = 1|C_{max}$ is $\mathcal{NP}$-complete. This result may also be derived from the $\mathcal{NP}$-completeness of $\bar{P}|prec, p_j = 1, c = 1|C_{max}$ (see Picouleau 1992). More recently, it has been shown by Hoogeveen et al. (1992) that deciding whether the problem $P|bipartite, p_j = 1, c = 1|C_{max}$ has a schedule satisfying $C_{max} \leq 4$ is an $\mathcal{NP}$-complete problem.

The same question for $C_{max} \leq 3$ and the problem $P|prec, c = 1, p_j = 1|C_{max}$ may be answered in polynomial time (see Picouleau 1992). So, unless $P = NP$, there is no polynomial approximation algorithm with worst case ratio smaller than $\frac{5}{4}$. It also may be proved from the preceding results that the problem $P|prec, c = 1, p_j = 1, dup|C_{max}$ is $\mathcal{NP}$-complete.

Some special cases of $P|prec, c = 1, p_j = 1|C_{max}$ have been studied by Picouleau (1992). When $G$ is an interval graph, a variant of the known list algorithm by Papadimitriou and Yannakakis (1979) for the problem without communication times solves the problem polynomially (see Picouleau 1992). The problem $P|tree, c = 1, p_j = 1|C_{max}$ has been proved to be $\mathcal{NP}$-hard by a reduction from SAT (see Veltman 1993). When the number $m$ of processors is fixed, a polynomial algorithm using dynamic programming has been developed by Varvarigou et al. (1993). Finally, if the tasks are a priori assigned to two processors, the problem is $\mathcal{NP}$-complete for an arbitrary precedence graph (see Veltman 1993). Note that the complexity of $2|prec, p_j = 1, c = 1|C_{max}$ is a quite motivating open question.

### 4.5.2 List scheduling for $P|prec, p_j = 1, c = 1|C_{max}$

We present in this section an extension of list scheduling developed by Rayward-Smith (1987). When adapted to $P|prec, p_j = 1, c = 1|C_{max}$, it is shown that any list scheduling algorithm **LS** satisfies

$$C_{max}(LS) \leq \left(\frac{3-2}{m}\right) C^*_{max} - \left(1 - \frac{1}{m}\right)$$

Assume as usual that time period $t$ is the time interval $[t-1..t]$. In classical list scheduling, a task that is ready at period $t$ may be processed by *any* free processor at that period. With communication delays, that property is no longer true. So a task is said to be *ready for a processor* $\pi$ at period $t$ if it may be scheduled on $\pi$ at period $t$. When applied to the $P|prec, p_j = 1, c = 1|C_{max}$ problem, a natural extension of classical list scheduling consists, in addition to the given tasks list, of fixing a priori a scanning order of the $m$ processors. Assuming a partial schedule is already built for the time periods $\{1, \ldots, t-1\}$, the algorithm scans in turn each processor to find a task that is ready for it at period $t$ and, if any, to assign it the first ready task in the list for the period $t$.

Note that any schedule provided by list scheduling is *active*, since if a processor is idle during the period $t$, no task scheduled after time $t$ is ready for it at period $t$. The quite general result proved by Rayward-Smith (1987) is that *any active schedule* satisfies the above performance ratio. The key idea of the proof is to show that with any active schedule is associated a subgraph of $G$, called a $(u,v)$-layered graph, whose structure allows to derive a lower bound of the makespan. A $(u,v)$-*layered graph* (Figure 4.18) is a graph whose nodes may be partitioned into $u$ layers $\{\mathscr{L}_1, \ldots, \mathscr{L}_u\}$ such that the terminal nodes are in $\mathscr{L}_u$, for any node $J_i$ in $\mathscr{L}_p$ ($p \neq 1$), $\emptyset \neq IN(i) \subset \mathscr{L}_{p-1}$, and for at least $v$ layers any node in these layers has at least two immediate predecessors. Clearly, from the definition, the makespan of any schedule of an $(u,v)$-layered graph is at least $u+v$.

Now let $S$ be a given active schedule. A unit period of this schedule is said to be *idle* if at least one processor is idle; it is said to be *dormant* if all the processors are idle. So, in a nondormant idle period, at least one processor is idle and at least one is busy. Dormant and nondormant idle periods provide the following information on the precedence graph: if the period $t$ is dormant, any task processed after time $t$ has at least two ancestors processed during period $t-1$; if the idle period $t$ is nondormant, any task scheduled after time $t$ has at least one ancestor processed during period $t-1$ or period $t$. From the above property, it may be derived that $G$ has a $(s + \max\{\lceil (r-2s-1)/2 \rceil + 1, s\})$-layered subgraph, where $s$ (respectively $r$) is the number of dormant (respectively idle) periods in $S$. As the subset of active schedules is dominant, we have $C^*_{max} \geq \frac{1}{2}(r+2s+1)$, and for the given active schedule $\hat{S}$ we know that the sum $mC(\hat{S})$ of the total active time and the total idle time is at most $(mC^*_{max}) + [sm + (r-s)(m-1)]$. Hence the desired performance ratio.

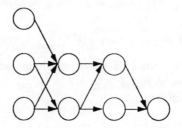

**Figure 4.18** A (4, 2)-layered graph.

## 4.6 MORE GENERAL PROBLEMS

Efe (1982) takes two criteria into account: the makespan and the load balance between the processors. In its first step the heuristic makes task clusters so that communication time between clusters is small and the number of clusters is at most $m$. The second step improves load balancing by task exchanges between clusters using a neighborhood search technique.

Kruatrachue and Lewis (1988) use task aggregation to make a set of longer tasks. The heuristic searches for a task set with not too many short tasks that would imply a large communication time and with not too many long tasks that would reduce the gain from parallelization. The heuristic computes first a good "grain size" by making a trade-off between processing time and communication time; it then uses task duplication to minimize the makespan. It is shown that, for a few small graphs, task duplication allows simultaneously a decrease in the delay due to communication times and better use of parallelism without increased makespan. The heuristic is an extension of that method to any graph. Once the heuristic has been applied, the processing times of the new tasks are greater than the communication delays. A multiprocessor scheduler is then used to cancel the communication delays. The grain size obtained by the heuristic is not optimal, but the results from several simulations are found to be better than those from list scheduling or load balancing.

Hwang, et al. (1989) investigated list algorithms for the problem where the communication times also depend on the processors assigned to the tasks. The **ELS** (for Extented List Schedule) algorithm computes in its first step a solution using classical list scheduling by ignoring the communication times. In its second step it adds the necessary communication delays to the schedule obtained in the first phase. Using Graham's (1969) bound for the first schedule and an upper bound for the communication delays, the following performance ratio is obtained:

$$C_{max}(ELS) \leq \left(2 - \frac{1}{m}\right) C'^{*}_{max} + \tau_{max} \sum_{(J_i, J_j) \in U} \eta(i,j)$$

where $\tau_{max}$ is the maximum interprocessor distance, $\eta(i,j)$ is the length of the message sent by $J_i$ to $J_j$ and $C'^{*}_{max}$ is the minimum makespan for the corresponding problem without communication times.

Because the performance of **ELS** is not so good, the **ETF** (earliest task first) heuristic is proposed. **ETF** uses the following greedy strategy: schedule first the earliest ready task. This strategy is further improved by the ability to postpone a scheduling decision to the next decision time if a task completion occurring between two successive decision times makes a more urgent task schedulable. From a sophisticated analysis of **ETF**, the following performance ratio is obtained:

$$C_{max}(ETF) \leq \left(2 - \frac{1}{m}\right) C'^{*}_{max} + \Lambda_G$$

where $\Lambda_G$ is the maximum sum of the communication times over all paths of $G$.

## 4.7 CONCLUSIONS

The main conclusion concerning the scheduling problems with communication delays is that unlike most classical scheduling problems, these problems are not really easier when the resource limitation is relaxed. Most of them remain $\mathcal{NP}$-hard even when strong restrictions are made on the precedence graph or on the values of the communication and/or processing times. Figure 4.19 summarizes these complexity results.

In this chapter a clear distinction between problems allowing, or not allowing, task duplication has been made. From a theoretical point of view, task duplication is an interesting hypothesis for computer scheduling over distributed memory multi-processor machines, but its implementation would certainly give rise to difficult operating system problems, since the number of copies induced by a minimal makespan schedule may be large. Moreover, task duplication is not a realistic assumption when dealing with shop problems or project management problems.

When duplication is allowed, quite interesting ideas on the way to design heuristics that really take communication times into account have been obtained. List scheduling does not seem to be fitted to these problems in the general case, since it has too strong a tendency to parallelize the executions of tasks.

It is also clear that the problems investigated up to now are far from real scheduling problems. Many constraints such as the communication channel capacity or unequal processor distances have not been taken into account. The results for the

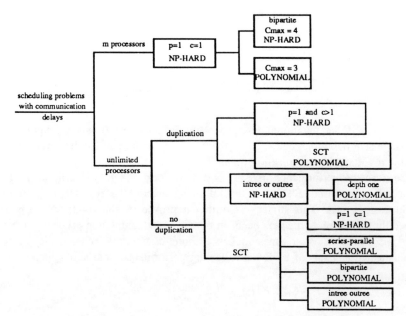

**Figure 4.19** Complexity of scheduling with communication times.

basic problems presented in this survey should provide good evaluation tools for exact (such as branch and bound) methods.

## REFERENCES

Chrétienne P. (1989) A polynomial algorithm to optimally schedule tasks over a virtual distributed system under tree-like precedence constraints *Eur. J. Oper. Res.*, **43**, 225–230.

Chrétienne P. (1992) Task scheduling with interprocessor communication delays *Eur. J. Oper. Res.*, **57**, 348–354.

Chrétienne P. (1994) Complexity of tree-scheduling with interprocessor communication delays *Discr. Appl. Math.*, **1**, 1–10.

Colin J.-Y. and Chrétienne P. (1991) CPM scheduling with small communication delays *Oper. Res.*, **39**, 680–684.

Efe K. (1982) Heuristic model of task assignment and scheduling in distributed systems. *IEEE Trans. Comput.*, **7**, 50–56.

Garey M. R. and Johnson D. S. (1979) *Computers and Intractability: A Guide to the Theory of NP-Completeness* Freeman, San Francisco.

Gerasoulis A. and Yang T. (1992) A comparison of clustering heuristics for scheduling dags on multiprocessors *J. Parallel Dist. Syst. Comput.*, **16**(4), 276–291.

Gerasoulis A. and Yang T. (1993) On the granularity and clustering of directed acyclic task graphs *IEEE Trans. Parallel Dist. Syst.*, **4**, 686–701.

Graham R. L. (1969) Bounds for certain multiprocessing anomalies *SIAM J. Appl. Math.*, **17**, 416–429.

Hoogeveen J. A., Lenstra J. K. and Veltman B. (1992) Three, four, five, six, or the complexity of scheduling with communication delays. Technical Report BS-R922, Center for Mathematics and Computer Science.

Hwang J. J., Chow Y. C., Anger F. D. and Lee C. Y. (1989) Scheduling precedence graphs in systems with interprocessor communication times *SIAM J. Comput.*, **18**, 244–257.

Jakoby A. and Reischuk R. (1992) The complexity of scheduling problems with communication delays for trees. In *Lecture Notes in Computer sciences*, No. 621, Vol. 3, pp 165–177 Springer, Berlin.

Jung H., Kirousis L. M. and Spirakis P. (1993) Lower bounds and efficient algorithms for multiprocessor scheduling of directed acyclic graphs with communication delays *Inf. Comput.*, **105**, 94–104.

Kruatrachue B. and Lewis T. (1988) Grain size determination for parallel processing *IEEE Soft.*, pp 23–32.

Papadimitriou C. H. and Yannakakis M. (1979) Scheduling interval-ordered tasks *SIAM J. Comput.*, **8**, 405–409.

Papadimitriou C. H. and Yannakakis M. (1990) Towards an architecture independent analysis of parallel algorithms *SIAM J. Comput.*, **19**, 322–328.

Picouleau C. (1992) Etude de problèms d'optimisation dans les systèmes distribués. Thèse, Université Pierre et Marie Curie

Picouleau C. (1995) Two new np-complete scheduling problems with communication delays and unlimited number of processors *Discr. Appl. Math.*, to appear.

Rayward-Smith V. J. (1987) Uet scheduling with unit interprocessor communication delays *Discr. Appl. Math.*, **18**, 55–71.

Sarkar V. (1989) *Partitioning and Scheduling Parallel Programs for Execution on Multiprocessors*. MIT Press, Cambridge, MA.

Valdes J, Tarjan R. E. and Lawler E. L. (1982) The recognition of series-parallel digraphs *SIAM J. Comput.*, **11**, 298–313.

Varvarigou T. A., Roychowdhury V. P. and Kailath T. (1993) Scheduling in and out forests in the presence of communication delays. In *Proceedings of 7th International Parallel Processing Symposium*, pp 222–229.

Veltman B. (1993) Multiprocessor scheduling with communication delays. PhD Thesis, CWI–Amsterdam.

Veltman B, Lageweg B. J. and Lenstra J. K. (1990) Multiprocessor scheduling with communication delays *Parallel Comput.*, **16**, 173–182.

*Institut Blaise Pascal, Laboratoire LITP, Université Pierre et Marie Curie, 4 Place Jussien, 75252 Paris Cedex 05, France*

CHAPTER 5

# Profile Scheduling by List Algorithms

**Zhen Liu**
*INRIA, Centre Sophia Antipolis*
**Eric Sanlaville**
*LITP Université Pierre et Marie Curie*

**Abstract**

The notion of profile scheduling was first introduced by Ullman in 1975 in the complexity analysis of deterministic scheduling algorithms. In such a model the number of processors available to a set of tasks may vary in time. Over the last decade, this model has been used to deal with systems subject to processor failures, multiprogrammed systems, or dynamically reconfigured systems. We give an overview of optimal polynomial solutions for scheduling a set of partially ordered tasks in these systems. Particular attention is given to a class of algorithms referred to as list scheduling algorithms. The objective of the scheduling problem is to minimize either the maximum lateness or the makespan. Results on preemptive and nonpreemptive deterministic scheduling and on preemptive stochastic scheduling are presented.

## 5.1 INTRODUCTION

Consider the problem of scheduling a set of partially ordered tasks represented by a directed acyclic graph, referred to as *task graph*, where vertices represent tasks and arcs represent precedence relations. Tasks are executed, subject to precedence constraints, on a set of parallel identical processors. The number of available processors, referred to as *profile*, may vary in time. Task and processor assignments must be nonredundant, i.e., at any time, a task can be assigned to at most one processor, and a processor can execute at most one task. Each task has a due date. The objective is to minimize the maximum lateness or, when due dates are not taken into consideration, the makespan.

---

*Scheduling Theory and its Applications* Edited by P. Chrétienne, E. G. Coffman, Jr., J. K. Lenstra and Z. Liu
© 1995 John Wiley & Sons Ltd

The notion of profile scheduling was first introduced by Ullman (1975) and later used by Garey *et al.* (1983) in the complexity analysis of deterministic scheduling algorithms. In this chapter we use the notion of profile to deal with systems subject to processor failures, multiprogrammed systems and dynamically reconfigured systems. In such cases the number of processors available to a set of tasks may vary in time.

The problem of minimizing the maximum lateness and the makespan is in general $\mathcal{NP}$-hard. We are interested in simple on-line or nearly on-line algorithms that yield optimal solutions under specific assumptions. Results on three problems will be presented here, with each result accounting for different task characteristics. The reader is referred to the survey paper by Lawler *et al.* (1989) for results on optimal scheduling under a constant profile, i.e. when the number of available processors is constant. The first results on optimal polynomial solutions for profile scheduling problems are due to Dolev and Warmuth (1984,1985a,b).

The chapter is organized as follows. Section 5.2 is devoted to notation. Section 5.3 deals with scheduling of nonpreemptive unit execution time (UET) tasks; optimality results concerning list schedules on a variable profile are surveyed. Section 5.4 is concerned with scheduling of preemptive real execution time (RET) tasks. We exhibit a tight relation between optimal list schedules and optimal priority schedules, which are counterparts of list schedules for preemptive scheduling. Section 5.5 focuses on tasks whose running times are independent and identically distributed random variables with a common exponential distribution. We prove the optimality of some list policies that stochastically minimize the makespan in several subproblems.

## 5.2 NOTATION

A task graph $G = (V, E)$ is a directed acyclic graph, where $V = \{1, 2, \ldots, |V|\}$ is the set of vertices representing the tasks, $E \subset V \times V$ is the set of arcs representing the precedence constraints: $(i,j) \in E$ if and only if task $i$ must be completed before task $j$ can start. Denote by $p_i$ and $d_i$ the respective processing time and due date of task $i \in V$.

Let $p(i)$ and $s(i)$ be the respective sets of immediate predecessors and successors of $i \in V$, i.e.

$$p(i) = \{j : (j,i) \in E\}, \quad s(i) = \{j : (i,j) \in E\}.$$

Let $S(i)$ be the set of (not necessarily immediate) successors of $i \in V$, i.e.

$$\forall i, \quad \text{if } s(i) = \emptyset \text{ then } S(i) = \emptyset; \quad \text{otherwise} \quad S(i) = s(i) \bigcup \left[ \bigcup_{j \in s(i)} S(j) \right].$$

A task without a predecessor (successor) is called an *initial* (*final*) task. Denote by $l_i$ the level of $i$, i.e. the number of arcs in a longest path from task $i$ to some final task. Denote by $h_i$ the height of $i$, i.e. the sum of the processing times of the tasks in a longest path from task $i$ to some final task, with final vertex included.

In the following, we shall consider different classes of precedence graphs. The three most interesting classes are the following.

- *Interval order* $G \in \mathcal{G}_{io}$: here each vertex $i$ corresponds to an interval $b_i$ of the real line such that $(i,j) \in E$ if and only if $x \in b_i$ and $y \in b_j$ imply $x < y$. These graphs have the following characteristic property:

$$\forall i,j \in V, \quad \text{either } S(i) \subseteq S(j), \quad \text{or } S(j) \subseteq S(i).$$

- *In-forest* $G \in \mathcal{G}_{if}$: here each vertex has at most one immediate successor: $|s(i)| \leq 1$, $i \in V$. A vertex $i \in V$ is called a leaf of in-forest $G$ if $p(i) = \emptyset$. A vertex $i \in V$ is called a root of in-forest $G$ if $s(i) = \emptyset$.
- *Out-forest* $G \in \mathcal{G}_{of}$: here each vertex has at most one immediate predecessor: $|p(i)| \leq 1$, $i \in V$. A vertex $i \in V$ is called a leaf of out-forest $G$ if $s(i) = \emptyset$. A vertex $i \in V$ is called a root of out-forest $G$ if $p(i) = \emptyset$.

There are $K \geq 1$ parallel and identical processors. The set of processors available to tasks varies in time, owing, for example, to failures of the processors or the execution of higher-priority tasks. The availability of the processors is referred to as the profile, and is specified by the sequence $M = \{a_n, m_n\}_{n=1}^{\infty}$, where $0 = a_1 < a_2 < \ldots < a_n < \ldots$ are the time epochs when the profile is changed, and $m_n$, $n \geq 1$, is the number of processors available during the time interval $[a_n, a_{n+1}[$. Without loss of generality, we assume that $m_n \geq 1$ for all $n \geq 1$. We shall assume that the profile is not changed infinitely often during any finite time interval: for all $x \in \mathbb{R}^+$, there is some finite $n \geq 1$ such that $a_n > x$.

The following three classes of profiles will often be referred to in this chapter.

- *Zigzag profiles* $M \in \mathcal{M}_z$: here the number of available processors is either $K$ or $K - 1$.
- *Increasing zigzag profiles* $M \in \mathcal{M}_{iz}$: here the number of available processors can decrease by at most one at any time. Between successive decrements, there must be at least one increase. That is, $\forall j$ and $\forall n \geq j, m_n \geq m_j - 1$.
- *Decreasing zigzag profiles* $M \in \mathcal{M}_{dz}$: here a symmetrical definition applies; that is, $\forall j$ and $\forall n \geq j, m_n \leq m_j + 1$. Such a profile is illustrated in Figure 5.1.

A scheduling algorithm (or policy) decides when an enabled task, i.e. an unassigned, unfinished task all of whose predecessors have finished, should be assigned to one of the available processors. A schedule is feasible if assignments are nonredundant and the constraints relative to the precedence relation and to the variable profile are respected. Scheduling can be either preemptive, i.e. the execution of a task can be stopped and later resumed on any processor without penalty, or nonpreemptive, i.e., once begun, the execution of a task continues on the same processor until its completion.

Let $S$ be an arbitrary feasible schedule of task graph $G$ under profile $M$. Let $C_i(S)$ be the completion time of task $i$ under $S$. The lateness of task $i$ is defined as

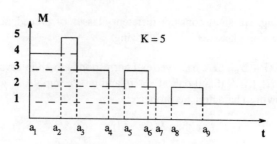

**Figure 5.1** Decreasing zigzag profile.

$L_i(S) = C_i(S) - d_i$. The maximum lateness of schedule $S$ is denoted by $L_S(G, M) = \max_{i \in V} L_i(S)$.

When all due dates are set to zero, the maximum lateness becomes the makespan. Denote by $C_S(G, M) = \max_{i \in V} C_i(S)$ the makespan of $(G, M)$ obtained by schedule $S$.

We extend the classical notation scheme of Graham et al. (1979) for scheduling problems to the case of variable profiles. We use $P(t)$ (respectively $Q(t)$, $R(t)$) to denote machine environment with identical (respectively uniform, unrelated) parallel processors whose number varies in time. For example, $P(t) \mid p_i = 1, prec \mid C_{max}$ denotes the nonpreemptive scheduling for makespan minimization of UET tasks subject to precedence constraints on identical parallel processors with variable profile.

## 5.3 NONPREEMPTIVE PROFILE SCHEDULING OF UET TASKS

In the framework of nonpreemptive profile scheduling, we shall consider UET tasks and integer profiles (those in which profiles change only at integer time epochs). The problems under consideration can be denoted by $P(t) \mid p_i = 1, prec \mid C_{max}$ or $P(t) \mid p_i = 1, prec \mid L_{max}$.

List algorithms are often used in nonpreemptive scheduling. These algorithms put the enabled tasks in an ordered list. Each time a processor becomes available, the task at the head of the list is assigned to that processor. The list can be dynamically updated. The highest level first (HLF) and the earliest due date (EDD) algorithms are well known examples. The reader is referred to Coffman (1976) for properties of list algorithms.

### 5.3.1 Complexity issues

Ullman (1975) exhibited a polynomial reduction from 3-SAT to $P(t) \mid p_i = 1, prec \mid C_{max}$, and then showed that $P \mid p_i = 1, prec \mid C_{max}$ had the same complexity (it suffices to "fill" the unavailable processors with dummy tasks). This is immediately generalized to the lateness minimization $L_{max}$.

Garey et al. (1987) also used profile scheduling for the study of opposing forests (union of in-forest and out-forest). They proved $\mathcal{NP}$-completeness for $P(t)decreasing \mid p_i = 1, intree \mid C_{max}$ (the decreasing profile refers to the case where the number of available processors is decreasing in time), and consequently for $P \mid p_i = 1, opposing\ forest \mid C_{max}$. For such precedence graphs, however, polynomial algorithms have been found for bounded profiles, i.e. $K$ is fixed (see discussion below).

### 5.3.2 Minimization of the makespan by list scheduling

Consider first the minimization of the makespan. We focus on the highest level first (HLF) algorithm: tasks are ordered by decreasing level. Note that HLF schedules differ only in the way ties are broken.

*Forests*

When the task graph is a forest, Hu (1961) proved that HLF yields an optimal schedule when the task graph is an in-forest and the profile is constant. Bruno (1981) extended this result to out-forests.

For variable profile, Dolev and Warmuth (1984,1985a,b) showed that HLF remains optimal in some special cases of profiles. For that, they first showed the so-called *elite theorem*. Define the height of a connected component of $G$ to be the highest level of its vertices. Let the components be ordered by decreasing height, and let $h(k)$ be the height of the $k$-th component. Suppose profile $M$ is bounded by $K$. The *median* of $(G, M)$ is defined as $\mu = h(K) + 1$. Consider

- $H(G)$: the set of components of $G$ whose heights are strictly greater than $\mu$ (high part);
- $E(G)$: the set of initial tasks of $H(G)$ whose levels are strictly greater than $\mu$ (elite);
- $L(G)$: the graph obtained from $G$ by removing the components of $H(G)$ (low part).

When the graph contains less than $K$ components, the median has value 0 and $G = H(G)$. These definitions are illustrated in Figure 5.2, where $K = 3$.

We now state the elite theorem of Doler and Warmuth (1985a).

**Theorem 5.1** *Consider a task graph $G$ and an integer profile $M$.*

(1) *If $|E(G)| > m_1$ then there is an optimal schedule such that $m_1$ tasks of $E(G)$ are executed during the first time unit.*
(2) *If $|E(G)| \leq m_1$ then for each set $E$ of $m_1$ tasks of maximum levels there exists an optimal schedule executing the tasks of $E$ during the first time unit.*
(3) *If $E(G) = \emptyset$ then any HLF schedule is optimal.*

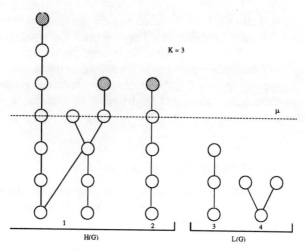

**Figure 5.2** Elite of a task graph.

As a consequence, we obtain the following.

**Corollary 5.1** *HLF minimizes the makespan*

- *when the task graph is an out-forest and the profile is decreasing zigzag;*
- *or when the task graph is an in-forest and the profile is increasing zigzag.*

In the first case the elite contains less than $K$ tasks, and $m_1 \geq K - 1$. Part (2) of Theorem 5.1 states that there exists an optimal schedule that coincides with any *HLF* schedule during the first time unit. The same reasoning is used during each time unit until the graph is empty. To prove the second case, it suffices to reverse the graph and the profile.

For bounded profiles, Dolev and Warmuth (1985b) provided a backward dynamic programming method whose time complexity is $O(n^{K-1} \cdot \log n)$ for scheduling an in-tree of size $n$ on an arbitrary profile bounded by $K$. This was based on two observations. First, if an optimal schedule is known for $H(G)$ then, according to the *merge theorem* of Dolev and Wrmuth (1984) an optimal schedule for $G$ may be obtained in linear time with respect to $|L(G)|$. Second, if we consider all subgraphs of some in-forest with at most $K - 1$ components, which could be obtained during the execution under any schedule, these subgraphs may be partitioned into at most $n^{K-1}$ equivalence classes. Here two graphs $G = (V, E)$ and $G' = (V', E')$ are equivalent if there is a bijection $\phi$ between $V$ and $V'$ such that the set of successors of vertex $\phi(v)$ is $\{\phi(v_1), \ldots, \phi(v_l)\}$, where $\{v_1, \ldots, v_l\}$ is the set of successors of $v$. The dynamic programming method decides whether a schedule of duration $D$ exists, building schedules backward from the $D$-th time unit. The $\log n$ factor comes from a bisection search to get the right value for $D$.

This method is then applied to out-forests under bounded profiles by reversing the directions of the precedence constraints and the profile. Further, such a dynamic programming algorithm is used to solve the scheduling problem of opposing forests as mentioned in Section 5.3.1.

In the special case of chains, i.e. $\mathcal{G}_{ch} \stackrel{def}{=} \mathcal{G}_{if} \cap \mathcal{G}_{of}$, Liu and Sanlaville (1992) showed the following.

**Theorem 5.2** *If G is a union of chains and M an arbitrary integer profile then any HLF minimizes the makespan.*

This was proved by an interchange argument, i.e. from an optimal non-HLF schedule, one can interchange the assignment decisions for two subchains so that the number of non-HLF decisions is decreased by at least one. Iterating this procedure for at most $n^2$ times yields an optimal HLF schedule. Note that all HLF schedules have the same makespan in that case.

*Interval order graphs*

Papadimitriou and Yannakakis (1978) have shown that task graphs with interval order structure have the following property: for any two vertices in the graph, their sets of (all) successors are comparable by inclusion relation: one set is included into the other. This property is essential in establishing the optimality of most successors first (MSF) schedules (forming a subset of HLF schedules), which respect the order of inclusion of the sets of successors. This result was obtained by Papadimitriou and Yannakakis (1978) for constant profiles, and extended to variable profiles in Sanlaville (1992).

**Theorem 5.3** *If G is a task graph with an interval order structure and M an arbitrary integer profile, then any MSF schedule minimizes the makespan.*

This result can again be proved using an interchange argument. Consider an optimal non-MSF schedule $\rho$ for graph $G$. Let $\tau$ be the first time when $\rho$ schedules some vertex $v$ instead of vertex $u$ if MSF rule is applied. Note that, by definition, all the predecessors of $u$ and $v$ have finished execution by time $\tau$. Let $\pi$ be the schedule obtained from $\rho$ by interchanging the execution times for $u$ and $v$. Since $G$ has an interval order structure, $S(v) \subset S(u)$, it is easy to see that $\pi$ satisfies the precedence constraints of $G$. Moreover, the number of non-MSF decisions in $\pi$ is decreased by one. Iterating this procedure for at most $n^2$ times yields an optimal MSF schedule.

*Arbitrary graphs with profiles bounded by 2*

Coffman and Graham (1972) proved that a subset of HLF schedules, referred to as lexicographic order schedules (LOS) in this chapter, minimizes the makespan when the profile is constant and equal to two. LOS is based on a static list of tasks defined by the lexicographic order as follows. Let there be $f$ final tasks. Assign labels

$1, \ldots, f$ to these final tasks in an arbitrary way. Suppose now that $k \geqslant f$ tasks have already been labeled by $1, 2, \ldots, k$. Consider all the tasks whose successors are all labeled. Assign label $k + 1$ to the task such that the decreasing sequence of the labels of its immediate successors is lexicographically minimal (ties are broken arbitrarily). Coffman and Graham (1972) showed that LOS schedules minimize the makespan of an arbitrary graph in a two-processor system. The main idea of their proof is to cut the Gantt chart of the LOS schedule into several segments such that each segment is composed of two sets of tasks $E_i$ and $F_i$, where $F_i$ is a singleton, and that all tasks of $E_i$ are predecessors of $E_{i+1}$. The optimality of LOS is extended in Sanlaville (1992) to variable profiles with at most two processors by a minor modification of the proof of Coffman and Graham (1972), which consists in assigning fictitious tasks to unavailable processors.

**Theorem 5.4** *Any LOS schedule is optimal for makespan minimization of any task graph G under any profile M bounded by 2.*

*Summary and discussion on makespan minimization*

These results are summarized in Figure 5.3, where "constant" and "arbitrary" stand for constant and arbitrary profiles. A gray area means that this particular subproblem is $\mathcal{NP}$-hard for arbitrary $K$. Since constant profiles are special cases of (increasing and decreasing) zigzag profiles, which are in turn special cases of arbitrary profiles, the implication relation for subproblems is clear.

Note that when the number of processors $K$ is fixed, the complexity remains an open problem, except for the case $K = 2$, for which the LOS algorithm of Coffman and Graham (1972) provides optimal schedules. However, for any fixed $K \geqslant 3$ the scheduling problem is still of unknown complexity, despite extensive research.

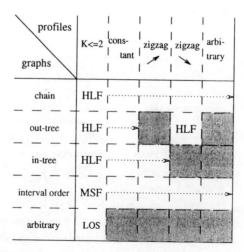

**Figure 5.3** Results on nonpreemptive profile scheduling of UET tasts ($C_{\max}$).

Bartusch et al. (1988) presented an interesting work on this problem, namely scheduling with a fixed number of processors. The authors provided a general two-phase solution scheme. In the first phase they tried to generalize the consistency notion of Brucker et al. (1977) and Garey and Johnson (1977). They computed a so-called "test choice" for the problem instance under consideration. This is derived from bounds on the execution periods of some special tasks. In the second phase, they constructed an optimal schedule in $O(n^K)$ time. This is based on sophisticated dominancy criteria and on the study of special partially ordered sets. The time complexity of the first phase is simple for the two-processor case, and is polynomial for all known polynomial cases (trees, interval-order graphs etc.). However, its complexity remains unknown in the general case. It seems that no further research results concerning this approach have been reported since 1989.

### 5.3.3 Minimization of the maximum lateness by list scheduling

Consider now the minimization of the maximum lateness. We analyze the earliest due date (EDD) algorithm. The optimality of the EDD schedules will rely on a two-step method: (1) modify the due dates so that they satisfy some consistency relation; (2) apply EDD to the modified due dates.

Roughly speaking, the consistency between due dates requires that whenever the due date of a task is met in some schedule, the due dates of its successors can be met too, provided there are enough processors. Such a consistency relation is not verified if, for instance, the due date of some task is earlier than the due date of one of its predecessors.

Initially, for any constant profile, Brucker et al. (1977) proved that if $G$ is an in-forest then EDD yields optimal schedules provided the modified due dates $d'_i$ are defined as follows:

$$d'_i = \begin{cases} d_i, & s(i) = \emptyset; \\ \min(d_i, d'_{s(i)} - 1), & s(i) \neq \emptyset, \end{cases} \quad (1)$$

where, by a harmless abuse of notation, $s(i)$ denotes the unique successor of task $i$. Note that such a modification has the purpose of obtaining consistent due dates. It was shown by Liu and Sanlaville (1992) that the optimality of EDD still holds for increasing zigzag profiles:

**Theorem 5.5** *For any in-forest $G \in \mathcal{G}_{if}$ and any increasing zigzag profile $M \in \mathcal{M}_{iz}$, the EDD schedule defined on modified due dates minimizes the maximum lateness.*

The proof proceeds by first establishing that EDD defined on the modified due dates meets all the original due dates if and only if such a feasible schedule exists, and then by using a standard argument to show the optimality of EDD schedules.

In a slightly more complicated way, Garey and Johnson (1976, 1977) showed the existence of modification schemes for the due dates, such that EDD applied to these due dates yields optimal schedules on two processors, even when release dates are associated with the tasks. In the latter case the algorithm is not on-line. These results can be extended to variable profiles (see Sanlaville 1992). However, even without release dates, the algorithms are then off-line, because the computation of the modified due dates depends not only on the release dates but also on the profile.

For profile of width $K \geqslant 3$ and general task graphs, there is no simple way to design some modification scheme so that EDD applied to the modified due dates becomes optimal. The general problem is $\mathcal{NP}$-hard. In fact, even for out-forests with constant profiles, the minimization of maximum lateness is $\mathcal{NP}$-hard, as was shown in Brucker *et al.* (1977). Unlike the makespan minimization, the scheduling of out-forests cannot be achieved by analyzing the "reverse" problem. Indeed, reversing the problem $P(t)dec.zig. \mid p_i = 1, out\ tree \mid L_{\max}$ yields the problem $P(t)inc.zig. \mid p_i = 1, r_i, in\ tree \mid C_{\max}$, where release dates $r_i$ are added to the tasks.

## 5.4 PREEMPTIVE PROFILE SCHEDULING

We now consider preemptive scheduling. The task processing times and profile change epochs are arbitrary real numbers. We are interested in optimal priority algorithms. In what follows, we first describe such algorithms, and then present a tight relation between optimal nonpreemptive list algorithms and optimal preemptive priority algorithms. Finally, we present simple optimal priority algorithms for specific problems.

### 5.4.1 Priority scheduling algorithms

Parallel to list algorithms, (dynamic) priority algorithms are used in preemptive scheduling. At any time, enabled tasks are assigned to available processors according to a priority list that can change in time and can depend on the partial schedule already constructed. A general description is as follows (see Muntz and Coffman 1970; Lawler 1982).

- At any time $t$, enabled tasks are ordered according to their priorities, thus forming subsets $V_1, \ldots, V_k$, where all tasks of $V_j$ have the same priority and greater priority than tasks in $V_{j+1}$.
- Suppose that tasks in $V_1, \ldots, V_{r-1}, r \leqslant k$, are assigned. Let $\tilde{m}_r(t)$ be the number of remaining free processors. If $\tilde{m}_r(t) \geqslant |V_r|$ then one processor is assigned to each of the tasks in $V_r$, and the algorithm deals with the next subset. Otherwise, the $\tilde{m}_r(t)$ processors are shared by the tasks of $V_r$ so that each task in $V_r$ is executed at speed $v_r = \tilde{m}_r(t)/|V_r|$.
- This assignment remains unchanged until one of the following events occurs:

(1) a task completes (or a new task is enabled, when release dates are considered);
(2) the priority order of tasks is changed;
(3) the profile changes.

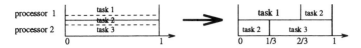

**Figure 5.4** An example of processor sharing with McNaughton's algorithm.

At such moments the processor assignment is recomputed.

In the above scheme the processor sharing can be achieved by McNaughton's (1959) wrap-around algorithm, which is linear in the number of tasks scheduled in each time interval. An example is illustrated in Figure 5.4, where three tasks are executed at speed $2/3$ on two processors during a unit length interval.

Note that other processor sharing schemes may be used. However, they will generate the same latenesses of the tasks provided the corresponding task processing speeds are the same in these processor sharing schemes. Therefore we shall not make any difference between them.

Denote by $p_i^S(t)$ the remaining processing requirement of task $i$ at time $t$ in schedule $S$. Define the laxity of task $i$ at time $t$ in this schedule as $b_i^S(t) = d_i - p_i^S(t)$. We shall consider SLF (smallest laxity first) algorithms. Figure 5.5 shows an SLF schedule for a set of independent tasks whose characteristics are indicated in Table 5.1. The schedule is optimal, which is not true in general when the maximum lateness is under consideration.

Priority schedules are also used for the minimization of makespan. Define the length of the remaining longest path of task $i$ at time $t$ in preemptive schedule $S$ as

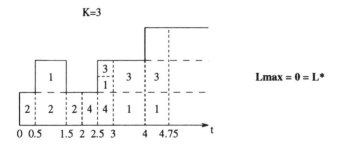

**Figure 5.5** An example of an SLF schedule.

**Table 5.1** A set of independent tasks

| $i$ | $r_i$ | $p_i$ | $d_i$ |
|---|---|---|---|
| 1 | 0 | 3 | 5 |
| 2 | 0 | 2 | 2 |
| 3 | 2 | 2 | 5 |
| 4 | 2 | 1 | 3 |

$r_i^S(t) = h_i + p_i^S(t)$. In an LRP (longest remaining path first) schedule, tasks are ordered by decreasing length of the remaining longest path. This corresponds to the HLF rule when tasks have unit execution times.

### 5.4.2 Relation between optimal nonpreemptive list algorithms and preemptive priority algorithms

It was shown in Liu and Sanlaville (1992) that there is a tight relation between the conditions under which nonpreemptive list algorithms (EDD, HLF) are optimal and those under which preemptive priority algorithms (SLF, LRP) are optimal. In order to state these results, we need the following notions of closure.

A class $\mathscr{G}$ of graphs is said to be closed under expansion if the following property is true for any graph $G = (V, E) \in \mathscr{G}$: for any vertex $i \in V$, if $G'$ is the graph obtained from $G$ by replacing vertex $i$ with a chain of two vertices $i_1$ and $i_2$ such that

$$p(i_1) = p(i), \quad s(i_1) = \{i_2\},$$
$$p(i_2) = \{i_1\}, \quad s(i_2) = s(i)$$

then $G'$ still belongs to the class $\mathscr{G}$.

A class $\mathscr{M}$ of profiles is said to be closed under translation if for any profile $M = \{a_r, m_r\}_{r=1}^{\infty}$ in $\mathscr{M}$, all profiles $M' = \{a'_r, m_r\}_{r=1}^{\infty}$ belong to $\mathscr{M}$, provided $\{a'_r\}_{r=1}^{\infty}$ is an increasing sequence of real numbers.

**Theorem 5.6** *Let $\mathscr{M}$ be a class of profiles that is closed under translation and $\mathscr{G}$ a class of graphs that is closed under expansion. If, for any integer profile $M \in \mathscr{M}$ and for any $G \in \mathscr{G}$ with UET tasks and integer due dates, there exists an EDD schedule minimizing the maximum lateness of $G$ within the class of nonpreemptive policies then, for any $M \in \mathscr{M}$ and any $G \in \mathscr{G}$, the SLF schedule minimizes the maximum lateness of $G$ within the class of preemptive schedules.*

**Proof** This proceeds in two steps. In the first we prove the result for the case where the pair $(G, M)$ has commensurable timing, i.e. the task processing times and due dates, and the profile change times are mutually commensurable. Real numbers $x_1, \ldots, x_r \in \mathbb{R}$ are said to be mutually commensurable if there exist $w \in \mathbb{R}$ and $r$ integers $\alpha_1, \ldots, \alpha_r$ such that $x_i = \alpha_i w$ for all $i = 1, \ldots, r$. In the second step, we extend the result to the general case with arbitrary real timing.

The scheme of the proof in the first step is similar to that of Muntz and Coffman (1970). Roughly speaking, we show that when the graph $G$ is sufficiently expanded, (i) an optimal preemptive solution for a pair $(G, M)$ may be approached arbitrarily closely by considering optimal nonpreemptive schedules, and, (ii) nonpreemptive EDD schedules coincide with the preemptive SLF schedule for $(G, M)$. Putting these two points together yields the desired result. Note that in an expansion, if vertex $i$ is split into two vertices $i_1$ and $i_2$ then their processing times and due dates are defined as follows:

$$p_{i_1} = p_{i_2} = \tfrac{1}{2} p_i,$$
$$d_{i_2} = d_i, \quad d_{i_1} = d_i - p_{i_2}.$$

In the second step we show that when $(G, M)$ has real timings, the absolute difference between the maximum lateness of SLF schedule and the optimal one is bounded by an arbitrarily small constant. This implies that SLF schedule does yield an optimal solution.

If the complements of the task heights are taken as due dates, i.e. $d_i = -h_i$ for all $i \in V$, then the EDD (respectively SLF) rule coincides with the HLF (respectively LRP) rule. It can also be shown that in such a case the maximum lateness coincides with the makespan (Liu and Sanlaville 1992). Therefore we have the following.

**Theorem 5.7** *Let $\mathscr{M}$ be a class of profiles that is closed under translation and $\mathscr{G}$ a class of graphs that is closed under expansion. If, for any integer profile $M \in \mathscr{M}$ and for any $G \in \mathscr{G}$ with UET tasks, there exists an HLF schedule minimizing the makespan of G within the class of nonpreemptive policies then, for any $M \in \mathscr{M}$ and any $G \in \mathscr{G}$, the LRP schedule minimizes the makespan of G within the class of preemptive schedules.*

Note that the above results actually hold in a more general case where the task executions are subject to release dates.

In the remainder of this section we apply these results together with the results of the previous section concerning optimal nonpreemptive scheduling in order to obtain optimal preemptive schedules.

### 5.4.3 Applications

We first consider the maximum lateness of in-forests. For a given in-forest $G \in \mathscr{G}_{\text{if}}$ with processing times $p_1, \ldots, p_n$ and due dates $d_1, \ldots, d_n$, we define an in-forest $G' \in \mathscr{G}_{\text{if}}$ such that $G'$ has the same set of tasks, the same precedence constraints and the same processing times. The due dates in $G'$ are modified as in (1). It can be shown (Liu and Sanlaville 1992) that such a modification on the due dates does not change the maximum lateness of any feasible schedule. The following then follows from Theorems 5.5 and 5.6.

**Corollary 5.2** *If $G \in \mathscr{G}_{\text{if}}$ is an in-forest and $M \in \mathscr{M}_{\text{iz}}$ is an increasing zigzag profile then the SLF schedule defined on the modified due dates minimizes the maximum lateness within the class of preemptive schedules.*

Note that this result extends Theorem 7.3 of Lawler (1982) to the case of increasing zigzag variable profile. It is possible to apply Theorem 5.6 to the case of arbitrary task graph and constant profile with two processors. In such a case a new proof of Theorem 8.3 of Lawler (1982) may be obtained for the case of identical processors (see Sanlaville 1992).

Consider now the makespan minimization problem. For the simplest case of the task graphs—the chains—the following follows from Theorems 5.3 and 5.7.

**Corollary 5.3** *For any graph consisting of chains and for any profile, the LRP schedule is an optimal preemptive schedule for makespan minimization.*

Note that in the preemptive case, scheduling problems for a union of disjoint chains and for a set of independent tasks are equivalent.

In the case of forests, as a consequence of Corollary 5.1 and Theorem 5.7, we obtain the following.

**Corollary 5.4** *If G is an in-forest and M is an increasing zigzag profile, or if G is an out-forest and M is a decreasing zigzag profile, then the LRP schedule minimizes the makespan within the class of preemptive schedules.*

Observe that this result extends a result of Muntz and Coffman (1970) to zigzag variable profiles.

For an arbitrary task graph, since the LOS schedule belongs to the class of HLF schedules, Theorems 5.4 and 5.7 allow us to conclude the following.

**Corollary 5.5** *LRP is an optimal preemptive schedule for makespan minimization of any task graph G under any profile M bounded by 2.*

This last result extends a result of Muntz and Coffman (1969) to variable profiles.

Note that in order to apply Theorems 5.6 and 5.7, the class of task graphs under consideration should be closed under expansion. Thus, we cannot apply Theorems 5.3 and 5.7 to obtain the optimality of LRP for graphs with interval order structure, since this class of graphs does not fulfill the condition.

## 5.5 STOCHASTIC PROFILE SCHEDULING

### 5.5.1 Problem description

In this section we consider the problem of stochastic scheduling under variable profile. We assume that the task processing times are independent and identically distributed random variables having a common exponential distribution. These processing times are independent of the profile $M = \{a_n, m_n\}_{n=1}^{\infty}$, which is a sequence of random vectors.

We assume that the scheduler has no information on the samples of the (remaining) processing times of the tasks. At any time $t$, $a_n \leqslant t < a_{n+1}$, the scheduler may not have any information on the truncated sequence $\{a_l, m_l\}_{l=n+1}^{\infty}$. In other words, the scheduler may not know either the future time epochs when the profile changes or the number of available processors at any future time. Within such a framework, dynamic preemptive scheduling is necessary.

We are interested in the stochastic minimization of makespan. A policy $\pi_*$ is said to stochastically minimize the makespan of $(G, M)$ within the above described class of policies if for any policy $\pi$ in that class the makespan of $\pi_*$ is stochastically smaller than that of $\pi$, where a random variable $X \in \mathbb{R}$ is said to be stochastically smaller than a random variable $Y \in \mathbb{R}$ if for all $x \in \mathbb{R}$, $P[X \leqslant x] \geqslant P[Y \leqslant x]$.

### 5.5.2 Optimal algorithms for constant profiles

When the task graph is an in-forest, and the profile is a constant 2, Chandy and Reynolds (1975) proved that the HLF policy minimizes the expected makespan. Bruno (1985) subsequently showed that HLF stochastically minimizes the makespan. Pinedo and Weiss (1985) extended this last result to the case where tasks at different levels may have different expected task running times. Frostig (1988) further generalized the result of Pinedo and Weiss to include increasing likelihood ratio distributions for the task running times. These results do not hold for systems with three processors; see the counterexamples in Chandy and Reynolds (1975). However, Papadimitriou and Tsitsiklis (1987) proved that for any arbitrarily fixed number of processors, HLF is asymptotically optimal as the number of tasks tends to infinity, provided the task processing times have a common exponential distribution.

Coffman and Liu (1992) investigated the stochastic scheduling of out-forests on identical parallel processors with constant profile. For the uniform out-forests where all the subtrees are ordered by an embedding relation (see the definition below) they showed that an intuitive priority scheduling policy induced by the embedding relation, referred to as the largest tree first (LTF) policy in this chapter, stochastically minimizes the makespan when there are two processors. If, in addition, the out-forests satisfy a uniform root-embedding constraint then the greedy policy stochastically minimizes the makespan for an arbitrary number of processors.

### 5.5.3 Optimal algorithms for variable profiles

Stochastic profile scheduling was first investigated by Liu and Sanlaville (1991). They considered three kinds of task graphs: interval-order graphs, in-forests and out-forests. The results we are going to present in the remainder of this section are due to Liu and Sanlaville (1991), and were actually obtained in a more general framework, namely that of uniform processors, where the processors may have different speeds.

*Interval order graphs*

As in the deterministic UET case, MSF (most successor first) is optimal when the task graph has an interval-order structure.

**Theorem 5.8** *For any interval-order graph $G \in \mathcal{G}_{io}$ and any profile $M$, MSF stochastically minimizes the makespan of $G$.*

**Proof** This uses the uniformization technique, i.e. we can consider a coupled processing model where all processors $1,\ldots,K$, whenever they are available, are continually executing tasks. When a completion occurs, and no task was assigned to some processor, it corresponds to the completion of a fictitious task on this processor. When a task is assigned to a processor, it is assigned a running time equal to the remainder of the running time already underway at that processor. Owing to the memoryless property of exponential distributions, we can see that this coupled model is equivalent in law to the initial one.

Let $G = (V, E)$ be an interval-order graph, and $T(G) = \{T_1, T_2, \ldots, T_g\}$ a partition of $V$ obtained by the equivalence relation on the sets of successors: for all $1 \leqslant i \leqslant g$, $u, v \in T_i$ if and only if $S(u) = S(v)$. The sets $T_1, \ldots, T_g$ are labeled in such a way that for all $1 \leqslant i < j \leqslant g$, $u \in T_i$ and $v \in T_j$ imply $S(u) \supset S(v)$. We define a majorization relation as follows. Let $G^1 = (V^1, E^1)$ and $G^2 = (V^2, E^2)$ be two subgraphs of $G$ obtained by successively deleting vertices of $G$ having no predecessor in $G$ or in the previously obtained subgraphs. It is easy to see that $G^1$ and $G^2$ are in $\mathscr{G}_{io}$. Let $T_i^j = T_i \cap V^j$, $j = 1, 2$, $i = 1, \ldots, g$. Graph $G^1$ is said to be majorized by $G^2$, referred to as $G^1 \prec_s G^2$, if and only if

$$\forall i, \ 1 \leqslant i \leqslant g, \ \sum_{k=1}^{i} |T_k^1| \leqslant \sum_{k=1}^{i} |T_k^2|,$$

Now using the uniformization technique and the above notion of majorization, one proves the following properties.

- Let $G^1 = (V^1, E^1)$ and $G^2 = (V^2, E^2)$ be two subgraphs of $G \in \mathscr{G}_{io}$ obtained by successively deleting vertices of $G$ having no predecessor in $G$ or in the previously obtained subgraphs. If $G^1 \prec_s G^2$ then, under MSF policy, the makespan of $G^1$ is stochastically smaller than that of $G^2$.
- Let $G \in \mathscr{G}_{io}$ be a task graph. Let $\pi$ be a policy that follows the MSF rule all the time except at the first decision epoch. Then the makespan of $G$ under MSF is stochastically smaller than that under $\pi$.

This last property together with a backward induction allow us to conclude Theorem 5.8.

*In-forests*

When the task graph is an in-forest, we have the following.

**Theorem 5.9** *For any in-forest $G \in \mathscr{G}_{if}$ and any profile $M$ bounded by 2, HLF stochastically minimizes the makespan of $G$.*

The scheme of the proof is similar. However, we have to use another majorization relation, referred to as "flatter than" by Chandy and Reynolds (1975). Let $G^1 = (V^1, E^1)$ and $G^2 = (V^2, E^2)$ be two in-forests. Forest $G^1$ is said to be flatter than $G^2$, denoted by $G^1 \prec_f G^2$, if and only if

$$\forall i, \ i \geqslant 0, \quad \sum_{k \geqslant i} N_k(G^1) \leqslant \sum_{k \geqslant i} N_k(G^2),$$

where $N_k(G)$ denotes the number of vertices at level $k$ of graph $G$.

*Out-forests*

Suppose now that the task graph is an out-forest. Even for a profile bounded by two, examples may easily be found for which the HLF policy is not optimal (even in term of expected makespan) (see Coffman and Liu 1992). Instead of HLF, the greedy policy LTF introduced in Coffman and Liu (1992) turns out to be optimal in a subclass of out-forests.

Let $G = (V, E) \in \mathscr{G}_{of}$ be an out-forest. A vertex $v \in V$ and all its successors form a subtree of $G$, denoted by $T_G(v)$ or simply $T(v)$ when there is no ambiguity. We denote by $|T(v)|$ the size of $T(v)$, i.e. its number of vertices.

The largest tree first (LTF) policy is defined as follows: at any decision epoch, LTF assigns the task $v$ whose subtree $T(v)$ is the largest among all subtrees of the enabled tasks to an available processor. In general, LTF policy is not optimal within the class of out-forests $\mathscr{G}_{of}$. Counterexamples were provided in Coffman and Liu (1992). However, within the classes of uniform and r-uniform out-forests (introduced in Coffman and Liu 1992), a policy is optimal if and only if it is LTF.

Let $T_1, T_2 \in \mathscr{G}_{of}$ be two out-trees. An out-tree $T_1$ is said to embed out-tree $T_2$, or $T_2$ is embedded in $T_1$, denoted by $T_1 \succ_e T_2$ or $T_2 \prec_e T_1$, if $T_2$ is isomorphic to a subgraph of $T_1$. Formally, $T_1$ embeds $T_2$ if there exists an injective function $f$ from $T_2$ into $T_1$ such that $\forall u, v \in T_2$, $v \in s(u)$ implies $f(v) \in s(f(u))$. The function $f$ is called an embedding function.

Let $r_1$ and $r_2$ be the roots of the out-trees $T_1$ and $T_2$ respectively. If $T_1 \succ_e T_2$ and if there is an embedding function $f$ such that $f(r_2) = r_1$ then $f$ is a root-embedding function, and we write $T_1 \succ_r T_2$ or $T_2 \prec_r T_1$.

An out-forest $G \in \mathscr{G}_{of}$ is said to be uniform (respectively r-uniform) if all its subtrees $\{T(v), v \in G\}$ can be ordered by the embedding (respectively root-embedding) relation. The class of uniform (respectively r-uniform) forests is denoted by $\mathscr{G}_{uof}$ (respectively $\mathscr{G}_{rof}$). It is clear that $\mathscr{G}_{rof} \subset \mathscr{G}_{uof} \subset \mathscr{G}_{of}$.

The graph illustrated in Figure 5.6 is a uniform out-forest. However, it is not r-uniform. An example of r-uniform out-forest is given in Figure 5.7.

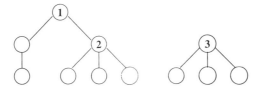

**Figure 5.6** An example of a uniform out-forest.

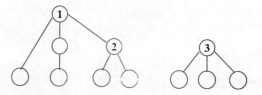

**Figure 5.7** An example of an r-uniform out-forest.

**Theorem 5.10** *LTF stochastically minimizes the makespan of an out-forest $G$ if*

- $G \in \mathscr{G}_{uof}$ is uniform and $M$ is bounded by 2; or
- $G \in \mathscr{G}_{rof}$ is r-uniform and $M$ is arbitrary.

The scheme of the proof is again similar to that of Theorem 5.8, with the majorization relation being defined as the embedding relation between uniform out-forests. Let $G^1 = (V^1, E^1)$ and $G^2 = (V^2, E^2)$ be two uniform out-forests. Assume that the vertices of $G^1$ and $G^2$ are indexed in such a way that

$$T_{G^1}(1) \succ_e T_{G^1}(2) \succ_e \cdots \succ_e T_{G^1}(|V^1|).$$
$$T_{G^2}(1) \succ_e T_{G^2}(2) \succ_e \cdots \succ_e T_{G^2}(|V^2|).$$

An out-forest $G^1$ is embedded in $G^2$, referred to as $G^1 \prec_e G^2$, if and only if

$$|V^1| \leq |V^2|$$

and

$$\forall i, \ 1 \leq i \leq |V^1|, \quad T_{G^1}(i) \prec_e T_{G^2}(i).$$

Similarly, $G^1 \prec_r G^2$ if and only if

$$T_{G^1}(1) \succ_r T_{G^1}(2) \succ_r \cdots \succ_r T_{G^1}(|V^1|),$$
$$T_{G^2}(1) \succ_r T_{G^2}(2) \succ_r \cdots \succ_r T_{G^2}(|V^2|),$$
$$|V^1| \leq |V^2|,$$

and

$$\forall i, \ 1 \leq i \leq |V^1|, \quad T_{G^1}(i) \prec_r T_{G^2}(i).$$

**ACKNOWLEDGMENT**

The work of Eric Sanlaville was partially supported by INRIA while visiting the GERAD Laboratory, Montréal, Canada.

## REFERENCES

Bartusch M., Mohring R. H. and Radermacher F. J. (1988) $M$-machine unit time scheduling: a report on ongoing research. In *Lecture Notes in Economics and Mathematical Systems*, Vol. 304 pp 165–212 Springer, Berlin.

Brucker P., Garey M. R. and Johnson D. S. (1977) Scheduling equal-length tasks under treelike precedence constraints to minimize maximum lateness *Math. Oper. Res.*, **2**, 275–284.

Bruno J. L. (1982) Deterministic and stochastic problems with tree-like precedence constraints. In: Dempster M. A. H. *et al.*, editors, *Deterministic and Stochastic Scheduling* pp 367–374, Reidel, Dordrecht.

Bruno J. L. (1985) On scheduling tasks with exponential service times and in-tree precedence constraints *Acta Informatica*, **22**, 139–148.

Chandy K. M. and Reynolds P. F. (1975) Scheduling partially ordered tasks with probabilistic execution times *Oper. Syst. Rev.*, **9**, 169–177.

Coffman E. G., Jr., editor, (1976) *Computer and Job-Shop Scheduling Theory* Wiley, New York.

Coffman E. G., Jr., and Graham R. L. (1972) Optimal scheduling for two-processor systems *Acta Informatica*, **1**, 200–213.

Coffman E. G., Jr., and Liu Z. (1992) On the optimal stochastic scheduling of out-forests *Oper Res.*, **40**, S67–S75.

Dolev D. and Warmuth M. K. (1984) Scheduling precedence graphs of bounded height *J. Algorithms*, **5**, 48–59.

Dolev D. and Warmuth M. K. (1985a) Scheduling flat graphs *SIAM J. Comput.*, **14**, 638–657.

Dolev D. and Warmuth M. K. (1985b) Profile scheduling of opposing forests and level orders *SIAM J. Alg. Disc. Meth.*, **6**, 665–687.

Frostig E. (1988) A stochastic scheduling problem with intree precedence constraints *Oper. Res.*, **36**, 937–943.

Garey M. R. and Johnson D. S. (1976) Scheduling tasks with nonuniform deadlines on two processors *J. Assoc. Comput. Mach.*, **23**, 461–467.

Garey M. R. and Johnson D. S. (1977) Two-processor scheduling with start-times and deadlines *SIAM J. Comput.*, **6**, 416–426.

Garey M. R., Johnson D. S., Tarjan R. E. and Yannakakis M. (1983) Scheduling opposite forests *SIAM J. Alg. Disc. Meth.*, **4**, 72–93.

Graham R. L., Lawler E. L., Lenstra J. K. and Rinnooy Kan A. H. G. (1979) Optimization and approximation in deterministic sequencing and scheduling: a survey *Ann. Discr. Math.*, **5**, 287–326.

Hu T. C. (1961) Parallel sequencing and assembly line problems *Oper. Res.*, **9**, 841–848.

Lawler E. L. (1982) Preemptive scheduling of precedence constrained jobs on parallel machines. In Dempster M. A. H. *et al.*, editors, *Deterministic and Stochastic Scheduling*, pp 101–123 Reidel, Dordrecht.

Lawler E. L., Lenstra J. K., Rinnooy Kan A. H. G. and Shmoys D. B. (1989) Sequencing and scheduling: algorithms and complexity. Report BS-R8909, CWI, Amsterdam.

Liu Z. and Sanlaville E. (1991) Stochastic scheduling with variable profile and precedence constraints. Rapport de Recherche INRIA, 1525. (Submitted for publication in *SIAM J. Comput.)*

Liu Z. and Sanlaville E. (1992) Preemptive scheduling with variable profile, precedence constraints and due dates. Rapport de Recherche MASI 92.5, Université P. et M. Curie (to appear in *D.A.M.*).

McNaughton R. (1959) Scheduling with deadlines and loss functions *Management Sci.*, **6**, 1–12.

Muntz R. R. and Coffman E. G., Jr. (1969) Optimal preemptive scheduling on two-processor systems *IEEE Trans. Comput.*, **18**, 1014–1020.

Muntz R. R. and Coffman E. G., Jr. (1970) Preemptive scheduling of real-time tasks on multiprocessor systems *J. Assoc. Comput. Mach.*, **17**, 325–338.

Papadimitriou C. H. and Tsitsiklis J. N. (1987) On stochastic scheduling with in-tree precedence constraints *SIAM J. Comput.*, **16**, 1–6.

Papadimitriou C. H. and Yannakakis M. (1978) Scheduling interval-ordered tasks. Report 11.78, Center for Research in Computer Technology, Harvard University.

Pinedo M. and Weiss G. (1985) Scheduling jobs with exponentially distributed processing times and intree precedence constraints on two parallel machines *Oper. Res.*, **33**, 1381–1388.

Sanlaville E. (1992) Conception et analyse d'algorithmes de liste en ordonnancement préemptif. Thèse de l'Université P. et M. Curie.

Ullman J. D. (1975) NP-complete scheduling problems *J. Comput. Syst. Sci.*, **10**, 384–393.

---

*INRIA, Centre Sophia Antipolis, 2004 Route des Lucioles, BP 93, 06902 Sophia Antipolis Cedex, France*

---

*Laboratoire LITP, Case 168, Université Pierre et Marie Curie, 4 Place Jussieu, 75252 Paris Cedex 05, France*

CHAPTER 6

# Efficient Algorithms and a Software Tool for Scheduling Parallel Computation

Apostolos Gerasoulis
*Rutgers University*

Tao Yang
*University of California at Santa Barbara*

**Abstract**

To parallelize an application program for a distributed memory architecture, we can use a precedence task graph to represent the parallelism of this program, schedule tasks onto the given physical processors and then distribute program and data accordingly. We discuss program partitioning techniques for constructing task graphs and present several static scheduling algorithms that consider the overhead of interprocessor communication. Finally, we give an overview of a software system PYRROS that uses scheduling algorithms to generate parallel code for distributed memory parallel machines.

## 6.1 INTRODUCTION

In this chapter we consider the partitioning and scheduling problem for directed acyclic program task graphs (DAG). We emphasize algorithms for scheduling parallel architectures based on the asynchronous message passing paradigm for communication. Such architectures are becoming increasingly popular, but programming them is very difficult since both the data and the program must be partitioned and distributed to the processors. The following problems are of major

---

*Scheduling Theory and its Applications* Edited by P. Chrétienne, E. G. Coffman, Jr., J. K. Lenstra and Z. Liu
© 1995 John Wiley & Sons Ltd

importance for distributed memory architectures:

(1) the program and data partitioning and the identification of parallelism;
(2) the mapping of the data and program onto an architecture;
(3) the scheduling and coordination of the task execution.

From a theoretical point of view, all the above problems are extremely difficult in the sense that finding the optimum scheduling solution is $\mathcal{NP}$-complete in general (Chrétienne 1989a; Coffman and Denning 1973; Lenstra and Rinnooy Kan 1978; Papadimitriou and Yannakakis 1990). In practice, however, parallel programs are written routinely for distributed memory architectures with excellent performance. Thus one of the grand challenges in parallel processing is to build a compiler that will *automatically* partition and parallelize a sequential program and then produce a schedule and generate the target code for a given architecture. For a specialized class of sequential program definitions, the identification of parallelism becomes simpler. However, choosing good partition even in this simple case is difficult, and requires the computation of a schedule.

We discuss the techniques of data and program partitioning, and present an overview of the scheduling problem. We emphasize static scheduling over dynamic, because it is still an open problem how to reduce the run-time overhead of dynamic scheduling for distributed memory architectures. We have addressed the issues of static scheduling and developed algorithms along with a software system named PYRROS (Yang and Gerasoulis 1992). PYRROS takes as an input a task graph and produces schedules for message passing architectures such as nCUBE-II. The current PYRROS prototype has a low complexity and can handle task graphs with millions of tasks.

An automatic system for scheduling and code generation is useful in many ways. If the scheduling is determined at compile time then the architecture can be utilized better. Also a programmer does not have to get involved in low-level programming and synchronization. The system can be used to determine a good program partitioning before actual execution. It can also be used as a testbed for comparing manually written scheduling with an automatically generated scheduling.

## 6.2 PROGRAM PARTITIONING AND DATA DEPENDENCE

We start with definitions of the task computation model and architecture:

*Directed acyclic weighted task graph (DAG)* This is defined by a tuple $G = (V, E, C, T)$, where $V = n_j, j = 1, \ldots, v$ is the set of task nodes and $v = |V|$ is the number of nodes, $E$ is the set of communication edges and $e = |E|$ is the number of edges, $C$ is the set of edge communication costs and $T$ is the set of node computation costs. The value $c_{i,j} \in C$ is the communication cost incurred along the edge $e_{i,j} = (n_i, n_j) \in E$, which is zero if both nodes are mapped in the same processor. The value $\tau_i \in T$ is the execution time of node $n_i \in V$.

*Task* This is an indivisible unit of computation, which may be an assignment statement, a subroutine or even an entire program. We assume that tasks are convex, which means that once a task starts its execution, it can run to completion without interrupting for communications (Sarkar 1989).

*Static macro-dataflow model of execution* (Sarkar 1989; Wu and Gajski 1988; El-Rewini and Lewis 1990) This is similar to the dataflow model. The data flow through the graph and a task waits to receive all data in parallel before it starts its execution. As soon as the task completes its execution, it sends the output data to all successors in parallel.

### 6.2.1 Program partitioning

A program partitioning is a mapping of program statements onto a set of tasks. Since tasks operate on data, their input data must be *gathered* from a data structure and *transmitted* to the task before execution, then operated by the task and finally transmitted and *scattered* back to the data structure. If the data structure is distributed amongst many processors then the *gather/scatter* and *transmission* operations are costly in terms of communication cost unless the data are *partitioned* properly.

We present an example for partitioning the Gaussian elimination (GE) algorithm without pivoting. Figure 6.1(a) shows a fine-grain partitioning where tasks are defined at the statement level

$$u_{ij,k}: \quad a(i,j) = a(i,j) - a(i,k) * a(k,j)/a(k,k).$$

This partitioning fully exposes the parallelism of the GE program, but a fine-grain machine architecture is required to exploit this parallelism. For coarse-grain architectures we need to use coarse-grain program partitionings. Figure 6.1(b) shows a coarse-grain partitioning where the interior loop is taken as one task $U_k^i$. Each task $U_k^i$ modifies row $i$ using row $k$.

### 6.2.2 Data dependence graph

Once a program is partitioned, data dependence analysis must be performed to determine the parallelism in the task graph. For $n = 4$ the fine- and coarse-grain dependence graphs corresponding to Figure 6.1 are depicted in Figure 6.2. The statement-level fine-grain graph has the dependence edges between node $u_{ij,k}$ for $k = 1, \ldots, 3$ and $i,j = k+1, \ldots, 4$. Notice that task $u_{33,2}$ must begin execution after $u_{22,1}$ is completed, since it uses the output of $u_{22,1}$. The direction of the dependence arrow shown in the DAG is determined by using the sequential execution of the *kij* program in Figure 6.1. However, there is no dependence between $u_{22,1}$ and $u_{23,1}$ and they may be executed in parallel. All transitive edges have been removed from the graph.

```
for k = 1 : n − 1                for k = 1 : n − 1
  for i = k + 1 : n                for i = k + 1 : n
    for j = k + 1 : n               U_k^i : { for j = k + 1 : n
       u_{ij,k}                              u_{ij,k}
    end                                   end}
  end                            end
end                            end
```

**Figure 6.1** The *kij* fine-grain (a) and coarse-grain (b) partitionings for GE.

The coarse-grain graph is shown in Figure 6.2 in ovals by aggregating several $u_{ij,k}$ into a coarser-grain task $U_k^i$. We combine the edges between two oval tasks, and a clear picture of the dependence task graph is shown as the U-DAG in Figure 6.4.

### 6.2.3 Algorithms for partitioning

Partitioning algorithms require a cost function to determine if a partitioning is good or not. One widely used cost function is the minimization of the parallel time. Unfortunately, for this cost function the partitioning problem is $\mathcal{NP}$-complete in most cases (Sarkar 1989). However, instead of searching for the optimum partitioning, we can search for a partitioning that has sufficient parallelism for the given architecture and also satisfies additional constraints. The additional constraints must be chosen so that the search space is reduced. An example of such a constraint is to search for tasks of a given maximum size that have no cycles. This is known as the *convexity constraint* in the literature (Sarkar 1989). A *convex* task is nonpreemptive in the sense that it receives all necessary data items before starting

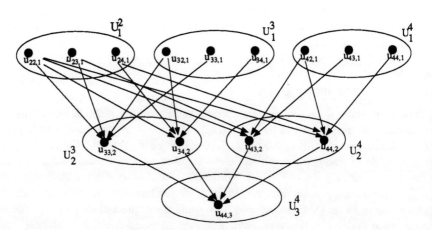

**Figure 6.2** The fine-grain DAG for GE and $n = 4$. Ovals show a coarse-grain partitioning by aggregating small computation units $u_{ij,k}$.

```
for k = 1 : n - 1
    for j = k + 1 : n
        T_k^j : { for i = k + 1 : n
                    u_{ij,k}
                  end }
    end
end
```

**Figure 6.3** The *kji* coarse-grain partitioning for GE.

execution, and completes its execution without any interruption. After that, it sends the data items to the successor tasks that need those data.

*Top-down*

One methodology for program partitioning is to start from the top level (the program) and go down the loop nesting levels until sufficient parallelism is discovered. At each loop level a partitioning is defined by mapping everything below that level in a task. Then a data dependence analysis is performed to find the parallelism at that level. If no sufficient parallelism is found at that level then program transformations such as loop interchange can be performed and the new loop tested again for parallelism. Incorporating this loop interchange program transformation technique can also change the data access pattern of each task.

We show how the top-down approach works for the GE example. There are three nesting loop levels in the program of Figure 6.1. Starting from the top (outer loop), we see that there is no parallelism. At the next level there is parallelism for some of the loops, but the task convexity constraint sequentializes the task graph, so we must go to the next level which is the interior loop level for our program. At the interior loop the tasks are convex, and there is sufficient parallelism for coarse-grain architectures as shown in the U-DAG in Figure 6.4.

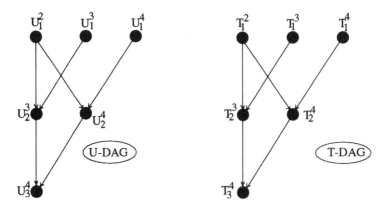

**Figure 6.4** Dependence task graphs corresponding to two coarse-grain partitionings: U-DAG with row data access pattern and T-DAG with column data access pattern.

By loop interchanging loops $j$ and $i$ in the $kij$ GE program and taking the interior loop as a task, the result is the $kji$ form of GE algorithm shown in Figure 6.3. The dependence graph is the T-DAG in Figure 6.4. Each task $T_k^j$ uses column $k$ to modify column $j$.

*Bottom-up*

One difficulty with the top-down approach is that this approach follows the program structure level to partition, and it is difficult to identify an appropriate level other than the statement level that has sufficient parallelism. Thus this approach will usually end up with a fine-grain statement-level task partitioning. If that is the case and we are interested in coarse-grain partitioning then we must go bottom-up to determine such partitioning. Finding an optimal partitioning is $\mathcal{NP}$-complete, and heuristics must be used.

We show an example of the bottom-up approach for Figure 6.2. Given the fine-grain DAG, the partitioning in the ovals is a mapping corresponding to U-DAG coarse-grain DAG. Another coarse-grain partitioning is to aggregate $u_{22,1}, u_{32,1}$ and $u_{42,1}$ into $T_1^2$ and so on; this results in the T-DAG shown in Figure 6.4. The T-DAG and U-DAG have the same dependence structure but different task definitions. The two partitionings are also known as row and column partitionings because of their particular data access patterns.

### 6.2.4 Data partitioning

For shared memory architectures, the data structure is kept in a common shared memory while for distributed memory architectures, the data structure must be partitioned into *data units* and assigned to the local memories of the processors. A data unit can be a scalar variable, a vector or a submatrix block. For distributed memory architectures coarse-grain data partitioning is preferred because there is a high communication startup overhead in transferring a small size data unit. If a task requires to access a large number of distinct data units and data units are evenly distributed among processors then there will be substantial communication overhead in fetching a large number of non-local data items for executing such task. Thus the following property can be used to determine program and data partitionings.

**Consistency** *The program partitioning and data partitioning are consistent if sufficient parallelism is provided by partitioning and at the same time the number of distinct units accessed by each task is minimized.*

Let us assume that the fine-grain task graph in Figure 6.2 is given and also that the data unit is a row of the matrix. Then the program partitioning shown in ovals is consistent with such a data partitioning, and it corresponds to the U-DAG in Figure 6.4. The resulting coarse-grain tasks $U_k^j$ access an extensive number of data elements of rows $k$ and $j$ in each update. Making the data access pattern of a task consistent with data partitioning results in efficient reuse of data that reside in the local cache or the local memory.

Let us now assume that the matrix is partitioned in column data units. Then each task $U_k^j$ needs to access $n$ columns for each update, which results in excessive data movement. On the other hand, T-DAG task partitioning in Figure 6.4 is consistent with column data partitioning, since each task $T_k^j$ only accesses two columns ($k$ and $j$) for each update.

### 6.2.5 Computing the weights for the DAG

Sarkar (1989, p. 139) has proposed a methodology for the estimation of the communication and computation cost for the macro dataflow task model. The computation cost is the time $E$ for a task to execute on a processor. The communication cost consists of two components:

(1) *the processor component*, which is the time that a processor participates in communication; the cost is expressed by the reading and writing functions $R$ and $W$;
(2) *transmission delay component*, which is the time $D$ for the transmission of the data between processors; during that time, the processors are free to execute other instructions.

The weights can be obtained from

$$\tau_i \approx E_i, \qquad c_{ij} \approx R_i + D_{i,j} + W_j.$$

The parameters $R_i, D_{i,j}$ and $W_j$ are functions of the *message size*, the *network load* and the *distance* between the processors. When there is no network contention, a very common approximation to $c_{i,j}$ is the *linear model*:

$$c_{i,j} \approx (\alpha + k\beta)d(i,j),$$

where $\alpha$ is known as the startup time, $\beta$ is the transmission rate, $k$ is the size of the message transmitted between tasks $n_i$ and $n_j$, and $d(i,j)$ is the processor distance between tasks $n_i$ and $n_j$. This linear communication model is a good approximation to most currently available message passing architectures (see Dunigan 1991). For the nCUBE-II hypercube we have $\alpha = 160$ $\mu$s and $\beta = 2.4$ $\mu$s per word transfer for single-precision arithmetic.

For the GE example, if $\omega$ is the time that it takes for each $u_{ij,k}$ operation then $\tau_{kj} = (n-k)\omega$ for task $T_k^j$ in the T-DAG (or $U_k^i$ in the U-DAG) of Figure 6.4. The communication weights are all equal to $[\alpha + (n-k)\beta]d(T_k^j, T_{k+1}^j)$, since only $n-k$ elements of the data unit are modified in $T_{k+1}^j$.

Of course, for some task graphs the computation and communication weights or even the dependence structure can only be determined at run time. For such cases run-time scheduling techniques are useful (see e.g. Saltz et al. 1990).

## 6.3 GRANULARITY AND THE IMPACT OF PARTITIONING ON SCHEDULING

### 6.3.1 Scheduling and clustering definitions

Scheduling is defined by a processor assignment mapping, $PA(n_j)$, of the tasks onto the $p$ processors, and by a starting times mapping, $ST(n_j)$, of all nodes onto the real positive numbers set. Figure 6.5(a) shows a weighted DAG with all computation weights assumed to be equal to 1. Figure 6.5(b) shows a processor assignment using two processors. Figure 6.5(c) shows a *Gantt chart* of a schedule for this DAG. The Gantt chart completely describes the schedule, since it defines both $PA(n_j)$ and $ST(n_j)$. The scheduling problem with communication delay has been shown to be $\mathcal{NP}$-complete for a general task graph in most cases (Sarkar 1989; Chrétienne 1989a; Papadimitriou and Yannakakis 1990).

*Clustering* is a mapping of the *tasks* onto *clusters*. A *cluster* is a set of tasks that will execute on the same processor. Clusters are not tasks, since tasks that belong to a cluster are permitted to communicate with the tasks of other clusters immediately after completion of their execution. The clustering problem is identical to processor assignment part of scheduling. Sarkar (1989) calls it an *internalization prepass*. Clustering is also $\mathcal{NP}$-complete for the minimization of the parallel time (Chrétienne 1989a; Sarkar 1989).

A clustering is called *nonlinear* if two independent tasks are mapped in the same cluster; otherwise it is called *linear*. In Figure 6.6(a) we give a weighted DAG, in Figure 6.6(b) a linear clustering with three clusters $\{n_1, n_2, n_7, n_3, n_4, n_6\}$ and $\{n_5\}$, and in Figure 6.6(c) a nonlinear clustering with clusters $\{n_1, n_2, n_3, n_4, n_5, n_6\}$ and $\{n_7\}$. Notice that for the nonlinear cluster independent tasks $n_4$ and $n_5$ are mapped in the same cluster.

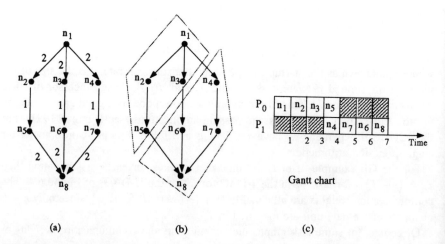

**Figure 6.5** (a) DAG with node weights equal to 1. (b) A processor assignment of nodes. (c) The Gantt chart of a schedule.

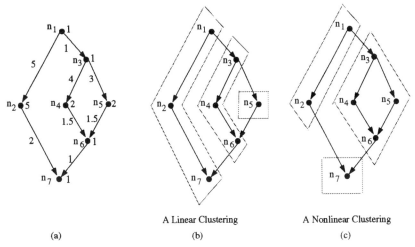

**Figure 6.6** (a) A weighted DAG. (b) A linear clustering. (c) A nonlinear clustering.

In Figure 6.7(a) we present the Gantt chart of a schedule for the nonlinear clustering of Figure 6.6(c). Processor $P_0$ has tasks $n_1$ and $n_2$ with starting times $ST(n_1) = 0$ and $ST(n_2) = 1$. If we modify the clustered DAG as in Sarkar (1989) by adding a zero-weighted pseudo edge between any pair of nodes $n_x$ and $n_y$ in a cluster, if $n_y$ is executed immediately after $n_x$ and there is no data dependence edge between $n_x$ and $n_y$, then we obtain what we call a *scheduled DAG*. Figure 6.7(b) is a scheduled DAG, and the dashed edge between $n_4$ and $n_5$ shows the pseudo-execution edge.

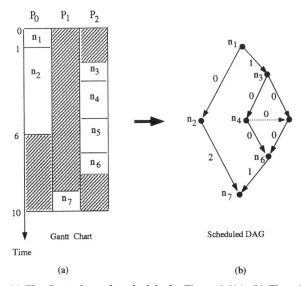

**Figure 6.7** (a) The Gantt chart of a schedule for Figure 6.6(c). (b) The scheduled DAG.

We call the longest path of the scheduled DAG the *dominant sequence* (DS) of the clustered DAG, to distinguish it from the *critical path* (CP) of a clustered but not scheduled DAG. For example, the clustered DAG in Figure 6.6(c) has the sequence $\langle n_1, n_2, n_7 \rangle$ as its CP with length 9, while a DS of this clustered DAG is $DS = \langle n_1, n_3, n_4, n_5, n_6, n_7 \rangle$ and has length 10 using the schedule of Figure 6.7(b). In the case of linear clustering the DS and CP of the clustered DAG are identical (see Figure 6.6b).

### 6.3.2 The granularity theory

One goal of partitioning is to produce a DAG that has sufficient parallelism for a given architecture. Another is to have a partition that minimizes the parallel time. These two goals are in conflict, because having a partitioning with a high degree of parallelism does not necessarily imply the minimization of the parallel time, unless communication cost is zero. It is therefore the communication and computation costs derived by a partitioning that will determine the "useful parallelism" that minimizes the parallel time. This has been recognized in the literature, as can be seen from the following quote from Heath and Romine (1988, p. 559):

> Another important characteristic determining the overall efficiency of parallel algorithms is the relative cost of communication and computation. Thus, for example, if communication is relatively slow, then coarse-grain algorithms in which relatively large amount of computation is done between communications will be more efficient than fine-grain algorithms.

Let us consider the task graph in Figure 6.8. If the computation cost $w$ is greater or equal to the communication cost $c$ then the parallel time is minimum when $n_2$ and $n_3$ are executed in two separate processors as shown in Figure 6.8(c). In this case all parallelism in this partitioned graph can be fully exploited, since it is "useful parallelism". If, on the other hand, we assume that $w < c$ then the parallelism is not "useful", since the minimum parallel time is derived by sequentializing the tasks $n_2$ and $n_3$ as shown in Figure 6.8(b).

Notice that linear clustering preserves the parallelism embedded in the DAG while nonlinear clustering does not. We make the following observation.

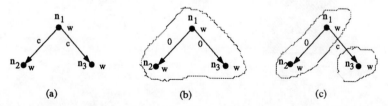

**Figure 6.8** Sequentialization versus parallelization: (a) a weighted DAG; (b) sequentialization using nonlinear clustering; (c) parallelization using linear clustering.

*If the execution of a DAG uses linear clustering and attains the optimal time then this indicates that the program partitioning is appropriate for the given architecture; otherwise the partitioning is too fine and the scheduling algorithm still needs to execute independent tasks together in the same processor using the nonlinear clustering strategy.*

It is therefore of interest to know when we can fully exploit the parallelism in a given task graph. In this section we assume that the architecture has an unbounded number of processors that are completely connected (clique).

In Figure 6.8 we saw the impact of the ratio $w/c$ on scheduling a simple DAG. The interesting question arises as to whether this analysis can be generalized to arbitrary DAGs. In Gerasoulis and Yang (1993) we have introduced a new notion of *granularity* using the ratio of the computation to communication costs taken over all fork-and-join subgraphs of a task graph. The importance of this choice of granularity definition will become clear later on.

A DAG consists of *fork* and/or *join* sets such as those shown in Figure 6.9. The join set $J_x$ consists of all immediate predecessors of node $n_x$. The fork set $F_x$ consists of all immediate successors of node $n_x$. Let $J_x = \{n_1, n_2, \ldots, n_m\}$ and $F_x = \{n_1, n_2, \ldots, n_m\}$ and define

$$g(J_x) = \frac{\min_{k=1,\ldots,m} \{\tau_k\}}{\max_{k=1,\ldots,m} \{c_{k,x}\}}, \quad g(F_x) = \frac{\min_{k=1,\ldots,m} \{\tau_k\}}{\max_{k=1,\ldots,m} \{c_{x,k}\}}.$$

We define the *grain* of a task $n_x$ and the *granularity* of a DAG as

$$g_x = \min\{g(G_x), g(J_x)\}, \quad g(G) = \min_{x=1,\ldots,v} \{g_x\}$$

respectively. We call a DAG *coarse-grain* if $g(G) \geq 1$; otherwise it is *fine-grain*. If all task weights are equal to $R$ and all edge weights are equal to $C$ then the granularity reduces to $R/C$, which is the same as that of Stone (1987). For coarse-grain DAGs each task receives or sends data with a small amount of communication cost compared with the computation cost.

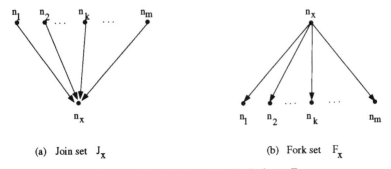

(a) Join set $J_x$          (b) Fork set $F_x$

**Figure 6.9** (a) Join set $J_x$. (b) Fork set $F_x$.

For example, the granularity of the graph in Figure 6.6(a) is $g = 1/5$, which is derived as follows. The node $n_1$ is a fork and its *grain* is $g_1 = 1/5$, the ratio of the minimum computation weights of its successors $n_2$ and $n_3$, and the maximum communication cost of the outgoing edges. The node $n_2$ is in both a fork and a join, and the grain for the join is 1/5, which is the computation weight of its only predecessor $n_1$ and the cost of the edge $(n_1, n_2)$, while the grain for the fork is the weight of $n_7$ over the weight of $(n_2, n_7)$, which is again 1/2. Continuing, we finally determine the granularity as the minimum grain over all nodes of the graph, which in our case is $g = 1/5$.

In Gerasoulis and Yang (1993) we prove the following theorems.

**Theorem 6.1** *For a coarse-grain task graph there exists a linear clustering that minimizes the parallel time.*

The above theorem is true only for our definition of granularity, which is why we have chosen it. We demonstrate the basic idea of the proof by using the example in Figure 6.8. We show in Gerasoutis and Yang (1993) that for any nonlinear clustering we can extract a linear clustering whose parallel time is less than or equal to the nonlinear clustering. If we assume that $w \geq c$ in Figure 6.8 then the parallel time of the nonlinear clustering in Figure 6.8(b) is $3w$. By extracting $n_3$ from the nonlinear clustering and making it a new cluster, we derive a linear clustering shown in Figure 6.8(c), whose parallel time is $2w + c \leq 3w$. We can always perform this extraction as long as the task graph is coarse-grain.

Theorem 6.1 shows that the problem of finding an optimal solution for a coarse-grain DAG is equivalent to that of finding an optimal linear clustering. Picouleau (1992) has shown that the scheduling problem for coarse grain DAGs is $\mathcal{NP}$-complete; therefore optimal linear clustering is $\mathcal{NP}$-complete.

**Theorem 6.2** *Determining the optimum linear clustering is $\mathcal{NP}$-complete.*

Thus, even though linear clustering is a nice property for task graphs, determining the optimum linear clustering is still a very difficult problem. Fortunately, for coarse-grain DAGs any linear clustering algorithm guarantees performance within a factor of two of the optimum, as the following theorem demonstrates.

**Theorem 6.3** *For any linear clustering algorithm we have*

$$PT_{opt} \leq PT_{lc} \leq \left[1 + \frac{1}{g(G)}\right] PT_{opt},$$

*where $PT_{opt}$ is the optimum parallel time and $PT_{lc}$ is the parallel time of the linear clustering. Moreover, for a coarse-grain DAG we have*

$$PT_{lc} \leq 2 PT_{opt}.$$

Notice that when communication tends to zero, we have $g(G) \to +\infty$ and $PT_{opt} = PT_{lc}$. The above theorems provide an explanation of the advantages of linear clustering, which has been widely used in the literature particularly for coarse grain dataflow graphs (see e.g. Geist and Heath 1986; Kim and Browne 1988; Kung 1988; Ortega 1988; Saad 1986). We present an example.

**Example** A widely used assumption for clustering is "the owner computes rule" (Callahan and Kennedy 1988), i.e. a processor executes a computation unit if this unit modifies the data that the processor owns. This rule can perform well for certain *regular* problems, but in general it could result in workload imbalances—especially for unstructured problems. The "owner computes rule" has been used to cluster both the U_DAG and the T_DAG in Figure 6.4, see Saad (1986), Geist and Heath (1986) and Ortega (1988). This assumption results in the following clusters for the U-DAG shown in Figure 6.10:

$$M_j = \{U_1^j, U_2^j, \ldots, U_k^j, \ldots, U_{j-1}^j\}, \quad j = 2, \ldots, n.$$

For each cluster $M_j$ row $j$ remains local in that cluster, while it is modified by rows $1, \ldots, j-1$ (similarly for columns in the T-DAG). The tasks in $M_j$ are chains in the task graph in Figure 6.10 that imply that linear clustering was the result of the "owner computes rule". We call this special clustering the *natural linear clustering*.

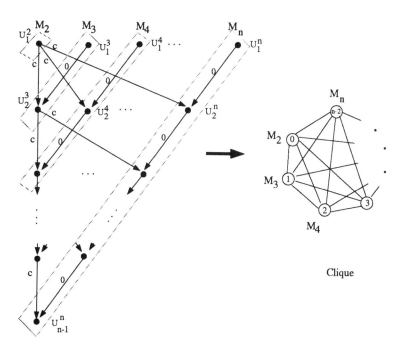

**Figure 6.10** The natural linear clustering for the U-DAG executed on a clique with $p = n - 1$ processors.

What is so interesting about the natural linear clustering? Let us assume that the computation size of all tasks is equal to $\tau$, and communication weights are equal to $c$ in the U-DAG. Then the following theorem holds.

**Theorem 6.4** *The natural linear clustering is optimal for executing the U-DAG on a clique architecture with $n - 1$ processors, provided the granularity $g = \tau/c \geq 1$.*

An application of Theorem 6.4 is the *kji* column partitioning form of the Gauss–Jordan (GJ) algorithm. At each step of the GJ algorithm all $n$ elements of a column are modified and then transmitted to the successor tasks. The weights are then given by

$$\tau = n\omega, \quad c = \alpha + n\beta$$

and, as long as $n\omega/(\alpha + n\beta) \geq 1$, the GJ natural linear clustering is optimum. For the GE DAG, the weight of a task $U_k^i$ in U-DAG or $T_k^j$ in T-DAG is $(n - k)\omega$ and its incoming edge weights are $\alpha + (n - k)\beta$. For large $n$ only a small portion in the bottom of the DAG is fine-grain, and the natural clustering is asymptotically optimal by ignoring the insignificant low-order computation cost in this bottom portion.

We summarize our conclusion of this section as follows. For a program with coarse-grain partitioning, linear clustering is sufficient to produce a good result. For a program with fine-grain partitioning, linear clustering that preserves the parallelism of a DAG could lead to high communication overhead.

The granularity theory is a characterization of the relationship between partitioning and scheduling. In a real situation some parts of a graph could be fine and others coarse. In such cases clustering and scheduling algorithms are needed to identify such parts and use proper clustering strategies to obtain the shortest parallel time. We consider these problems next.

## 6.4 SCHEDULING ALGORITHMS FOR MIMD ARCHITECTURES

We distinguish between two classes of scheduling algorithms. The one-step methods schedule a DAG directly on the $p$ processors. The multistep methods perform a clustering step first, under the assumption that there is an unlimited number of completely connected processors, and then in the following steps the clusters are merged and scheduled on the $p$ available processors. We consider heuristics that have the following properties:

(1) they do not duplicate the same tasks in two different processors;
(2) backtracking is allowed only if the cost is small.

### 6.4.1 One-step scheduling methods

We present two methods. One is the classical list scheduling and the other is the modified critical path (MCP) heuristic proposed by Wu and Gajski (1988).

### 6.4.1.1 The classical list scheduling heuristic

The classical list scheduling schedules free[†] tasks by scanning a priority list from left to right. More specifically, the following steps are performed.

(1) Determine a priority list.
(2) When a processor is *available* for execution, scan the list from left to right and schedule the first free task. If two processors are available at the same time, break the tie by scheduling the task in the processor with the smallest processor number.

When communication cost is zero, a good choice for a priority list is the critical path (CP) priority list. The priority of a task is its *bottom-up level*, the length of the longest path from it to an exit node. The CP list scheduling possesses many nice properties when communication cost is zero. For example, it is optimum for tree DAGs with equal weights and for any arbitrary DAG with equal weights on two processors (Coffman and Denning 1973). For arbitrary DAGs and $p$ processors any list scheduling including CP is within 50% of the optimum. Moreover, the experimental results of Adam et al. (1974) show that CP is near-optimum in practice in the sense that it is within 5% of the optimum in 90% of randomly generated DAGs. Unfortunately, these nice properties do not carry over to the case of nonzero communication cost.

In the presence of communication, it is extremely difficult to identify a good priority list. This is because the communication edge weight becomes zero when its end nodes are scheduled in the same processor, and this makes the computation of the level priority information nondeterministic.

Let us consider the CP algorithm in the case where the level computation includes both edge communication and node computation. For example, a task graph is shown in Figure 6.11(a) along with a list schedule based on the highest-level first-priority list. The level of $n_6$ is 2 and the level of $n_3$ is 4, which is equal to the maximum level of all successor tasks, which is 2, plus the communication cost in the edge $(n_3, n_6)$, which is 1, plus the computation cost of $n_3$, which is 1. The resulting priority list is $\{n_1, n_2, n_5, n_4, n_3, n_6, n_7\}$. Both $n_1$ and $n_2$ are free and the processors $P_0$ and $P_1$ available. At time 0, $n_1$ is scheduled in $P_0$ first, and in the next step $n_2$ is scheduled in the only available processor $P_1$. At time 1, the tasks $n_3, n_4$ and $n_5$ are free, and, since $n_5$ has the highest priority, it is scheduled in processor $P_0$, while the next highest priority $n_4$ is scheduled in processor $P_1$. Even if $n_4$ was scheduled in $P_1$, it needs to wait 4 unit times to receive the data from $P_0$, and thus $n_4$ is ready to start its execution at time 5. The task $n_5$ scheduled in $P_0$ can start execution immediately, since the data are local in that processor. Continuing in a similar manner, we get the final schedule shown in Figure 6.11(b) with $PT = 10$.

---

[†] A task is *free* if all of its predecessors have completed execution. A task is *ready* if it is free and all of the data needed to start its execution are available locally in the processor where the task has been scheduled.

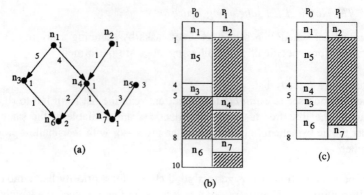

**Figure 6.11** (a) A DAG. (b) The schedule by CP. (c) The schedule by MCP.

One problem with the CP heuristic in the presence of communication is that it schedules a *free* task when a processor becomes available, even though this task is not *ready* to start execution yet. This could result in poor performance, as shown in Figure 6.12(b). Task $n_3$ is scheduled in $P_1$, since it becomes free at time $w$. When $cw$ a better solution is to schedule $n_3$ to $P_0$, as shown in Figure 6.12(c). We now present a modification of the CP heuristic.

### 6.4.1.2 The modified critical path (MCP) heuristic

Wu and Gajski (1988) have proposed a modification of the CP heuristic. Instead of scheduling a free task in an available processor, the free task is scheduled in the available processor that allows the task to start its execution at the *earliest* possible time. The computation of the priorities again uses the highest bottom-up level including both communication and computation costs. For the example in Figure 6.12(a) the priority list is $\{n_1, n_2, n_3\}$. The schedule is shown in Figure 6.12(c). The task $n_3$ becomes free at time $w$, and it is scheduled in processor $P_0$ because it can start its execution at time $2w$, which is earlier than the time $w + c$, since $c > w$.

For the example in Figure 6.11(a) the priority list is the same as in CP: $\{n_1, n_2, n_5, n_4, n_3, n_6, n_7\}$. After $n_1, n_2$ and $n_5$ are scheduled, task $n_4$ has the highest priority and is free at time 2, but is not ready at that time unless it is scheduled at $P_0$.

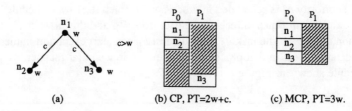

**Figure 6.12** (a) A fork DAG. (b) The schedule by CP. (c) The schedule by MCP.

Now $n_4$ is picked up for scheduling, and it is scheduled in processor $P_0$ because it can start executing at time 4, which is earlier than time 5 if it was scheduled in $P_1$. The parallel time reduces to $PT = 8$, as depicted in Figure 6.11(c).

Even though the MCP performs better than CP, it could still perform poorly, as can be seen in the scheduling of a join DAG shown in Figure 6.13. MCP gives the same schedule as CP, and if the communication cost is greater than the computation cost then the optimum schedule executes all tasks in one processor. The MCP cannot recognize this, since it uses the earliest starting time principle and it starts both $n_2$ and $n_3$ at time 0. One weakness of such one-pass scheduling is that the task priority information is nondeterministic because the communication cost between tasks will become zero if they are allocated in the same processor.

It has been argued by Sarkar (1989) and Kim and Browne (1988) that a better approach to scheduling when communication is present is to perform scheduling in more than one step. We discuss this approach next.

### 6.4.2 Multistep scheduling methods

#### 6.4.2.1 Sarkar's approach

Sarkar's (1989) heuristic is based on the assumption that a scheduling prepass is needed to cluster tasks with high communication between them. Then the clusters are scheduled on $p$ available processors. More specifically, Sarkar advocates the following two-step method.

(1) Determine a clustering of the task graph by using scheduling on an unbounded number of processors and a clique architecture.
(2) Schedule the clusters on the given architecture with a bounded number of processors.

Sarkar uses the following heuristics for the above two steps.

(1) Zero the edge with the highest communication cost. If the parallel time does not increase then accept this zeroing. Continue with the next highest edge until all edges have been visited.
(2) After $u$ clusters are derived, schedule those clusters to $p$ processors by using a priority list. Assuming that the $v$ task nodes are sorted in a descending order of

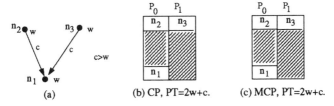

**Figure 6.13** (a) A join DAG. (b) The schedule by CP. (c) The schedule by MCP.

their priorities and the nodes are scanned from left to right. The scanned node, along with the cluster to which it belongs, is mapped on one of the $p$ processors that results in *the minimum increases in parallel time*. The parallel time is determined by executing the examined clusters in the physical processors and the unexamined clusters in virtual processors.

Let us see how this two-step method works for the example in Figure 6.11(a). Initially, the parallel time is 10. Sarkar's first clustering step zeroes the highest communication edge $(n_1, n_3)$, and the parallel time does not increase and this zeroing is accepted. The next highest edge $(n_1, n_4)$ is zeroed, and the parallel time reduces, by executing $n_3$ either before or after $n_4$, so that this zeroing is also accepted. Assume that $n_3$ is executed before $n_4$. Next, the edge $(n_5, n_7)$ is zeroed, and after that the edge $(n_4, n_6)$, and the parallel time reduces to 5, which is determined by a $DS = \langle n_1, n_3, n_4, n_6 \rangle$. By zeroing both $(n_2, n_4)$ or $(n_4, n_7)$, the parallel time increases and these zeroings are not accepted. The final result is three clusters: $M_1 = \{n_1, n_3, n_4, n_6\}, M_2 = \{n_2\}$ and $M_3 = \{n_5, n_7\}$, shown in Figure 6.14(a).

Assume there are two processor $P_0$ and $P_1$ available. The second step in Sarkar's algorithm determines a priority list based on the highest-level first principle. The initial list is $\{n_2, n_1, n_5, n_3, n_4, n_6, n_7\}$, because the level of $n_2$ is 5 while the level of $n_1$ is 4, and so on. The algorithm first picks $n_2$ to schedule; let us assume that it is scheduled in processor $P_1$. Next, the task $n_1$ is chosen to be scheduled. If it is scheduled to $P_0$ then all nodes in $M_1$ are scheduled to $P_0$, and $PT$ is 5. If it is scheduled to $P_1$ then $PT$ becomes 6, since now $n_1$ and $n_2$ must be sequentialized. Thus we assign $M_1$ to $P_0$. Next, $n_5$ is scanned, and it is scheduled to $P_0$, otherwise scheduling to $P_1$ will make $PT = 9$. Next, $n_3$ is scanned; if it is assigned to $P_1$ then all other nodes in $M_1$ will be reassigned to $P_1$, and $PT = 10$. Thus $n_3$ remains in $P_0$. Finally, we have the schedule shown in Figure 6.14(c).

### 6.4.2.2 PYRROS's multistep scheduling algorithms

The PYRROS tool (Yang and Gerasoulis 1992) uses a multistep approach to scheduling:

(1) perform clustering using the dominant sequence algorithm (DSC);

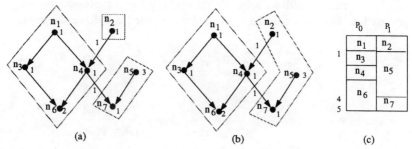

**Figure 6.14** (a) The clustering result. (b) Clusters after merging. (c) The schedule with $PT = 5$.

(2) merge the $u$ clusters into $p$ completely connected virtual processors if $u > p$;
(3) map the $p$ virtual processors into $p$ physical processors;
(4) order the execution of tasks in each processor.

This approach has similarities to Sarkar's two-step method. There is, however, a major difference. The algorithms used here are faster in terms of complexity. This is because we should like to test the multistep method on real applications and parallel architectures, and higher-complexity algorithms offer very little performance gains, especially for coarse-grain parallelism.

*The DSC clustering algorithm*

Sarkar's clustering algorithm has a complexity $O(e(v+e))$. Furthermore, zeroing the highest communication edges is not the best approach, since this edge might not belong in the DS, and as a result the parallel time cannot be reduced. In Yang and Gerasoulis (1991) and Gerasoulis and Yang (1992) we have proposed a new clustering algorithm called the DSC algorithm, which has been shown to outperform other algorithms from the literature—both in terms of complexity and parallel time. The DSC algorithm is based on the following heuristic.

- The parallel time is determined by the DS. Therefore if we want to reduce it, we must zero at least one edge in the DS.
- A DS zeroing-based algorithm could zero one or more edges in DS at a time. This zeroing can be done incrementally in a sequence of steps.
- A zeroing should be accepted if the parallel time reduces from one step to the next.

We show how DSC works for the example of Figure 6.11(a). Figure 6.15(a) is the initial clustering. The DS is shown by thick arrows. There are two dominant sequences in Figure 6.15(a) with $PT = 10$. In the first step the edge $(n_1, n_3)$ in one DS is zeroed as shown in Figure 6.15(b). The new DS is $\langle n_1, n_4, n_6 \rangle$ and $PT = 10$. This zeroing is accepted, since $PT$ does not increase. In the second step $(n_1, n_4)$ is zeroed and $n_4$ is added as the last task of cluster $n_1, n_3$, which results in two new DS, $\langle n_1, n_3, n_4, n_6 \rangle$ and $\langle n_5, n_7 \rangle$ (shown in Figure 6.15c) with $PT = 7$, and this zeroing is also accepted. In the third step $(n_4, n_6)$ is zeroed as shown in Figure 6.15(d), and this zeroing is accepted, since $PT = 7$—determined by the DS $\langle n_5, n_7 \rangle$. Next, $(n_5, n_7)$ is zeroed, and the $PT$ is reduced to 5. Finally, $(n_2, n_4)$ and $(n_4, n_7)$ cannot be zeroed, because zeroing them will increase the parallel time. Thus three clusters are produced.

Notice that in the third step shown in Figure 6.15(c) an ordering algorithm is needed to order the tasks in the nonlinear cluster, and then the parallel time must be computed to get the new DS. One of the key ideas in the DSC algorithms is that it computes the schedule and parallel time incrementally from one step to the next in $O(\log v)$ time. Thus the total complexity is $O((v+e) \log v)$. If the parallel time were not computed incrementally, then the total cost would be greater than $O(v^2)$,

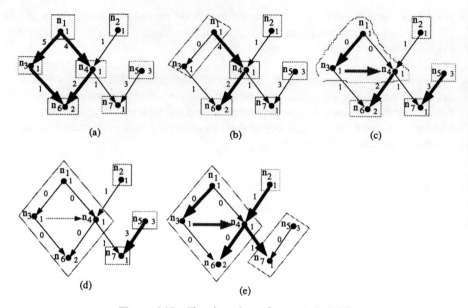

**Figure 6.15** The clustering refinements in DSC.

which is not practical for large graphs. More details can be found in Yang and Gerasoulis (1991).

The $\mathcal{NP}$-completeness of clustering for parallel time minimization has been shown by Sarkar (1989), Chrétienne (1989a) and Papadimitriou and Yannakakis (1990). Chrétienne (1989b) shows that the problems of scheduling a join-and-fork DAG or coarse-grain tree DAG are solvable in polynomial time, but the complexity jumps to $\mathcal{NP}$-complete for scheduling fine-grain tree DAGs and a DAG structure obtained by concatenating a fork and a join together. In Yang and Gerasoulis (1991) we show that DSC performs well for general DAGs by examining a set of randomly generated DAGs, but also produces the following optimal solutions.

**Theorem 6.5** *DSC is optimal for fork-and-join and coarse-grain tree DAGs.*

*Cluster merging*

The cost of the Sarkar cluster merging and scheduling algorithm is $O(pv(v + e))$ which is time-consuming for a large graph. PYRROS uses a variation of the *work profiling method* suggested by George *et al.* (1986) for cluster merging. This method is simple, and has been shown to work well in practice by, for example, Saad (1986), Geist and Heath (1986), Ortega (1988) and Gerasoulis and Nelken (1989). The complexity of this algorithm is $O(u \log u + v)$, which is less than $O(v \log v)$. The procedure is as follows.

(1) Compute the arithmetic load $LM_j$ for each cluster.

(2) Sort the clusters in an increasing order of their loads.
(3) Use a load balancing algorithm so that each processor has approximately the same load.

Let us consider an example. For the GE U-DAG in Figure 6.10 there are $n-1$ clusters $M_2, M_3, \ldots, M_n$. We have

$$LM_j = \sum_{i=1}^{j}(n-i)\omega \approx (nj - \tfrac{1}{2}j^2)\omega.$$

These clusters can be approximately load-balanced by using the wrap or reflection mapping, $VP(j) = (j-2) \bmod p$ (Geist and Heath 1986).

For the example in Figure 6.14(a) with three clusters and two processors, the result of merging is two clusters, as shown in Figure 6.14(b).

*Physical mapping*

We now have $p$ virtually processors (or clusters) and $p$ physical processors. Since the physical processors are not completely connected, we must take the processor distance into account. Determining the optimum mapping of the virtual to physical processors is a very difficult problem, since it can be instantiated as a graph isomorphism problem.

Let us define $TC_{i,j}$ to be the total communication, which is the sum of the costs of all edges between virtual processors $i$ and $j$. $CC = \{TC_{i,j} : TC_{i,j} \neq 0\}$ and $m = |CC|$. In general, we expect that $m \ll e$.

The goal of the physical mapping is to determine the physical processor number $P(V_i)$ for each virtual processor $V_i$ that minimizes the following cost function:

$$F(CC) = \sum_{TC_{i,j} \in CC} distance(P(V_i), P(V_j)) TC_{i,j}.$$

Figure 6.16 is an example of a physical mapping for a T-DAG. A clustering for this DAG is shown in Figure 6.16(a). The total communication between four virtual processors (clusters) is shown in Figure 6.16(b). Figure 6.16(c) shows one physical mapping to a 4-node hypercube with $F(CC) = 24$, and Figure 6.16(d) shows another mapping, with $F(CC) = 21$.

Currently we use a heuristic algorithm due to Bokhari (1990). This algorithm starts from an initial assignment, and then performs a series of pairwise interchanges so that the $F(CC)$ reduces monotonically as shown in the example above.

*Task ordering*

Once the physical mapping has been decided, a task ordering is needed to define the scheduling. Since we no longer move tasks between processors, the communication cost between tasks becomes deterministic. We show how important task ordering is with an example. The processor assignment along with communication and

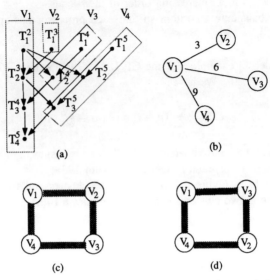

**Figure 6.16** An example of physical mapping. Each nonzero edge cost is 3 time units. (a) a T-DAG linear clustering. (b) Virtual cluster graph. (c) A mapping to a hypercube. (d) A better mapping.

computation weights are shown in Figure 6.17(a). In Figure 6.17(b) we show one ordering with $PT = 12$ and in Figure 6.17(c) another ordering, in which the parallel time increases to $PT = 15$.

Finding a task ordering that minimizes the parallel time is $\mathcal{NP}$-hard (Hoogeveen et al. (1992). We have proposed a modification of the CP heuristic for the ordering problem in Yang and Gerasoulis (1995). This heuristic, ready critical path (RCP), costs $O(v \log v + e)$, and is as follows.

(1) Adjust the communication edges of the DAG based on the processor assignment and physical distance.
(2) Determine a global priority list based on the highest-level first principle. The level computation includes both communication and computation cost in a path.
(3) In addition to the global priority list, each processor maintains a priority list of *ready* tasks for each processor. The ready task with the highest priority is executed as soon as this processor becomes free.

Let us consider the processor assignment in Figure 6.17(a). The level priorities of tasks are $L(n_1) = 12, L(n_2) = 7, L(n_3) = 1, L(n_4) = 1, L(n_5) = 2$ and $L(n_6) = 2$. The priority list is $\{n_1, n_2, n_5, n_6, n_3, n_4\}$. Initially, $n_1$ is ready and is scheduled on the first processor. At time 5, $n_2$ and $n_5$ are ready in the second processor, and $n_2$ is scheduled because of higher priority. The case is similar in the third processor for

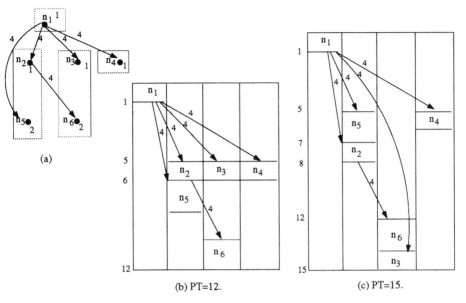

**Figure 6.17** (a) A physical mapping of a DAG. (b) The RCP ordering. (c) Another ordering.

scheduling $n_3$ and $n_6$. The resulting schedule is shown in Figure 6.17(b), and its parallel time is $PT = 12$.

The task-ordering problem is $\mathcal{NP}$-hard even for chains of tasks (Hoogeveen *et al.* 1992); however, in Yang and Gerasoulis (1995) we prove that fork-and-join DAGs are tractable.

**Theorem 6.6** *RCP is optimal for fork-and-join DAGs.*

### 6.4.3 Load balancing versus Sarkar's cluster merging algorithms

As discussed above, PYRROS uses a simple heuristic based on load balancing for merging clusters. This heuristic uses only the cluster load information, and completely ignores task precedences and intercluster communication. It is of interest to see how such a simple heuristic will perform in comparison with a more sophisticated, but more expensive in terms of complexity, heuristic such as Sarkar's cluster merging algorithm. To make a fair comparison, we use the same clustering algorithm for both cases—the DSC algorithm. We next merge the clusters using (1) the load balancing heuristic and (2) Sarkar's merging algorithm. We assume a clique architecture to avoid any mapping effects, and use the RCP ordering in both cases to order tasks.

We randomly generate 100 DAGs and weights as follows. The number of tasks and edges are randomly generated, and then computation and communication weights are randomly assigned. The size of the graphs varies from a minimum average of 143 nodes and 264 edges to a maximum average of 354 nodes and 2620 edges. In our experiments the number of processors is chosen based on the widths of the graphs. The widths and depths of graphs very from 8 to 20, and thus we choose $p = 2, 4$ and 8. Also, to see the performance for both fine- and coarse-grain graphs, we vary the granularity by varying the ratio of average computation over communication weights from 0.1 to 10.

Figure 6.18 shows that the average improvement ratio of Sarkar's algorithm over the load balancing heuristic is between 10 and 35%. When the width of the graph is small compared with the number of processors (e.g. $p = 8$), Sarkar's algorithm is better than load balancing by about 30%. On the other hand, when the width is much larger than the number of processors then the performance differences become smaller, especially for coarse-grain graphs; for example, for $p = 2$ the improvement ratio reduces to about 10% for coarse-grain graphs. Intuitively, this is expected, since each processor is assigned a larger number of tasks when the width-to-processor ratio increases and the RCP ordering heuristic can better overlap the computation and communication.

With respect to the execution time of the heuristics, for a Sun Sparcstation computer the load balancing heuristic takes about 0.1 s to produce a solution for graphs with average $v = 200$ and $e = 400$, while Sarkar's algorithm takes about 40 s. When we double the graph size, the load balancing heuristic takes 0.2 s while

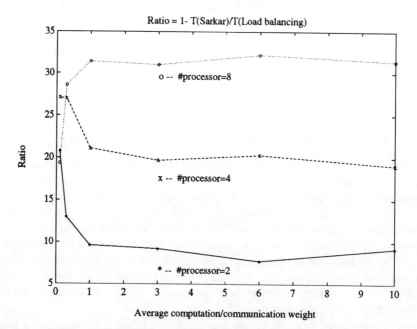

**Figure 6.18** The performance of Sarkar's merging algorithm compared with the load balancing algorithm. The graph width and depth are between 8 and 20.

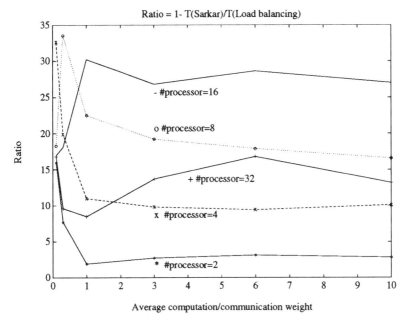

**Figure 6.19** The performance of two merging algorithms for graphs with widths between 30 and 40 and depths between 5 and 8.

Sarkar's needs 160 s. For the above graphs and $p$ the time spent for each graph varied from 0.05 to 0.3 s for the load balancing heuristic and from 9.8 to 725 s for Sarkar's. On the average, the load balancing heuristic was 1000 times faster than Sarkar's for those cases.

To verify our conclusions, we increased the width of the graphs from 8–20 to 30–40, but then reduced the depth of graphs to between 5–8 to keep the number of tasks sufficiently small for the complexity of Sarkar's algorithm. The results are shown in Figure 6.19, and are consistent with our previous conclusions. The performance of Sarkar's algorithm becomes better as the number of processors increases from $p = 2$ to $p = 16$, but then the performance reverses for $p = 32$, as expected, since $p$ approaches the width of the graph.

Our experiments show that, on average, the performance of the load balancing algorithm is within 35% of Sarkar's algorithm for those random graphs. This is very encouraging for the widely used load balancing heuristic. However, more experiments are needed to verify this result.

## 6.5 THE PYRROS SOFTWARE TOOL

The input of PYRROS is a weighted task graph and the associated sequential C or Fortran code. The output is a static schedule and parallel code for a given

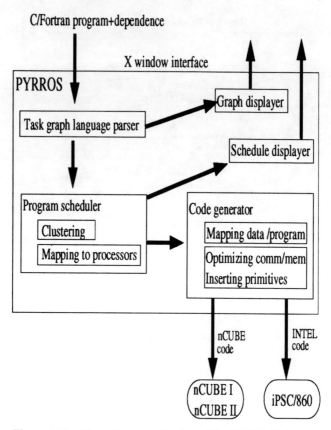

**Figure 6.20** The system organization of the PYRROS prototype.

architecture. The function modules of PYRROS are shown in Figure 6.20. The current PYRROS tool has the following components:

- a task graph language with an interface to C and Fortran, allowing users to define partitioned programs and data;
- a scheduling system for clustering the graph, load balancing and physical mapping, and communication/computation ordering;
- a graphic displayer for displaying task graphs and scheduling results;
- a code generator that inserts synchronization primitives and performs code optimization for nCUBE-I, nCUBE-II and INTEL iPSC/860 hypercube machines.

A more detailed description of PYRROS is given in Yang and Gerasoulis (1992).

There are several other systems related to PYRROS. PARAFRASE-2 (Polychronopoulos et al. 1990) is a parallelizing compiler system that performs dependence analysis, partitioning and dynamic scheduling on shared memory machines. SCHEDULER (Dongarra and Sorensen 1987) uses centralized dynamic scheduling for a shared memory machine. KALI (Koelbel and Mehrota 1990) addresses code generation and is currently targeted at DOALL parallelism. Kennedy's group (Hiranandani *et al.* 1991) is also working on code generation for Fortran D for distributed memory machines. PARTI (Saltz *et al.* 1990) focuses on run-time DOALL parallelism with irregular distribution of data, and optimizes performance by precomputing data accessing patterns. HYPERTOOL (Wu and Gajski 1988) and TASKGRAPHER (El-Rewini and Lewis 1990) use the same task model as PYRROS. The time complexity of these two systems is over $O(v^2)$.

### 6.5.1 Task graph language

The PYRROS system uses a simple language for defining task graphs. For example, the program code in Figure 6.21 is a description of the T-DAG partitioning shown in Figures 6.3 and 6.4 in terms of PYRROS task graph language. The key words are boldfaced. The semantic of the loop is the same as that in Figure 6.3. The interior loop body contains the data dependence and weight information for a task $T_k^j$ along with the specification of task computation. Task $T_k^j$ reads column $k$ and $j$ if $k > 1$. The *c-update* is an external C function that defines the updating of column $j$ using column $k$ for task $T_k^j$ corresponding to the interior loop in the GE program in Figure

```
struct Dataitem column[n];
task T[k][j]{
   int b;
   set_weight(n-k);
   if(k>1){
      read(&column[k], (n-k+1)*ELESIZE);
      read(&column[j], (n-k+1)*ELESIZE);
   }
   c_update(&column[k], &column[j],k,j);
   if(k<n-1){
      if(k!=j-1)
         write(&column[j], T[k+1][j], (n-k)*ELESIZE);
      else for(b=j+1; b<=n; b=b+1)
         write(&column[j],T[k+1][b], (n-k)*ELESIZE);
   }
}
dag Tdag{
   int k, j;
   for(k=1; k<n; k=k+1)
      for(j=k+1; j<=n; j=j+1)
         eval_task(T[k][j] );
}
```

**Figure 6.21** PYRROS task specification for the T-DAG.

6.3. After the *c-update* is executed, if $k < n$ then this task writes column $j$ to be used by $T_{k+1}^j$ and also performs a broadcast if $k = j - 1$.

PYRROS will read this program and perform lexical and semantic analysis to generate an internal representation of the DAG. Then, using the X-window DAG displayer, we can verify whether the definition of the task graph is correct.

### 6.5.2 A demonstration of PYRROS usage

In this section we demonstrate one usage of PYRROS. For GE T-DAG we choose $\alpha = 10, \beta = \omega = 1$ and $n = 5$, and PYRROS displays the dependence graph in the screen as shown in the left part of Figure 6.22. Task $T(1,2)$ corresponds to task $T_1^2$ in the T-DAG of Figure 6.4, and has an internal task number 1 written to its right. The edges of the DAG show the columns sent from one task to the successors.

As mentioned above, when a program is manually written for a library such as LINPACK, the clustering must be given in advance. Let us assume that the widely used natural linear clustering $M_j$ defined previously is used. This implies that $M_1 = T(1,2) = T1, M_2 = \{T(1,3), T(2,3)\} = \{T2, T5\}$ and so on. At this point, the user, executing the program with natural linear clustering, cannot determine how many processors to choose so that the parallel time is minimized. If the user chooses $p = 4$ because the width of the graph is 4 parallel tasks, the parallel time will be 75 time units, as shown in the right part of Figure 6.22, after mapping clusters to processors. The striped lines in this Gantt chart represent communication delay on the hypercube with $p = 4$ processors. The internal numbers of tasks are used in the Gantt chart.

On the other hand, if the scheduling is determined automatically by PYRROS, a better utilization of the architecture and shorter parallel time can be accomplished. In Figure 6.23 PYRROS using the DSC algorithm determines that $p = 2$ processors are sufficient for scheduling this task graph, and the parallel time is reduced to 26 time units. The reason that natural clustering performs poorly here is that the graph is fine-grain. Thus PYRROS is useful in determining the number of processors suitable for executing a task graph. This demonstrates one advantage of an automatic scheduling system.

### 6.5.3 Experiments with PYRROS

We report here our experiments on the BLAS-3 GE program in nCUBE-II. The dependence graph is similar to that in Figure 6.4, except that tasks operate on submatrices instead of array elements. The handwritten program uses the data column block partitioning with cyclic wrap mapping along the gray code of a hypercube following the algorithm of Moler (1986) and Saad (1986). Tasks that modify the same column block are mapped in the same processor. The broadcasting uses a function provided by the nCUBE-II library. The extra memory storage optimization for the handwritten program is not used to avoid the management

A. GERAOULIS AND T. YANG 139

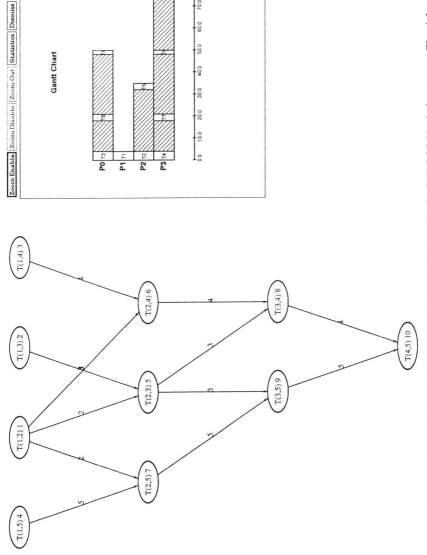

**Figure 6.22** The left part is a GE DAG with $n = 5$ displayed in the PYRROS X-window screen. The right p using natural clustering.

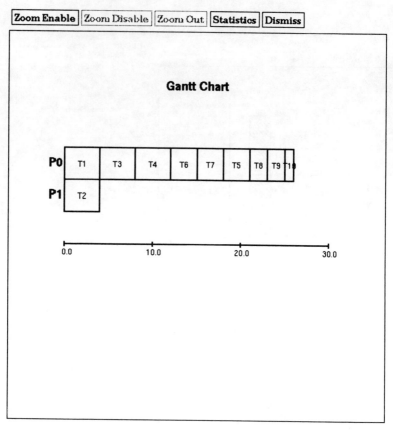

**Figure 6.23** The automatic scheduling result by PYRROS.

overhead, and consequently the maximum matrix size that this simple program can handle is $n = 450$.

The performance improvement of PYRROS code over this hand-written program, $1 - PT(\text{hand})/PT(\text{PYRROS})$, for block sizes 5 and 10 is shown in Figure 6.24. We can see the improvement is small for $p = 2$ because each processor has enough work to do. When $p$ increases, the PYRROS optimization plays an important role, resulting in 5–40% improvement. The speedup ratio of PYRROS over the sequential program for matrix sizes of 450 and 1000 is shown in Table 6.1.

## 6.6 CONCLUSIONS

Scheduling program task graphs is an important optimization technique for scalable MIMD architectures. Our study of the granularity theory shows that scheduling needs to take communication overhead into account, especially for message passing architectures. We have described several scheduling heuristic algorithms that achieve

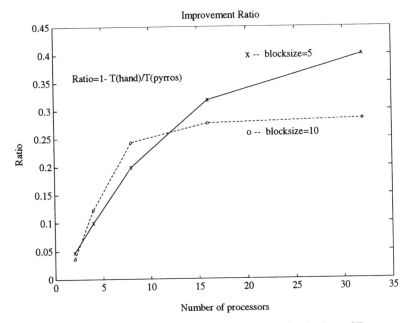

**Figure 6.24** The improvement ratio of PYRROS over a handwritten GE program on nCUBE-II.

**Table 6.1** The speedup ratio of PYRROS over the sequential GE program. The block size is the dimension of a submatrix.

| $p$ | $n = 450$, Block size = 5 | $n = 450$, Block size = 10 | $n = 1000$, Block size = 10 |
|---|---|---|---|
| 2  | 1.97 | 1.9  | 1.99 |
| 4  | 3.8  | 3.7  | 3.9  |
| 8  | 7.3  | 6.9  | 7.8  |
| 16 | 12.8 | 11.9 | 14.4 |
| 32 | 19.0 | 12.9 | 25.7 |

good performance in solving the scheduling problem. Those scheduling techniques are shown to be practical in PYRROS, which integrates scheduling optimization with other compiler techniques to generate efficient parallel code for arbitrary task graphs.

## ACKNOWLEDGMENTS

We thank Weining Wang for developing the task graph language parser, Milind Deshpande for the X window schedule displayer, Probal Bhattacharjya for the graph generator of sparse matrix solver, and Ye Li for programming the INTEL i860 communication routines.

The work presented here was in part supported by ARPA Contract DABT-63-93-C-0064 and by the Office of Naval Research under Grant N000149310114, and Sophia Shen Fellowship and Excellence Fellowship from

Rutgers University, and in part by startup funds from University of California at Santa Barbara. The content of the information herein does not necessarily reflect the position of the US Government, and official endorsement should not be inferred.

# REFERENCES

Adam T., Chandy K. M. and Dickson J. R. (1974) A comparison of list schedules for parallel processing systems *Commun. Assoc. Comput. Mach.*, **17**, 685–690.

Bokhari S. H. (1990) *Assignment Problems in Parallel and Distributed Computing* Kluwer, Dordrecht.

Callahan D. and Kennedy K. (1988) Compiling programs for distributed-memory multiprocessors *J. Supercomput.*, **2**, 151–169.

Chrétienne P. (1989a) Task scheduling over distribute memory machines. *Workshop on Parallel and Distributed Algorithms* North-Holland, Amsterdam.

Chrétienne P. (1989b) A polynomial algorithm to optimally schedule tasks over an ideal distributed system under tree-like precedence constraints *Eur. J. Oper. Res.*, **2**, 225–230.

Coffman E. G. and Denning P. J. (1973) *Operating Systems Theory*, Prentice-Hall, Englewood Cliffs, NJ.

Cosnard M., Marrakchi M., Robert Y. and Trystram D. (1988) Parallel Gaussian elimination on an MIMD computer *Parallel Comput.*, **6**, 275–296.

Dongarra J. J. and Sorensen D. C. (1987) SCHEDULE: tools for developing and analyzing parallel Fortran programs. In Gannon, D. B., Jamieson L. H. and Douglass R. J., editors, *The Characteristics of Parallel Algorithms*, pp 363–394 MIT Press, Cambridge, MA.

Dunigan T. H. (1991) Performance of the INTEL iPSC/860 and nCUBE 6400 hypercube. Oak Ridge National Laboratory Report ORNL/TM-11790.

El-Rewini H. and Lewis T. G. (1990) Scheduling parallel program tasks onto arbitrary target machines *J. Parallel Distrib. Comput.*, **9**, 138–153.

Geist G. A. and Heath M. T. (1986) Matrix factorization on a hypercube multiprocessor. In *Hypercube Multiprocessors*, pp 161–180 SIAM.

George A., Heath M. T. and Liu J. (1986) Parallel Cholesky factorization on a shared memory processor, *Lin. Algebra Applic.*, **77** 165–187.

Gerasoulis A. and Nelken I. (1989) Static scheduling for linear algebra DAGs. In *Proceedings of HCCA 4*, pp 671–674.

Gerasoulis A. and Yang T. (1992) A comparison of clustering heuristics for scheduling DAGs on multiprocessors *J. Parallel Distrib. Comput.*, **16**, 276–291.

Gerasoulis A. and Yang T. (1993) On the granularity and clustering of directed acyclic task graphs *IEEE Trans. Parallel Distrib. Syst.*, **4**, 686–701.

Girkar M. and Polychronopoulos C. (1988) Partitioning programs for parallel execution. In *Proceedings of ACM International Conference on Supercomputing, St Malo*.

Heath M. T. and Romine C. H. (1988) Parallel solution of triangular systems on distributed memory multiprocessors *SIAM J. Sci. Statist. Comput.*, **9** 558–588.

Hiranandani S., Kennedy K. and Tseng C. W. (1991) Compiler optimizations for Fortran D on MIMD distributed-memory machines. *Proceedings of Supercomputing '91, pp 86–100* IEEE, New York.

Hoogeveen J. A., Van de Velde S. L. and Veltman B. (1992) Complexity of scheduling multiprocessor tasks with prespecified processor allocations. CWI Report BS-R9211, Amsterdam.

Koelbel C. and Mehrotra P. (1990) Supporting shared data structures on distributed memory architecture. In *Proceedings of ACM SIGPLAN Symposium on Principles and Practice of Parallel Programming* pp 177–186.

Kim S. J. and Browne J. C. (1988) A general approach to mapping of parallel computation upon multiprocessor architectures. In *Proceedings of International Conference on Parallel Processing*, Vol. 3, pp 1-8.

Lenstra J. K. and Rinnooy Kan A. H. G. (1978) Complexity of scheduling under precedence constraints *Oper. Res.*, **26**.

Kung S. Y. (1988) *VLSI Array Processors* Prentice-Hall, Englewood Cliffs, NJ.

Moler C. (1986) Matrix computation on distributed memory multiprocessors In *Hypercube Multiprocessors* pp 181–195 SIAM.

Ortega J. M. (1988) *Introduction to Parallel and Vector Solution of Linear Systems* Plenum, New York.

Papadimitriou C. and Yannakakis M. (1990) Towards on an architecture-independent analysis of parallel algorithms *SIAM J. Comput.*, **19** 322–328.

Picouleau C. (1992) New complexity results on the UET–UCT scheduling algorithms. In *Proceedings of Summer School on Scheduling Theory and its Applications, Chateau De Bonas, France*, pp 487–502.

Polychronopoulos C., Girkar M., Haghighat M., Lee C., Leung B. and Schouten D. (1990) The structure of Parafrase-2: an advanced parallelizing compiler for C and Fortran. In Gelernter D., Nicolau A. and Padua D. editors, *Languages and Compilers for Parallel Computing*.

Saad Y. (1986) Gaussian elimination on hypercubes. In Cosnard M. *et al.*, editors *Parallel Algorithms and Architectures*, North-Holland, Amsterdam.

Saltz J., Crowley K., Mirchandaney R. and Berryman H. (1990) Run-time scheduling and execution of loops on message passing machines *J. Parallel Distrib. Comput.*, **8**, 303–312.

Sarkar V. (1989) *Partitioning and Scheduling Parallel Programs for Execution on Multiprocessors* MIT Press, Cambridge, MA.

Stone H. (1987) *High-Performance Computer Architectures*, Addison-Wesley, Reading, MA.

Varvarigou T., Roychowdhury V. and Kailath T. Scheduling in and out forests in the presence of communication delays *IEEE Trans. Parallel Distrib. Syst.*, to appear. (A shorter version appeared in *Proceedings of International Parallel Processing Symposium, April 1993, CA.*)

Wu M. Y. and Gajski D. (1988) A programming aid for hypercube architectures *J. Supercomput.*, **2**, 349–372.

Yang T. and Gerasoulis A. (1991) A fast static scheduling algorithm for DAGs on an unbounded number of processors. In *Proceedings of Supercomputing '91*, pp 633–642 IEEE, New York.

Yang T. and Gerasoulis A. (1992) PYRROS: static task scheduling and code generation for message-passing multiprocessors. In *Proceedings of 6th ACM International Conference on Supercomputing, Washington, DC*, pp 428–437.

Yang T. and Gerasoulis A. (1993) List scheduling with and without communication delay *Parallel Comput.*, **19**, 1321–1344.

*Department of Computer Science, Rutgers University, New Brunswick, NJ 08903, USA*
gerasoulis@cs.rutgers.edu

*Department of Computer Science, University of California, Santa Barbara, CA 93106, USA*
tyang@cs.ucsb.edu

CHAPTER 7

# Scheduling Parallel Programs Assuming Preallocation

**V. J. Rayward-Smith**
*School of Information Systems, University of East Anglia*

**F. Warren Burton**
*School of Computing Science, Simon Fraser University, Burnaby, BC, Canada*

**G. J. Janacek**
*School of Mathematics, University of East Anglia*

**Abstract**

Parallel programs comprise a number of processes that may synchronize with each other. Each process is a piece of software or, equivalently, the set of actions (or tasks) generated by that software. In this study we assume that a process is generally pre-allocated to a processor and that thereafter it cannot be moved. We show that the order in which the actions of the processes are scheduled may make a spectacular difference in the performance of a program. In particular, for sufficiently large problems we can construct a program that can have a speed-up of $m$ where $m$ is the number of processors, or a speed-up arbitrarily close to 1 (i.e. no effective parallelism), depending on the order in which processes are scheduled. We argue that, in practice, a system is not likely to have sufficient information to be able to determine which action to schedule, so it is not possible to avoid bad schedules. Moreover, these bad schedules can arise in situations where the process allocation has balanced the total load on each process. In general, the problem of determining a good assignment of processes to processors is shown to be impossible with the information that is likely to be available to a system at the time a decision must be made. Even with complete information, finding an optimal schedule is shown to be $\mathcal{NP}$-hard, even for tree-structured problems and $m = 2$. Obvious heuristics are also shown to fail. We also consider the role of the machine allocation function of a scheduler, and, in particular, the expected optimal schedule length achieved by a random allocation. Our conclusion is that fully automatic scheduling of arbitrary parallel programs on systems that do not allow processes to move once started cannot be done in a way that will ensure good performance.

---

*Scheduling Theory and its Applications* Edited by P. Chrétienne, E. G. Coffman, Jr., J. K. Lenstra and Z. Liu
© 1995 John Wiley & Sons Ltd

## 7.1 INTRODUCTION

Let us first consider a deterministic parallel program where the work that is to be done by the program is fixed and where all the results are uniquely defined. However, the order in which things are to be done may be nondeterministic.

Traditional scheduling theory applies to the case where any processor may, at any time, be running any runnable process (Burton et al. 1990). Hence any action (or task) generated may be run on any processor. If we ignore overheads and assume that the scheduler never allows a processor to be idle when there is a process that can run then the worst possible schedule takes less than twice as long as the best possible schedule. In fact, the schedule length can be bounded in terms of the average parallelism in the problem (i.e. the speed-up that would be possible with an unbounded number of processors), and the number of processors. For $p$ the average parallelism, $T_1$ the time required by the program on one processor, $T_m$ the time required on $m$ processors and $S_m = T_1/T_m$ the speed-up on $m$ processors, it was shown (Eager et al. 1989) that

$$\frac{mp}{m+p-1} \leqslant S_m \leqslant \min(p, m).$$

If $m = p$ then $S_m \geqslant \frac{1}{2}m$. If either $m$ or $p$ is very large compared with the other value then the speed-up approaches the smaller of these two values, and is always at least half the smaller value. For example, if the average parallelism in a program exceeds the number of processors then a speed-up equal to at least half the number of processors is guaranteed, and the guaranteed speed-up must approach the number of processors as the average parallelism becomes large. On the other hand, the speed-up cannot be greater than the minimum of the average parallelism and the number of processors. This result suggests that, in practice, we do not need to worry much about scheduling in these situations.

While the choice of a schedule may make only a relatively minor difference in the time requirements of such parallel programs, it may make spectacular differences in the space requirements (Burton 1988).

If a program is nondeterministic then any difference in the schedule can completely alter the program behavior. For example, in combinatorial search algorithms, such as branch-and-bound, minor scheduling differences may make major differences in the time required by a program (Burton et al. 1981; Lai and Sprague 1985). Since we are demonstrating negative results, it does not matter that things can get even worse in a more general setting. We shall limit our attention to deterministic parallel programs and consider only the time required to execute a program.

We shall consider what can happen in a system where all processing actions (or tasks) are preallocated to processors. We call a set of actions that must be processed together on the same processor a *process*. Thus we are concerned with systems where, once a process starts to execute on a given processor, it must remain on that processor. Such systems are sometimes called *distributed systems*, but a preallocation of processes to processors is common in many forms of parallel programming (see e.g. CSP and occam). In practice, some distributed systems may allow processes to migrate to other processors, but this can cause communication problems. For example, if a process uses a large data structure, and the process is moved frequently, then the communication cost of moving the data structure, or accessing the data structure remotely, may dominate the computational costs.

In Section 7.2 we define our model, state our assumptions more precisely and introduce terminology to be used in the remainder of the chapter. In Section 7.3 we give an example of a program that can have a speed-up arbitrarily close to $m$ if scheduling is done in an optimal order, but will have a speed-up arbitrarily close to 1 (i.e. no effective parallelism), depending on the problem size, if scheduling is done in the worst possible order. We further argue that it is not possible in practice to avoid bad schedules. We briefly consider the problem of deciding which processor should execute a given process, and see that a system is not likely to have sufficient information to make a good choice.

In Section 7.4 we show that computing the optimal schedule given complete information is $\mathcal{NP}$-hard, even for very restricted problem classes. An obvious heuristic, level scheduling, which might be expected to give reasonable results with complete information and to suggest possible heuristics for practical systems, is shown in Section 7.5 to fail completely.

In Sections 7.6 and 7.7 we consider the preallocation function. In particular, we consider "balanced" preallocations and random preallocation, determining bounds on (the expected length) of optimal schedules. Section 7.8 gives our conclusions.

## 7.2 MODEL, ASSUMPTIONS AND TERMINOLOGY

We model a deterministic distributed program by an ordered triple $(A, \sqsupset, f)$, where $A$ is a set of atomic actions, simply called actions hereinafter, $\sqsupset$ is a partial order defined on those actions, and $f$ is a total function $A \rightarrow 1, 2, \ldots, m$.

If, for actions $a_1$ and $a_2$, $a_1 \sqsupset a_2$ then $a_1$ must be performed before $a_2$. The function $f$ is an assignment of actions to processors, referred to as the *preallocation function*.

In the remainder of this chapter we shall assume that each action can be performed by a processor in one unit of time. This is not a serious restriction in our model, but simplifies the presentation. A process in a high-level programming language corresponds to a collection of actions, which for the majority of this study must all be assigned to the same processor by the preallocation function.

A *schedule* is a total function $s : A \rightarrow \{1, 2, \ldots\}$ subject to the restrictions that

(1) if $f(a_1) = f(a_2)$ and $a_1 \neq a_2$ then $s(a_1) \neq s(a_2)$;
(2) if $a_1 \sqsupset a_2$ then $s(a_1) < s(a_2)$.

A schedule assigns actions to times when they are to be performed, subject to the restriction that no two actions are performed at the same time on the same processor, and the partial order $\sqsupset$ is respected. If other than unit execution time tasks are allowed, this definition must be generalized.

We define the length of a schedule by

$$length(s, (A, \sqsupset, f)) = \max \{s(a) : a \in A\}.$$

When $(A, \sqsupset, f)$ is clear from the context, instead of writing $lengths(s, (A, \sqsupset, f))$, we shall write $length(s)$. An optimal schedule of $(A, \sqsupset, f)$ is a schedule $s$ such that $length(s) \leq length(t)$ for all other schedules $t$ of $(A, \sqsupset, f)$.

We call a schedule *work-preserving* if a processor never idles when it could work, i.e. $s$ is work-preserving iff

(3) if processor $k$ idles at time $t$ and $f(a) = k$ and $s(a) > t$ then there exists $b$ such that $b \sqsupset a$ and $s(b) \geq t$.

It is easy to show that there is a work-preserving schedule of optimal length, so we shall restrict all our discussion to such schedules. A schedule will be called *valid* only if it satisfies conditions (1)–(3).

For modeling purposes, we could assume that the number of processes in a distributed program equals the number of processors, with one process per processor. Then, we could use the term process to mean precisely the set of actions assigned to a given processor. Such a process will generally contain internal parallelism, i.e. the actions within a process need not be totally ordered with respect to $\sqsupset$. If a reader likes to think of a process as a totally ordered set of actions (i.e. a sequential process) then he or she may wish to regard the collection of actions assigned to a given processor as a collection of processes that happen to be assigned to the same processor.

In a practical system for distributed computing a scheduler is likely to have only limited information. If a system could determine how much time a process will require, it could solve the halting problem. Therefore a processor is unlikely to know how many actions are assigned to it, although it may know how many runnable actions it has at any one time. Similarly, if semaphore operations are contained in conditionals then a system cannot predict how actions will synchronize. In many practical applications it is fair to assume that a scheduler does not know what is yet to come. That is, given a collection of runnable actions, a processor has no way to distinguish between them.

## 7.3 WORST CASE SCHEDULING

Consider the deterministic, distributed program whose actions are illustrated in the directed acyclic graph (dag) of Figure 7.1(a). Each subscripted variable represents an action. For actions $a$ and $b$, $a \sqsupset b$ iff there is a downwards-directed path from $a$ to $b$. The second subscript of an action indicates the processor to which an action is assigned, i.e. $f(a_{ij}) = f(b_{ij}) = j$.

There are $m$ processors, and $k$ is an arbitrary, large value. The $r$th chain ($1 \leqslant r \leqslant m$) comprises $a_{r1}, a_{r2}, ..., a_{rr}$, with the remaining actions $b_{r+1r}, b_{r+2r}, ..., b_{kr}$ all assigned to $r$. This example is similar to one cited by Jaffe (1980). In Figure 7.1(b) we given an optimal schedule for this program on $m$ processors. Each row represents one processor, each column represents one unit of time, and the entry in a given row and column gives the action that the processor performs at that time. On the other hand, Figure 7.1(c) describes a poorly performing

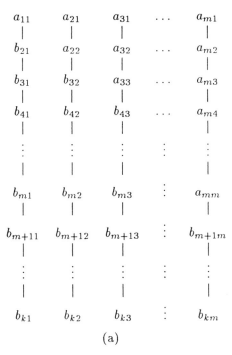

**Figure 7.1** (a) A directed acyclic graph. (b) An optimal schedule. (c) A depth-first schedule.

(depth-first) schedule. While the schedule of Figure 7.1(b) is of length $k + m - 1$, the poor schedule of Figure 7.1(c) is of length $mk - \frac{1}{2}(m-1)(m-2)$. There are $mk$ actions, so, as $k \to \infty$, the first schedule has a speed-up tending to $m$ while the second has a speed-up tending to 1.

It is clear that the assignment of processes to processors is critical. It is generally assumed that an allocation where each processor is allocated the same total workload will work well. It is important to realize that scheduling is still critical—the above example is reasonably balanced, since, as $k \to \infty$, each processor is allocated the same share of activities.

In many practical situations attempting any sort of balancing is very difficult, since it is not possible to attempt to allocate processes to processors wisely without detailed knowledge of what is yet to happen. We illustrate this with some examples.

Consider the collection of $m^2$ processes, to be run on $m$ processors, illustrated in Figure 7.2. Each process consists of two actions: an $a$ action and a corresponding $b$ action. No $b$ action can be performed until all $a$ actions are finished. We shall assume that each $a$ action requires one unit of time, but $b$ actions are of two types. Nice $b$ actions require one unit of time, but nasty $b$ actions require $k$ units of time, for some arbitrary, large, value $k$. (If the reader wishes to continue to think in terms of unit execution time tasks, replace each nasty $b$ action with a chain of $k$ unit execution time action.) A process containing a nice or nasty action will be called nice or nasty respectively. There are $m$ nasty processes and $m(m-1)$ nice processes.

We shall assume that a computer system must assign each process to a processor before that process starts execution. Furthermore, we shall assume that the system cannot distinguish a nasty process from a nice one before the $b$ action starts executing. (Again, if we could tell how much time a computation was going to require in advance, we could solve the halting problem.) Since all $a$ actions must be performed before any $b$ action, all processes must be assigned to processors before the character of any process is known.

In the best case each processor will be assigned 1 nasty process and $m - 1$ nice processes. The resulting schedule will require $2m + k - 1$ units of time (2 units for

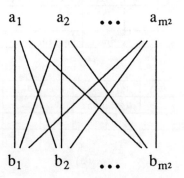

**Figure 7.2** Synchronizing processes.

each of $m-1$ nice processes and $k+1$ units for the nasty process). All processes will be busy until all work is done, so the speed-up will be $m$.

On the other hand, with $m^2$ processes, at least one processor must have at least $m$ processes, no matter how processes are assigned to processors. If one processor happens to be assigned all $m$ nasty processes then the schedule length will be $m(k+1)$, giving a speed-up of $m(2m+k-1)/m(k+1)$. As $k$ becomes large, nasty processes become increasingly dominant, and the speed-up drops to arbitrarily close to 1. Hence, with process placement, as with scheduling, there are cases where programs may have a speed-up equal to the number of processors, virtually no speed-up, or anything in between. In this case, if processes were assigned to processors in a completely random fashion, we should not expect to do too badly.

The problem of assigning processes to processors is considered in more detail in Sections 7.6 and 7.7, and the implications of scheduling to parallel functional programming in Burton and Rayward-Smith (1993).

## 7.4 OPTIMAL SCHEDULING

In this section we show the problem of finding an optimal schedule for an arbitrary $(A, \sqsupset, f)$ on an arbitrary number $m$ of processors, is $\mathcal{NP}$-hard. The decision problem associated with our scheduling problem is as follows.

*Optimal preallocated schedule (OPS)*

*Instance* A partially ordered set of unit execution time (UET) actions, $(A, \sqsupset)$, a (total) preallocation function $f : A \to 1, 2, \ldots, m$, a number of processors, $m \in Z^+$, and a time limit, $L \in Z^+$.

*Question* Is there a valid schedule of $(A, \sqsupset, f)$ on $m$ processors of length $\leqslant L$?

This decision problem was proved $\mathcal{NP}$-complete for $m=2$ and $\sqsupset$ arbitrary and for $m$ arbitrary and $(A, \sqsupset)$ a forest by Goyal (1976). We shall show that the problem remains $\mathcal{NP}$-complete for any fixed $m \geqslant 2$ and even when $(A, \sqsupset)$ is a set of chains of actions.

**Theorem 7.1** *OPS is $\mathcal{NP}$-complete even when $m$ is fixed at any value $\geqslant 2$ and $(A, \sqsupset)$ is constrained to be a set of chains.*

**Proof** The problem is similar to flowshop scheduling (Gonzalez and Sahni, 1978) and the proof of its $\mathcal{NP}$-completeness is similar though simpler.

The following problem, 3-PARTITION, is known to be strong $\mathcal{NP}$-complete (Garey and Johnson 1979), i.e. $\mathcal{NP}$-complete even if unary notation is used to represent the integers. To show that OPS is $\mathcal{NP}$-complete, we show in Lemma 7.1 that there is a polynomial transformation from 3-PARTITION to OPS, and then that OPS$\in \mathcal{NP}$.

## 3-PARTITION

**Instance** A bag of $s$ positive integers, $S = \{a_1, a_2, \ldots, a_s\}$ with $s = 3n$,

$$\sum \{a_i : 1 \leqslant i \leqslant S\} = nB, \quad \tfrac{1}{4}B < a_i < \tfrac{1}{2}B \text{ for } 1 \leqslant i \leqslant s.$$

**Question** Does there exist a partition of $S$ into $n$ bags, $S_1, S_2, \ldots, S_n$, such that $\sum \{a : a \in S_i\} = B$ for $i = 1, 2, \ldots, n$?

Note that each such subset must contain precisely three elements.

**Lemma 7.1** *3-PARTITION $\propto$ OPS*

**Proof** Given an instance, $I = (a_1, a_2, \ldots, a_{3n}, B)$, of 3-PARTITION, construct $g(I) = ((A, \sqsupset, f), m, L)$, an instance of OPS, as follows.

$A$ comprises $3n + 1$ chains. For each $a_i$ we have a chain $c_i$ comprising $a_i$ tasks assigned to 1 followed by $a_i$ tasks assigned to 2. We also have a chain $d$ of length $2nB$ comprising $n$ repeats of a pattern consisting of $B$ tasks assigned to 2 followed by $B$ tasks assigned to 1. Then we set $m = 2$ and $L = 2nB$.

Since 3-PARTITION is strong $\mathcal{NP}$-complete, we can assume all integers are represented in unary notation, and thus $g$ is a polynomial time transformation. We need to prove $I \in Y_{3-\text{PARTITION}}$ iff $g(I) \in Y_{\text{OPS}}$.

($\Rightarrow$) Say $S_i = \{a_{i1}, a_{i2}, a_{i3}\} (1 \leqslant i \leqslant n)$ is the partition of $A$ satisfying $\sum \{a : a \in S_i\} = B$ for $1 \leqslant i \leqslant n$. Then a valid schedule of $A$ is given in Figure 7.3. In this schedule $S_i^1$ denotes the tasks assigned to 1 in chains $c_{i1}, c_{i2}$ and $c_{i3}$, and $S_i^2$ denotes the tasks assigned to 2 in those chains.

($\Leftarrow$) Since $L = 2nB$ and we have $2 \sum \{a_i : a_i \in A\} + 2nB = 4nB$ tasks allocated to only two processors, the schedule can have no idle time whatsoever. Also since $d$ is a chain of $2nB$ tasks, the schedule of these tasks is fixed as in Figure 7.3. Now consider the tasks scheduled on 2 in $[B, 2B[$. They belong to chains $c_i$, but since there must be $B$ such tasks, we must have at least three chains contributing tasks. These chains must all have had all their first halves processed on machine 1 in time $[0, B[$. At most three such top halves could have been processed in time $[0, B[$. Hence we deduce that exactly three top halves are processed in $[0, B[$ and their corresponding bottom halves in $[B, 2B[$. We can continue this reasoning and deduce that for any $[(2k-2)B, (2k-1)B[, 1 \leqslant k \leqslant n$, the top halves of precisely three chains are

| $S_1^1$ | next $B$ tasks of $d$ | $S_2^1$ | next $B$ tasks of $d$ | $\ldots$ | $S_n^1$ | last $B$ tasks of $d$ |
|---|---|---|---|---|---|---|
| first $B$ tasks of $d$ | $S_1^2$ | next $B$ tasks of $d$ | $S_2^2$ | $\ldots$ | next $B$ tasks of $d$ | $S_n^2$ |

**Figure 7.3** A schedule for OPS.

processed and in $\lfloor(2k-1)B, 2kB\rfloor$ the corresponding bottom halves are processed. Suppose that $c_{i1}, c_{i2}, c_{i3}$ are the three chains processed in $\lfloor(2k-2)B, 2kB\lfloor, 1 \leqslant k \leqslant n$; then the sets $S_k = \{a_{i1}, a_{i2}, a_{i3}\}$ form a 3-partition of $S$. The proof of Lemma 7.1 is thus completed. □

The proof of Theorem 7.1 then follows from the easily verified observation that $OPS \in NP$. □

## 7.5 LEVEL SCHEDULING

In job shop scheduling (see e.g. Coffman 1976), partially ordered UET sets of tasks (actions) are assigned to processors. The basic model differs from our model in the sense that there is no predetermined assignment; any task can be processed on any machine. Jaffe (1980) developed this basic model to include types of task and corresponding types of processor. He showed that with $k$ types of task and $m_i$ identical processors for tasks of type $i$, the length $\omega$ of any work-preserving schedule satisfies

$$\omega \leqslant \left[(k+1) - \frac{1}{\max\{m_1, m_2, \ldots, m_k\}}\right]\omega_{\text{opt}},$$

where $\omega_{\text{opt}}$ is the length of the optimal schedule. Our model thus corresponds to a special case of Jaffe's model where $k = m$ and $m_1 = m_2 = \ldots = m_k = 1$.

A successful heuristic in job shop scheduling is the level scheduling strategy (Lam and Sethi 1977). In the terminology of this chapter, an action $a \in A$ is a *terminal action* iff there is no $b \in B$ such that $a \sqsupset b$. A chain of length $n$ in $A$ from $a$ to $b$ is a sequence of actions in $A$, $a = a_0 \sqsupset a_1 \ldots \sqsupset a_{n-1} = b$. The *level* of $a \in A$ is the length of the longest chain from $a$ (to a terminal action). We have argued that at the time $a$ is scheduled we cannot know the descendants of $a$. Hence the level of $a$ can only be finally determined when the whole program is completed. However, it may be possible to estimate the level.

A schedule $s$ of $(A, \sqsupset, f)$ is called a *level schedule* iff each processor, given a choice of possible actions to process at any time, always selects that action at the greatest level.

Consider the deterministic distributed program illustrated in Figure 7.4(a). As before, each subscripted variable represents an action, and, for actions $x$ and $y$, $x \sqsupset y$ iff there is a downwards-directed path from $x$ to $y$. The second subscript indicates the processor to which an action is assigned. That is, $f(x_{ij}) = j$, where $x$ may be $a, b$ or $c$. There are $m$ processors, and $k$ is an arbitrary, large value.

At each level there are $k + m$ actions. Note how the number of $b$ actions decreases by one and the number of $c$ actions increases by one each time we go down one level. Note also how each process of the form $b_{ii}$ is a critical process, with all actions of type $a$ or $b$ at the next level down having to wait for the critical process to be performed. Whenever we have a critical action $b_{ii}$, we also have $k$ actions of type $a$ assigned to the same processor.

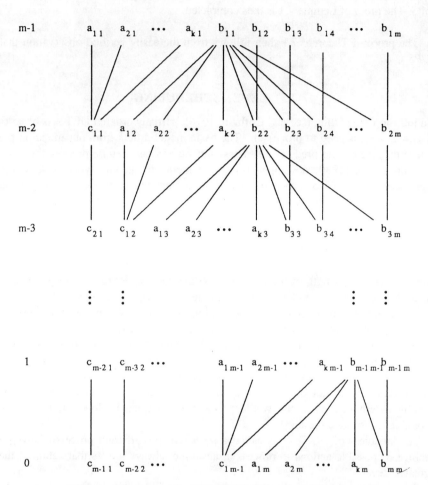

**Figure 7.4** (a) A collection of partially ordered actions.

There are a total of $m(m + k)$ actions in the $m$ rows of Figure 7.4(a). Hence, if all actions were moved to a single processor, the length of a schedule would be $m(m + k)$, assuming the processor is not idle when there is work to be done.

Figure 7.4(b) gives an optimal schedule for this program on $m$ processors. Whenever a processor has a choice as to which action to perform next, it chooses the critical action, $b_{ii}$, before doing any type-$a$ action. The length of this schedule is $m + k$, giving a speed-up of $m$, which is the best possible speed-up on $m$ processors.

| $b_{11}$ | $a_{11}$ | $a_{21}$ | ... | $a_{k1}$ | $c_{11}$ | ... | $c_{m-1,1}$ |
|---|---|---|---|---|---|---|---|
| $b_{12}$ | $b_{22}$ | $a_{12}$ | ... | | $a_{k2}$ | $c_{12}$ | ... | $c_{m-2,2}$ |
| ⋮ | ⋮ | ⋱ | | | | | ⋮ |
| $b_{1m}$ | $b_{2m}$ | ... | $b_{mm}$ | $a_{1m}$ | $a_{2m}$ | ... | $a_{km}$ |

**Figure 7.4** *(cont.)* (b) An optimal schedule.

On the other hand, consider a level schedule where each processor always performs type-*a* actions before type-*b* actions, when given the choice. For example, processor 1 spends $k$ units of time performing *a*s and then one unit of time performing $b_{11}$. During this time, the other processors each perform one type-*b* action and then sit idle for $k$ units of time. Next, processor 2 spends $k$ units of time on *a*s, followed by one unit of time performing a critical *b*, while processors $3, \ldots, m$ perform only one more noncritical *b* and processor 1 finishes up its *c*s and then sits idle for evermore. The schedule length is $m(k+1)$, giving a speed-up of $m(m+k)/m(k+1)$, which can be made arbitrarily close to 1 by making $k$ large enough. Note that processors are idle only when they have no runnable actions.

It is worth noting that if scheduling were done in a completely random way (which is the best we can hope for if we cannot distinguish critical actions from other actions) then we should expect to do half of the *a*s before doing the critical *b*, which would improve things by only a factor of 2. That is, our speed-up in this example would approach 2 rather than 1, no matter how large $m$ is.

Level scheduling can thus do as badly as any work-preserving scheduling strategy on certain programs. If the program is not distributed, so any processor can perform any action and $(A, \sqsupset)$ is a tree then level scheduling produces an optimal job shop schedule (Hu 1961). The fact that this result cannot hold for our model follows from Theorem 7.1.

## 7.6 BALANCING THE ALLOCATION FUNCTION

As we have explained, part of the implementation of a parallel program involves preassigning processes to the various processors. In this section we consider the role of the preallocation function.

In the worst case all processes are allocated to the same processor and all potential parallelism is lost. In practice, preallocation attempts to spread the likely work load between processors, usually based on records of previous program runs. We know that we can still get arbitrarily bad schedules even with such load balancing.

However, in certain circumstances, load balancing is a useful first step in obtaining a good schedule. We explore this claim in this section.

Up to now, we have assumed that each activity within the program is associated with a particular process and hence must be processed on the processor to which that process is assigned. However, in many practical situations, such as when implementing divide-and-conquer or branch-and-bound algorithms, there are activities that can be processed on any processor. The implementor then contrives an allocation of activities that shares out the workload between processors as evenly as possible—this is called *load balancing*. In this section we study such balanced allocations. We assume each activity is preassigned to some processor in such a way that the total workload is as balanced as possible.

It may be thought that if the preallocation function $f$ is balanced, i.e. if each processor gets either $\lceil n/m \rceil$ or $\lfloor n/m \rfloor$ of the activities, then the very bad schedules of Section 7.3 may no longer arise. In this section we show this to be false even under the stringent balancing condition defined below.

An activity $a$ is at depth 0 iff $\sim \exists b(b \sqsupset a)$. Such activities are called *root activities*. If an activity $a$ is not a root activity then $depth(a) = \max \{depth(b) : b \sqsupset a\} + 1$.

The depth of an activity $a$ in $(A, \sqsupset)$ is thus the length of the longest path from a root activity to $a$ in the graphical representation of $(A, \sqsupset)$.

Let $d_i$ denote the number of activities at depth $i$. A schedule is *depth-balanced* if each processor is preassigned either $\lceil d_i/m \rceil$ or $\lfloor d_i/m \rfloor$ of the activities at depth $i$ for all $i$.

If $D_i = \sum_{j \leq i} d_j$ then the schedule is *sum–depth-balanced* if each processor is preassigned either $\lceil D_i/m \rceil$ or $\lfloor D_i/m \rfloor$ of the activities at depth $\leq i$ for all $i$.

Note that a sum–depth-balanced schedule is necessarily balanced, but this is not necessarily the case for a depth-balanced schedule.

**Theorem 7.2** *For any depth-balanced or sum–depth-balanced work-preserving schedule $s$,*

$$length(s) \leq m\omega_{opt},$$

*where $\omega_{opt}$ is the optimal schedule length subject to the same balanced preallocation function. Moreover, this bound is tight.*

**Proof** Figure 7.5 describes a program for which the bound is tight. In this example $f$ is defined by $f(a_{ij}) = f(b_{ij}) = j$, and is thus both depth- and sum–depth-balanced. At each depth there is a critical value $b_{ii}$, which can be significantly delayed in a bad schedule by doing all the activities $a_{1i}, a_{2i}, \ldots, a_{k-i+1\,i}$. The worst schedule thus has length $\frac{1}{2}m(2k - m + 3)$ while $\omega_{opt} = m + k$. As $k \to \infty$, the ratio tends to $m$, establishing the tightness of the bound. □

There is a clear message from Theorem 7.2. Balancing the load is not sufficient; attention must still be paid to the scheduling of the activities.

A schedule is said to be *local breadth-first* if each processor always selects the next activity to be processed from those available by choosing the one at least depth.

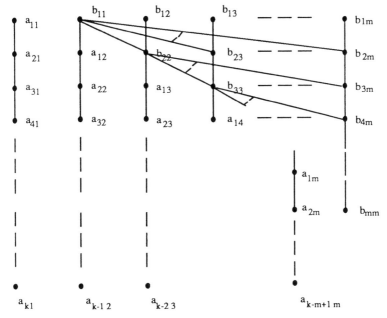

**Figure 7.5**

**Theorem 7.3** *If a schedule is depth-balanced and local breadth-first then*
$$length(s) \leq 2\omega_{opt},$$
*Moreover, the bound is tight as $m \to \infty$.*

**Proof** Every node at depth $i$ must be processed by time
$$\lceil d_0/m \rceil + \lceil d_1/m \rceil + \ldots + \lceil d_i/m \rceil.$$
Thus if $l$ is the maximum depth of any node then the total schedule has
$$\begin{aligned} length(s) &\leq \lceil d_0/m \rceil + \lceil d_1/m \rceil + \cdots + \lceil d_l/m \rceil \\ &\leq \lceil (d_0 + d_1 + \cdots + d_l)/m \rceil + l \\ &= \lceil n/m \rceil + l \\ &\leq \omega_{opt} + \omega_{opt} = 2\omega_{opt}. \end{aligned}$$

Figure 7.6 describes a program with $\omega_{opt} = m + 1$ and for which the worst schedule has length $2m$. As $m \to \infty$, the ratio between these tends to 2. As before, $f(a_{ij}) = j$. □

A sum–depth-balanced schedule is not necessarily depth-balanced. For example, with $m = 2, d_0 = 3$ could be split 2 to $P_1$ and 1 to $P_2$, and $d_1 = 6$ could be split 2 to $P_1$ and 4 to $P_2$. This would be a sum–depth-balanced schedule but not a depth-balanced schedule. However, we do have the following result.

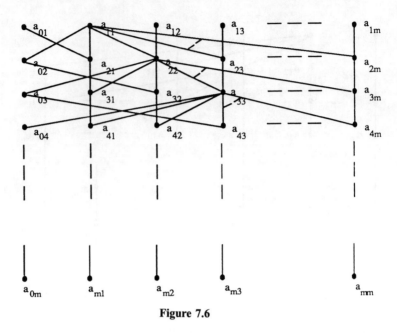

Figure 7.6

**Lemma 7.2** *In a sum–depth-balanced schedule each processor is assigned at least $\lfloor d_i/m \rfloor - 1$ and at most $\lceil d_i/m \rceil + 1$ of the activities at depth i.*

**Proof** $d_i = D_i - D_{i-1}$. The minimum number of activities of $d_i$ given to a processor is thus $\lfloor D_i/m \rfloor - \lceil D_{i-1}/m \rceil$. Say $D_i = am + b (0 \leq b < m)$ and $D_{i-1} = cm + d (0 \leq d < m)$. Then $d_i = (a-c)m + (b-d)$, so $\lfloor d_i/m \rfloor \leq a - c$. But $\lfloor D_i/m \rfloor = a$ and $\lceil D_{i-1}/m \rceil \leq c + 1$. Hence $\lfloor D_i/m \rfloor - \lceil D_{i-1}/m \rceil \geq a - c - 1 \geq \lfloor d_i/m \rfloor - 1$. The maximum value follows similarly. □

**Theorem 7.4** *If the schedule is sum–depth-balanced and is local breadth-first then*
$$length(s) \leq 3\omega_{opt}$$

**Proof** From Lemma 7.2, every node at depth $i$ must be processed by time
$$(\lceil d_0/m \rceil) + (\lceil d_1/m \rceil + 1) + \ldots + (\lceil d_i/m \rceil + 1).$$
The result then follows using similar arguments to that of Theorem 7.3. □

We have proved in Theorems 7.3 and 7.4 that, providing the schedule is depth- or sum–depth-balanced, a reasonable schedule can be obtained using breadth-first. Breadth-first is not sufficient by itself though. Figure 7.7 describes a program for

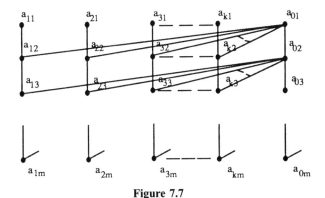

**Figure 7.7**

which a balanced (but neither depth- nor sum–depth-balanced) assignment is given by $f(a_{ij}) = j$. Using a local breadth-first strategy is not sufficient to distinguish critical tasks $a_{0i}$ from other tasks. The worst schedule using a breadth-first strategy has length $m(k + 1)$. The optimal schedule has length only $m + k$. Again, as $k \to \infty$, the ratio tends to $m$.

Similar results to Theorems 7.2–7.4 can be established for level balanced assignments and local level schedules. Since the level of an activity, i.e. the length of a longest path to a terminal activity, is difficult to know in advance, these results have less practical significance.

## 7.7 USING A RANDOM ALLOCATION FUNCTION

Since, in general, we cannot know the action dag in advance and it is difficult to decide on the preallocation function, it is no surprise that we can get very poor schedules. The previous section considered balanced schedules where the preallocation function achieved a very even distribution of the workload. In this section we consider the case where activities are preassigned randomly to processors. In the case where any activity can be assigned to any processor, this models random assignment.

In the case where each activity belongs to a process preassigned to a processor, the results also have applications. If we assume that the allocation of processes to processors has been "fair", then, taken over a large range of program structures, we might assume that each action in the dag is thereby assigned to a random processor.

The activities at a given depth in the dag form an independent set, so our first task is to consider the random allocation of $n$ independent activities to $m$ processors and the likely loads on individual processors.

Let $f$ be some processor assignment. Then the number of activities assigned by $f$ to processor $p$ is denoted by $r(f,p)$.

An *occupancy sequence* (defined by $n$ and $m$) is any sequence of nonnegative integers of the form

$$r_{11}, r_{12} \cdots r_{1\lambda_1}, r_{21}, r_{22} \cdots r_{2\lambda_2}, \ldots r_{k1}, r_{k2} \cdots r_{k\lambda_k}$$

where $\sum \lambda_i = m$ and $\sum r_{ij} = n$.

An occupancy sequence $\mathbf{r}$ is the occupancy vector for $f$ if $\mathbf{r}$ is a sorted sequence comprising all the integers in the multiset $\{r(f,p) : p \text{ is a processor}\}$.

There are $n^m$ possible distinct allocations of $n$ activities to $m$ processors. The probability of getting an occupancy sequence

$$\mathbf{r} = (r_{11}, r_{12} \cdots r_{1\lambda_1}, r_{21}, r_{22} \cdots r_{2\lambda_2}, \ldots r_{k1}, r_{k2} \cdots r_{k\lambda_k})$$

from a random processor allocation is

$$P(r) = \frac{n!}{r_{11}! r_{12}! \ldots r_{l\lambda_k}!} \frac{m!}{\lambda_1! \lambda_2! \ldots \lambda_k!} \frac{1}{n^m}$$

(Feller 1967).

Thus the probability of getting a schedule of length $t$ when randomly allocating $n$ independent activities on $m$ machines is

$$P(t) = \sum \{P(\mathbf{r}) : r_{11} = t \text{ and } r \text{ is an occupancy sequence defined by } n, m\}.$$

The expected length of a random schedule of $n$ independent activities on $m$ machines is

$$E(n,m) = \sum \{tP(t) : 1 \leqslant t \leqslant n\}.$$

The variance of a random schedule of $n$ independent activities on $m$ machines is

$$\mathrm{var}(n,m) = \sum \{t^2 P(t) : 1 \leqslant t \leqslant n\} - E(n,m)^2.$$

Table 7.1 shows these values for a range of $m$ and $n$. It is encouraging since it shows that $E(n,m)$ is reasonably close to $n/m$ even for quite moderate values of $n$ and $m$. The variance, which is a measure of the "spread" of values about the mean, is also not too great.

Now consider an arbitrary processor $p$. Regard an allocation of an activity to $p$ as a "success". Let $S_n$ denote the number of successes in $n$ allocations and let

$$S_n^* = \frac{S_n - n/m}{\sqrt{(n/m)(1 - 1/m)}}.$$

The Laplace limit theorem asserts that

$$P\{S_n^* > x\} \sim 1 - \Phi(x)$$

where $\Phi$ is the c.d.f. of the normal distribution. Thus, for any fixed $n$ and $m$, it is improbable to have a large $S_n^*$. As $n$ increases, one would expect the probability that $S_n^*$ exceeds some given value also to increase. However, by the law of the iterated

**Table 7.1** $E(n, m)/\text{var}(n, m)$.

| n/m | 2 | 3 | 4 |
|---|---|---|---|
| 2  | 1.5/0.25  | 2.25/0.19 | 2.75/0.44 |
| 3  | 1.33/0.22 | 1.89/0.32 | 2.37/0.31 |
| 4  | 1.25/0.19 | 1.69/0.34 | 2.12/0.33 |
| 10 | 1.10/0.09 | 1.29/0.23 | 1.53/0.27 |
| 20 | 1.05/0.05 | 1.15/0.13 | 1.28/0.22 |
| 30 | 1.03/0.03 | 1.10/0.09 | 1.19/0.16 |

| n/m | 10 | 20 | 50 | 100 |
|---|---|---|---|---|
| 2  | 6.23/0.99 | 11.76/1.90 | 27.81/4.62 | 53.98/9.10 |
| 3  | 4.93/2.83 | 8.89/1.69  | 20.16/3.46 | 38.26/7.80 |
| 4  | 4.20/0.80 | 7.38/1.45  | 16.23/3.40 | 30.23/6.77 |
| 10 | 2.75/0.52 | 4.41/0.85  | 8.69/1.76  | 15.11/3.21 |
| 20 | 2.18/0.32 | 3.23/0.55  | 5.81/1.05  | 9.59/1.74  |
| 30 | 1.93/0.31 | 2.75/0.47  | 4.77/0.79  | 7.55/1.29  |

algorithm (Feller 1967), we have

$$\limsup \frac{S_n^*}{\sqrt{2 \log \log n)}} \to +1 \quad \text{as } n \to \infty,$$

$$\liminf \frac{S_n^*}{\sqrt{2 \log \log n)}} \to -1 \quad \text{as } n \to \infty$$

Alternatively, the result can be expressed as follows.

**Theorem 7.5** *For $\lambda > 1$, with probability one, only finitely many of the events*

$$S_n > \frac{n}{m} + \lambda \sqrt{\frac{2n}{m}\left(1 - \frac{1}{m}\log\log n\right)}$$

*occur.*

*For $\lambda < 1$, with probability one, this inequality holds for infinitely many n.*

As we have seen, exact calculations are available for small $m$ and $n$, and there are asymptotics for large $n$. In the intermediate range we can devise some useful approximations. David and Barton (1962) obtained an approximation to $F(k)$, the cumulative distribution function of the length, i.e. $F(k) = P(length \leq k)$, namely

$$F(k) = e^{-\lambda}, \text{ where } \lambda = \frac{m\theta^{k+1}e^{-\theta}}{(k+1)!} \qquad \theta = \frac{n}{m}.$$

While this gives the whole probability structure, it is in a form that makes computation of the moments rather difficult. A simpler approach is based on the quantities $Q_q$ defined for given $q$ by

$$F(Q_q) = q, \quad 0 \leq q \leq 1.$$

The natural counterpart for the mean is the median $Q_{0.5}$, while $Q_{0.05}$ and $Q_{0.95}$ give the likely range of lengths, since $P(Q_{0.05} \leq length \leq Q_{0.95}) = 0.9$.

Rather than use the David and Barton approximation directly, we can approximate using the central limit theorem to give

$$F(Q_q) \approx \exp\left\{-m\left[1 - \Phi\left(\frac{Q_q - \theta}{\sqrt{\theta}}\right)\right]\right\}.$$

Taking logarithms, we deduce the following.

**Theorem 7.6**

$$Q_q = \frac{n}{m} + \sqrt{\frac{n}{m}}\,\Phi^{-1}\left(1 + \frac{\ln q}{m}\right), \text{ provided that } \frac{\ln q}{m} < 1.$$

Table 7.2 gives values of $Q_{0.5}, Q_{0.05}$ and $Q_{0.95}$ for a range of values of $m$ for $n = 50$ and $n = 500$. We can see that the range of possible values of the length is quite narrow, and is reasonably approximated as a constant multiple of $\sqrt{n}/m$.

These results suggest that a random processor allocation of $n$ independent activities on $m$ machines is going to give a reasonable approximation to the optimal schedule of length $\lceil n/m \rceil$.

**Table 7.2**

(a) $n = 50$

| $m$ | Lower ($Q_{0.05}$) | Median ($Q_{0.5}$) | Upper ($Q_{0.95}$) |
|---|---|---|---|
| 4 | 10 | 16 | 20 |
| 5 | 9 | 13 | 17 |
| 10 | 6 | 8 | 11 |
| 15 | 5 | 6 | 8 |
| 20 | 4 | 5 | 7 |
| 30 | 3 | 4 | 5 |
| 40 | 3 | 3 | 5 |

(b) $n = 500$

| $m$ | Lower ($Q_{0.05}$) | Median ($Q_{0.5}$) | Upper ($Q_{0.95}$) |
|---|---|---|---|
| 100 | 9 | 11 | 12 |
| 200 | 6 | 7 | 8 |
| 300 | 5 | 5 | 6 |
| 400 | 4 | 5 | 5 |

We now consider the case where the $n$ activities are a poset presented as a dag. Let $\omega_0$ be the length of the shortest schedule of the activities in this dag when executed on $m$ processors, assuming that any activity can be processed on any processor. This provides a lower bound on the length of an optimal schedule with preallocation. If $\omega(f)$ denotes the length of the optimal schedule with machine allocation function $f$ then $\omega_0 \leqslant \omega(f)$ for any $f$. In fact,

$$\omega_0 = \min \{\omega(f) : f \in A \to M\},$$

where $A \to M$ denotes the set of all $n^m$ possible machine allocation functions.

The *expected optimal schedule length* using a random machine allocation function is

$$E_0 = \frac{\sum \{\omega(f) : f \in A \to M\}}{n^m}$$

**Theorem 7.7**  $\max(\omega_0, E(n,m)) \leqslant E_0 \leqslant \min(E(m,m)\omega_0, n)$.

**Proof**  The lower bound is obvious and tight when all activities from a single chain (are independent). The upper bound of $n$ is obvious, and that of $E(m,m)\omega_0$ can be proved by a simple induction on $\omega_0$. It is also a tight bound—consider $\omega_0$ layers of $m$ activities such that every activity in layer $l$ is a parent of every activity in layer $l+1$, $1 \leqslant l < \omega_0$. □

$E(m,m)$ is a small-valued function even for quite large values of $m$, so Theorem 7.7 shows that random machine allocation is only likely to introduce a small multiplicative increase ($< 3.5$ for $m \leqslant 30$) in expected schedule length, provided the optimal time schedule is used. Unfortunately, as previously stated, finding such an optimal schedule is $\mathcal{NP}$-hard. This appears to suggest that simple "load balancing" heuristics for machine allocations are a reasonable approach for average case performance but that scheduling remains a critical problem.

## 7.8  CONCLUSIONS

We have seen that it is not safe to rely on automatic process placement or scheduling for arbitrary programs in a distributed system where processes cannot be moved. There are several possible ways around this problem.

The most obvious solution is to use a system where processes can be freely moved. We believe that trying to do this in a distributed system will introduce bad worst case communication costs. A more practical approach might be to use a shared memory system—or at least a system where remote memory accesses are affordable.

However, there are problems similar to those described here with shared memory. With such architectures and with fine-grain parallelism, it is memory contention that causes the problem. If we view actions as accessing a particular memory module rather than being performed on a given processor, the scheduling of activities to access such memory modules becomes critical. Strategies used can introduce dramatic differences to the timing of a program.

On the other hand, the class of programs can be limited. This could either be done by constraints in the language or through good programming practices. We do not have a good characterization of the class of programs that are safe, but even simple chains are causing significant problems, so this is an unpromising avenue of research.

Finally, letting the programmer manage process placement and scheduling appears to be another solution. This is probably practical only in applications where there are a fixed number of processes, probably with this number depending on the number of processors.

If the goal for a particular computer system is to support general purpose parallel processing at a high level, we believe that a system allowing cheap process migration is essential.

## ACKNOWLEDGMENTS

This work was sponsored by the Natural Science and Engineering Research Council of Canada, the United Kingdom Science and Engineering Research Council and Ministry of Defence and the Simon Fraser University Centre for System Science.

## REFERENCES

Burton F. W. (1988) Storage management in virtual tree machines *IEEE Trans. Comput.*, **37**, 322–328.

Burton F. W. and Rayward-Smith V. J. (1994) Worst case scheduling for parallel functional programming *J. Funct. Programming*, **4** (1), 65–75.

Burton F. W., McKeown G. P., Rayward-Smith V. J. and Sleep M. R. (1981) Parallel processing and combinatorial optimization. In *Proceedings of Conference on Combinatorial Optimization, Stirling, Scotland, August 1981*, University of Stirling, pp 19–36.

Burton F. W., McKeown G. P. and Rayward-Smith V. J. (1990) Applications of UET scheduling theory to the implementation of declarative languages *Comput. J.*, **33**, 330–336.

Coffman E. G., Jr., editor (1976) *Computer and Job-Shop Scheduling Theory* Wiley, New York.

David F. N. and Barton D. E. (1962) *Combinatorial Chance* Griffin, London.

Eager D. L., Zahorjan J. and Lazowska E. D. (1989) Speed-up versus efficiency in parallel systems *IEEE Trans. Comput.*, **38**, 408–423.

Feller W. (1957) *An Introduction to Probability Theory and its Applications*, Vol. 1 Wiley, New York.

Feller W. (1967) *An Introduction to Probability Theory and its Applications*, Vol. 2 Wiley, New York.

Garey M. R. and Johnson D. S. (1979) *Computers and Intractability: Guide to the Theory of NP-Completeness* W. H. Freeman, San Francisco.

Gonzalez T. and Sahni S. (1978) Flowshop and jobshop schedules: complexity and approximation *Oper. Res.*, **26**, 36–52.

Goyal D. K. (1976) Scheduling processor bound systems. Report No CS-7-036, Computer Science Department, Washington State University, Pullman.
Hu T. C. (1961) Parallel sequencing and assembly line problems *Oper. Res.*, **9**, 841–848.
Jaffe J. M. (1980) Bounds on the scheduling of typed task systems *SIAM J. Comput.*, **9**, 541–551.
Lai T. H. and Sprague A. (1985) Performance of parallel branch-and-bound algorithms *IEEE Trans. Comput.*, **34**, 962–964.
Lam S. and Sethi R. (1977) Worst case analysis of two scheduling algorithms *SIAM J. Comput.*, **6**, 518–536.

*School of Information Systems, University of East Anglia, Norwich NR4 7TJ, UK*

*School of Computing Science, Simon Fraser University, Burnaby, BC, Canada V5A 1SJ*

*School of Mathematics, University of East Anglia, Norwich NR4 7TJ, UK*

CHAPTER 8

# Real-Time Scheduling of Periodic Tasks

Claudine Chaouiya, Sophie Lefebvre-Barbaroux and Alain Jean-Marie
*INRIA, Centre Sophia Antipolis*

**Abstract**

We study queuing systems with periodic arrivals, submitted to a hard real-time constraint. This constraint specifies that each customer must leave the system before some fixed deadline. It must be satisfied for any possible lag between the customer types. We first analyze the basic properties of a queue with preemptions, whose arrival process is the superposition of periodic processes. We then deduce *feasibility conditions* under which the real-time constraint is satisfied—first for a general work conserving service discipline, then for several classical policies: first come first served, preemptive and nonpreemptive static priorities. The case of a network of stations submitted to periodic arrivals is then considered. We describe an optimal policy in the case of a ring topology, and discuss feasibility conditions for the FIFO policy in the case of a series network. For this, we develop an original approach based on the computation of bounds for the workload of the stations and the delays of customers.

## 8.1 INTRODUCTION

The study of hard real-time systems is mainly concerned with *feasibility*. Given a set of tasks, described by their arrival dates, their service times and deadlines, the problem is to decide whether or not every task will meet its deadline when served according to a given policy. A related question is whether, given a set of tasks, it is possible to construct a service policy that makes this set feasible. These problems are generally NP-hard with respect to the number of tasks: see Lawler *et al.* (1989) for the general case and Leung and Merril (1980) for the case where tasks arrive periodically. However, particular (but still of practical interest) occurrences turn out to be easier to solve.

We are interested in systems where tasks of different types arrive at fixed intervals $R_i$ and have constant service times $L_i$ and deadlines $\sigma_i$. The only unknown of the

---

*Scheduling Theory and its Applications* Edited by P. Chrétienne, E. G. Coffman, Jr., J. K. Lenstra and Z. Liu
© 1995 John Wiley & Sons Ltd

system is the lag (or phase) between different types. A set of tasks is said to be feasible if every task is completed before its deadline, for all possible phases.
One may envision several approaches to feasibility:

- *Tests with trajectories* reduce the problem to the study of certain trajectories obtained with particular phases. They consist in verifying the feasibility of task sets on these trajectories, which may involve the simulation of the system.
- *Tests on parameters* of the system, such as message lengths and periods, or load factors: these conditions are much easier to compute but are often only necessary or only sufficient.
- *Tests on optimality*: in the context of real-time scheduling, a policy **P** is said to be optimal in some class of policies if for any other policy **Q** in this class all task sets that are feasible by **Q** are feasible by **P**. Therefore if a task set is not feasible for an optimal policy, it is not feasible for any other policy of the class.

Table 8.1 (see also Silly 1984) summarizes some of the feasibility results for various scheduling policies. Here $M$ is the number of tasks types and $\rho = \sum_{i=1}^{M} L_i/R_i$.

Liu and Layland (1973) give a test of feasibility with static preemptive priorities using a special trajectory of the system. They show that the *rate monotonic priority*, which assigns higher priority to the smaller periods, is optimal when the deadline for each type is exactly its period (Serlin 1972; Liu and Layland 1973).

A dynamic policy that has received much attention is the so-called *relative urgency* (RU) or *deadline driven* priority (Serlin 1972; Liu and Layland 1973). When the deadlines are equal to the periods, this policy is optimal, and any task set for which $\rho < 1$ is feasible. Labetoulle (1974) proves that this policy is also optimal when deadlines are different from periods. In this case, however, there is no known characterization of feasibility.

Serlin (1972) introduced the *infinite time slicing* (*ITS*) policy. It consists in assigning to each type a fixed ratio of the server power. If deadlines and periods are the same then ITS is also optimal. It is no longer optimal for general deadlines.

Labetoulle (1974) proposed the improved *ITS generalized* policy, which is actually a weighted processor sharing policy. When present, each type shares the processing power to a level that depends on the type and on the server occupation. ITSG is less efficient than RU for general deadlines, but it has the significant advantage that there exists a feasibility test based on trajectories.

All the policies discussed so far are preemptive. Nonpreemptive priorities are more difficult to analyze and have received less attention, although they are much easier to implement. Kim and Naghibzadeh (1980) showed the nonpreemptive version of the RU policy to be optimal in the class of nonpreemptive policies.

In this chapter, we propose a characterization of feasibility for any fixed nonpreemptive priority, based on a test on special trajectories (Section 8.3.3).

We also develop a new approach to feasibility based on bounds on workload processes and on task delays. Such an approach can be found in Lefebvre-Barbaroux (1992) and Cruz (1991a,b). It allows one to find necessary conditions for a task set

**Table 8.1** Scheduling algorithms and their properties.

| Name | Type[a] | Deadline[b] | Feasibility condition[c] | Optimality | Reference |
|---|---|---|---|---|---|
| Rate monotonic (RM) | s.p. | $R$ | NSC: $\Delta_i^0 \leq R_i \forall i$ <br> SC: $\rho < M(2^{1/M} - 1)$ <br> NSC: $\Delta_i^0 \leq \sigma_i \forall i$ | Optimal <br><br> in class s.p. <br> < Inversed deadline | Serlin (1972), Liu and Layland (1973) <br><br> Dhall (1977) |
| Inversed RM | s.p. | $R$ | NSC: $\sum_i L_i \leq \inf_i R_i$ | Worst in class s.p. | Dhall (1977) |
| Inversed deadline | s.p. | $\sigma$ | NSC: $\Delta_i^0 \geq \sigma_i \forall i$ | Optimal in class s.p. | Leung and Whitehead (1982) |
| Fixed priority | s.p. | $\sigma$ | SC on parameters | | Section 8.3.1 |
| Relative urgency | d.p. | $R$ | NSC: $\rho \leq 1$ | Optimal | Lin and Layland (1973) |
| | | $\sigma$ | ? | | Labetoulle (1974) |
| Infinite time slicing (ITS) | p.s. | $R$ | NSC: $\rho \leq 1$ <br> SC: $\sum_i L_i/\sigma_i \leq 1$ | Optimal <br> < Inversed deadline | Serlin (1972), Labetoulle (1974) |
| ITS generalized | p.s. | $\sigma$ | NSC: $\Delta_i^0 \geq \sigma_i \forall i$ | <relative urgency | Labetoulle (1974) |
| FIFO | s.n.p. | $\sigma$ | NSC: $\sum_i L_i \leq \inf_i \sigma_i$ | ? | Section 8.3.2 |
| Fixed priority | s.n.p. | $\sigma$ | NSC: $\Delta_i \leq \sigma_i \forall i$ | | Section 8.3.3 |
| Relative urgency nonpreemptive | d.n.p. | $R$ | SC on parameters | Optimal in class n.p. | Kim and Naghibzadeh (1980) |

[a] s.p., static preemptive; d.p, dynamic preemptive; n.p., nonpreemptive; s.n.p., static nonpreemptive; d.n.p., dynamic nonpreemptive; p.s., processor sharing.
[b] $R$, the deadline of each task is its period; $\sigma$, general deadline.
[c] $\Delta_i^0$ is the delay of the first task of type $i$ for synchronous arrivals; for $\Delta_i$ see Section 8.3.3.

to be feasible in the case of a server with preemptions (or "vacancies"), operating under the FIFO service discipline (Section 8.3.2) or under a fixed preemptive priority (Section 8.3.1).

We also apply this method to the case of a communication network with real-time constraints, for which tests based on trajectories seem difficult to find (Section 8.4.2).

In the first part of the chapter, which is devoted to a single server with periodic arrivals and vacations, we establish the existence of a stationary (periodic) regime. We then provide bounds on the load of the queue and on the length of activity periods, which remain satisfied for any work-conserving service discipline.

In the second part we analyze the real-time properties of this server.

In the third part (Section 8.4), we study a communication network with hard real-time constraints. In the case of a ring topology we propose an optimal policy such that every message set for which the network is stable (in a sense we shall define in Section 8.2) is feasible. We then use the analysis of the previous sections to obtain a feasibility test when the stations of the network are connected in series and send messages according to the FIFO discipline. A feasibility condition, which does not depend on the service discipline, is also provided.

## 8.2 A SINGLE-SERVER SYSTEM WITH PERIODIC ARRIVALS AND PREEMPTION

In this section we study the basic properties of a single queue where the arrival process is defined as a superposition of deterministic processes. We assume that at any moment the server may be "on vacation": at such an instant it does not work, even if there are customers waiting in the queue. In our applications we shall use these vacations to model periods when the server is busy working on "other" tasks with higher priority. For this reason, we shall speak of *preemption* of the server. The service policy for customers in the queue may itself be preemptive or not. We discuss the periodicity of the system, and give bounds on its workload and on the length of activity periods.

Let $M$ be the number of types of customers. For each type $i \in \{1, \ldots, M\}$, we define:

- $L_i$, the (constant) service time of a customer (or *length* of a task);
- $R_i$, the period of the arrival process, which is assumed to be an integer,
- $\Phi_i$, the instant of the first arrival (or initial *phase*).

The quantities $L_i, R_i$ and $\Phi_i$ will be expressed in the same unit. A "task set" is defined by the two vectors $\mathbf{L} = (L_1, \ldots, L_M)$ and $\mathbf{R} = (R_1, \ldots, R_M)$. The vector $\Phi = (\Phi_1, \ldots, \Phi_M)$, together with the initial load $W(0)$, defines a particular configuration of this task set.

We assume that the server is preempted according to some process described by the indicator function $\pi$ : $\pi(t) = 0$ if the server is preempted at time $t$, and $\pi(t) = 1$ if the server is free.

The following lemma states a sample path property of the single-server queue, and does not depend on the periodic structure of our problem. Let $W(t)$ be the workload (quantity of work waiting for service) in queue at time $t$. By convention, we consider that arrivals are taken into account "just after" time $t$. Recall that a service policy is said to be *non-idling* (or *work-conserving*) if it is such that the server works as long as the queue is not empty, unless it is preempted. For any $a \leqslant b$, let $S(a, b)$ be the quantity of work arrived in the interval $[a, b)$.

**Lemma 8.1** *For all non-idling service policies, $W$ is characterized by the functional equation*

$$\forall b \geqslant a \geqslant 0, \quad W(b) = W(a) + S(a,b) - \int_a^b \mathbf{1}_{\{W(u)>0\}} \pi(u) \, du. \tag{8.1}$$

**Proof** The workload at an instant $b$ equals the load at an instant $a$ ($a \leqslant b$), increased by the quantity of work arrived in the interval $[a, b)$, minus the work done in the same interval. Since the policy is non-idling, the server works if and only if the workload is positive and it is not preempted.

Conversely, this equation can be shown to have a single solution for any given $W(0)$. □

From this lemma, it is easily seen that $W$ is a left-continuous function that admits limits from the right: it is the sum of the increasing step function $S(a, .)$ which has these properties, and of an integral function that is continuous. We shall use the notation $W(t^+)$ for the right-hand limit of $W$ at $t$.

### 8.2.1 Periodicity

Let us now assume that the arrival process and the preemption process of the queue are periodic. Let $T$ be the common period, that is, the least common multiple of the $R_i$s and of $T_P$, the period of the preemption process. Let $S = S(0, T)$ be the amount of work arriving in any period. Let $\rho = S/T$ be the load factor of the system. Likewise, let $P = \int_t^{t+T} [1 - \pi(u)] \, du$ be the quantity of preemption during any period and let $\rho_P = P/T$.

The following lemma is an intermediate coupling result. It will be useful for proving the periodicity of the load process.

**Lemma 8.2** *Consider two load processes $W_1$ and $W_2$ with the same arrival and preemption processes, which satisfies $\rho + \rho_P < 1$. Assume that $W_1(0) > W_2(0)$.*

*Then the following holds.*

(i) *The two processes couple at a certain finite instant, given by* $\tau = \inf \{t : W_1(t) = 0\}$. *If* $t < \tau$ *then* $W_1(t) > W_2(t)$, *and if* $t \geq \tau$ *then* $W_1(t) = W_2(t)$.

(ii) *The coupling instant satisfies the bound*

$$\left\lfloor \frac{\tau}{T} \right\rfloor \leq \frac{1}{T} \frac{W_1(0)}{1 - \rho - \rho_P}.$$

**Proof** Let $V = W_1 - W_2$ and $X = \mathbf{1}_{\{W_2 > 0\}} - \mathbf{1}_{\{W_1 > 0\}}$. By Lemma 8.1, we have

$$V(t) = V(0) + \int_0^t X(u)\pi(u)\,du \quad \forall t \geq 0.$$

Hence $V$ is continuous on $R_+$. Let $b = \inf \{t : V(t) = 0\}$. It is clear that $V(t) = 0\ \forall t \geq b$, and (because $V(0) > 0$), $V(t) > 0\ \forall t < b$: this follows from the uniqueness of the solution of (8.1).

Let $\tau = \inf \{t : W_1(t) = 0\}$. We first prove that $b = \tau$, which gives (i). By the continuity of $V$, we know that for $t \leq b$, $0 < V(t) = W_1(t) - W_2(t)$, so that $W_1(t) > W_2(t) \geq 0$. Therefore $b \leq \tau$. Assume that $b < \tau$. This means in particular that $W_1(b) > 0$. Then $W_2(b) > 0$ (since $V(b) = 0$), and using the left continuity of $W_1$ and $W_2$, there exists $a < b$ such that $W_1$ and $W_2$ are positive in $[a, b]$. $X$ is therefore equal to 0 in this interval, and $V(a) = V(a) - V(b) = \int_a^b X(u)\pi(u)\,du = 0$. This contradicts the definition of $b$. Hence $b = \tau$.

Finally, we show that $\tau$ is finite. If $k$ is such that $kT \leq \tau$ then $\int_0^{kT} \mathbf{1}_{\{W_1(u) = 0\}}\,du = 0$, and with (8.1), we obtain $W_1(0) + kS + kP - kT = W_1(kT) \geq 0$. Owing to the assumption $\rho + \rho_P < 1$ (i.e. $S + P < T$) we deduce the bound $k \leq W_1(0)/(T - S - P)$. This proves (ii). □

The following theorem establishes the periodic behavior of the system.

**Theorem 8.1** *If* $\rho + \rho_P < 1$, $W$ *is periodic with period* $T$ *after a time* $\tau$ *determined by the initial load:*

- *overload: if* $W(0) > W(T)$ *then* $\tau = \inf \{t : W(t) = 0\}$;
- *equilibrium: if* $W(0) = W(T)$ *then* $\tau = 0$;
- *underload: if* $W(0) < W(T)$ *then* $\tau = -T + \inf \{t \geq T : W(t) = 0\}$.

**Proof** Let $W_1 = W(. + T)$. The load processes $W$ and $W_1$ have the same arrival process, because of the $T$-periodicity. The "equilibrium" case follows easily from the uniqueness of the solution of (8.1). The "overload" and "underload" cases are corollaries of Lemma 8.2 (i). □

By analogy with the vocabulary of stochastic processes, we shall say that a system is in a *stationary regime* after it has become periodic. If the system is periodic from instant $t = 0$ (the "equilibrium" case), it is said to be stationary. Otherwise, it is called transient. Theorem 8.1 states that any transient system (with $\rho + \rho_P < 1$) couples with its *stationary version* after a finite time $\tau$. The stationary version may be constructed by taking an initial load equal to $W(kT)$, with $kT \geqslant \tau$. If $W(0) = 0$, the coupling occurs during the first period, and $W$ is always less than its stationary version (from Lemma 8.2).

When $\rho + \rho_P < 1$, a stationary system empties in every period: this is easily seen from (8.1), for instance. Such a system is called *stable*.

If $\rho + \rho_P = 1$, the queue also becomes stationary, but the stationary regime depends on the initial load. The queue does not necessarily become empty. When $\rho + \rho_P > 1$, there is a "pseudo-periodicity" with a linear increasing of the load. Details can be found in Lefebvre-Barbaroux et al. (1992) and Chaouiya (1992).

When the system is stationary, it is sometimes convenient to extend $W(t)$ by periodicity to negative values of $t$.

It is interesting to note that the construction of the stationary regime requires the observation of at least a complete period of the system, which may be infeasible in practical cases. On the other hand, it can be seen that the "overload" case of theorem 8.1 yields the fastest determination of the coupling instant $\tau$.

### 8.2.2 Bounds

In this subsection we use the specific structure of the arrival process (i.e. a superposition of deterministic processes) to derive an upper bound for $W(t)$. This result is completed by a study of the activity periods.

We assume that the queue is stable, i.e. $\rho + \rho_P < 1$, and that the preemption process satisfies the property

$$\exists \Pi > 0, \quad \forall a < b, \quad \int_a^b \pi(u)du \leqslant \Pi + \rho_P(b - a). \tag{8.2}$$

It is important to note that the results of this section depend only on the existence of such a bound. The preemption process need not be periodic.

Observe that results obtained for stationary systems apply to transient ones after they have coupled with their stationary regime.

In addition to the previous notation, let, for any set $I$ of task types, $\Lambda_I = \sum_{i \in I} L_i$. The quantity $\Lambda = \sum_{i=1}^M L_i$ is the maximum quantity of work that can arrive instantaneously to the system (when all types have a synchronous arrival). Also let $\rho_I = \sum_{i \in I} L_i / R_i$.

An *activity period* of the system is defined as an interval $[a,b]$ such that $\int_a^b \mathbf{1}_{\{W(u)=0\}} du = 0$ and such that there is no interval including $[a,b]$ and satisfying this property. When there is no preemption, this corresponds to the classical notion of a *busy period* during which the queue is not empty, with the small difference that we allow $W(t) = 0$ in such a busy period if there is an arrival at $t$ (i.e. $W(t^+) > 0$). In other words, idle periods have a strictly positive duration.

We first characterize the quantity of work arriving during any time interval.

**Lemma 8.3** *For any $b > a \geq 0$, $S(a,b)$ is bounded by*
$$S(a,b) \leq \rho(b-a) + \Lambda.$$

**Proof** The number of arrivals of type $i$ in the interval $[a,b[$ is at most $\lceil (b-a)/R_i \rceil \leq (b-a)/R_i + 1$. Therefore $S(a,b) \leq \sum_i (b-a)L_i/R_i + \sum_i L_i = \rho(b-a) + \Lambda$. □

The key result for obtaining bounds on the workload of the server is the following bound on the quantity of work done by the server. Define the function $f$ by $f(t) = 0$ if the server is not working (either because $W(t) = 0$ or because of preemption), and $f(t) = i$ if the server is working on a task of type $i$.

**Lemma 8.4** *Assume the system is stationary. For any subset of tasks $I \subset \{1,\ldots,M\}$ and any $a < b$,*
$$\int_a^b \mathbf{1}_{\{f(u)\}} du \leq \Lambda_I + \rho_I \frac{\Pi + \Lambda - \Lambda_I}{1 - \rho_P - (\rho - \rho_I)} + \rho_I(b-a).$$

**Proof** Consider first the case where $W(a) > 0$. Let $c = \sup\{t \leq a : W(t^+) = 0\}$ be the instant at which the *activity period* containing $a$ begins. We have

$$\int_a^b \mathbf{1}_{\{f(u)\in I\}} du = \int_c^b \mathbf{1}_{\{f(u)\in I\}} du - \int_c^a \mathbf{1}_{\{f(u)\in I\}} du$$

$$= \int_c^b \mathbf{1}_{\{f(u)\in I\}} du - (a-c) + \int_c^a \mathbf{1}_{\{f(u)\notin I\}} du$$

$$= \int_c^b \mathbf{1}_{\{f(u)\in I\}} du - (a-c) + \int_c^a \mathbf{1}_{\{f(u)\in \bar{I}\}} du + \int_c^a (1-\pi(u)) du,$$

where $\bar{I}$ is the complement of $I$. This follows from the fact that $[c,a]$ is included in an activity period, so that if the server is not working on tasks of $I$ then it is either preempted or working on a task of $\bar{I}$. Observe that the sum of the two last integrals is less than $a - c$.

Given that $W(c) = 0$, we have $\int_c^b \mathbf{1}_{\{f(u)\in I\}} du \leq S_I(c,b)$, where $S_I(c,b)$ is the quantity of work of type in $I$ arrived in the interval $[c,b[$. The same holds for $\bar{I}$. Using Lemma 8.3 and the property (8.2), we obtain:

$$\int_a^b \mathbf{1}_{\{f(u)\in I\}} du$$
$$\leq \Lambda_I + \rho_I(b-a+a-c) - (a-c) + \min\{a-c, \Lambda_{\bar{I}} + \rho_{\bar{I}}(a-c) + \Pi + \rho_P(a-c)\}$$
$$= \Lambda_I + \rho_I(b-a) + \min\{\rho_I(a-c), \Lambda - \Lambda_I + \Pi + (\rho - \rho_I + \rho_P - 1)(a-c)\}.$$

It is easily checked that under the stability condition,

$$\max_{x \geqslant 0} \min \{\rho_I x, \Lambda - \Lambda_I + \Pi + (\rho + \rho_P - 1)x\} = \frac{\rho_I (\Pi + \Lambda - \Lambda_I)}{1 - \rho - \rho_P + \rho_I}.$$

The lemma follows in this case.

If now $W(a) = 0$, $\int_a^b \mathbf{1}_{\{f(u) \in I\}} \, du = \int_d^b \mathbf{1}_{f(u) \in I} \, du$, where $d$ is the start of the next activity period. The previous analysis is then easily adapted. □

A similar approach provides a bound on the workload of the server.

**Theorem 8.2** *If the system is stationary, or if $W(0) = 0$, then*

$$\forall t \geqslant 0, W(t) \leqslant \Lambda + \rho \frac{\Pi}{1 - \rho_P}.$$

**Proof** Assume first that the system is stationary. Without loss of generality, let $t = 0$ be the beginning of an activity period. Using equation (8.1), we have, for any $t$ in the first activity period,

$$W(t) = S(0, t) - t + \int_0^t \mathbf{1}_{\{W(u) > 0\}} [1 - \pi(u)] du \leqslant \Lambda + \rho t - t$$

$$+ \int_0^t [1 - \pi(u)] du \leqslant \Lambda + (\rho - 1)t + \min\{t, \Pi + \rho_P t\}.$$

Hence the result follows.

Assume now that $W(0) = 0$. Then (as remarked above), $W$ is less than its stationary version. The result follows from the first case. □

Note that if the server is not preempted, the bound reduces to $W(t) \leqslant \Lambda$. This is clearly true if there is at most one customer of each type in the system, which is the case if the task set is feasible (see Section 8.3). It is important to note that even when it is not the case, this inequality remains satisfied. If, for instance, $\mathbf{L} = (5, 1, 1)$, then a consequence of Theorem 8.2 is that there is always at most one customer of type 1 in queue (if $\rho < 1$).

**Theorem 8.3** *If $\rho + \rho_P < 1$, then the duration $\delta$ of any activity period satisfies*

$$\delta \leqslant \frac{\Lambda + \Pi}{1 - \rho - \rho_P}. \tag{1}$$

*If the preemption process is periodic then, in addition,*

$$\delta \leqslant T(\rho + \rho_P).$$

**Proof** Assume that the interval $[a, b]$ is an activity period. In particular, $W(a) = W(b) = 0$, and $\mathbf{1}_{\{W(u) < 0\}} = 1$ almost everywhere on $[a, b]$. Equation (8.1) gives $S(a,b) = \int_a^b \pi(u)\,du$, or equivalently $\delta = b - a = S(a,b) + \int_a^b [1 - \pi(u)]\,du$. With Lemma 8.4 and the property (8.2), we obtain $\delta \leqslant \Lambda + \rho\delta + \Pi + \rho_P\delta$, hence the result.

The second inequality is a consequence of the fact that on every period of the system, the total amount of time the server is preempted or working is $P + S$. □

## 8.3  A SINGLE SERVER UNDER HARD REAL-TIME CONSTRAINTS

For a given task set $\mathscr{M}$, assume that customers of type $i$ have a constant *deadline* $\sigma_i$, such that $L_i \leqslant \sigma_i \leqslant R_i$.

We shall say that a task set is *feasible* if, starting from an empty system ($W(0) = 0$), the delay of every customer is less than its deadline, for any initial phase. Equivalently, one may require that every customer arriving at a stationary system leave before its deadline. The assumption of an empty initial load is sensible in practice, and is implicitly made in the literature. Note, however, that it introduces some pitfalls in reasonings when it comes to change the initial phase of a system while trying to keep the same trajectories. For instance, the proof of Theorem 1 of Liu and Layland (1973) is imprecise in that respect.

Much attention has been devoted to preemptive disciplines to solve real-time problems (Liu and Layland 1973; Serlin 1972; Labetoulle 1974). In the particular high-speed networking application that prompted our study, preemptive policies may be difficult or inefficient to implement, owing to either hardware constraints or the cost of overhead. This motivates the analysis of real-time properties of nonpreemptive policies, which are simpler to implement and have a fixed overhead (each task enters service once).

It will be assumed throughout this section that $\rho + \rho_P < 1$: the stability of the system is clearly a necessary condition for the feasibility of the task set.

Theorem 8.3 provides a sufficient condition for the feasibility of a task set under any work-conserving policy, be it preemptive or nonpreemptive:

$$\frac{\Lambda + \Pi}{1 - \rho - \rho_P} \leqslant \min_i \sigma_i. \tag{8.4}$$

Indeed, the delay of any task is clearly bounded by the length of the activity period to which it belongs.

Because of its general nature, this condition is very weak, and better conditions are known for some classes of service disciplines (see Table 8.1).

In the following we give new feasibility tests for static preemptive policies, for the FIFO queueing discipline and for static nonpreemptive policies using the analysis of Section 8.2.

Using an approach involving trajectory transformations, we also provide a necessary and sufficient condition for the feasibility of task sets under a general fixed nonpreemptive policy.

### 8.3.1 Preemptive priorities

We assume in this subsection that each task has a fixed priority and can preempt tasks of lower priority. Assume that task types are ordered according to decreasing priority. The delay of a task of type $i$ is not influenced by tasks of type $j > i$. This allows one to refine the condition (8.4).

**Theorem 8.4** *A sufficient condition for a task set to be feasible under the fixed preemptive priority is*

$$\forall i, \quad \frac{\sum_{j=1}^{i} L_i + \Pi}{1 - \sum_{j=1}^{i} L_i/R_i - \rho_P} \leq \sigma_i.$$

### 8.3.2 FIFO service discipline

In this subsection we assume that the server operates under the FIFO (first in first out) service policy. In this case the delay of a task depends only on the workload found upon its arrival, and on the preemption process (and not on future arrivals). This allows us to derive a feasibility condition that is better than (8.4).

**Theorem 8.5** *A task set $\mathcal{M}$ with $\rho < 1$ is feasible by FIFO if*

$$\frac{\Lambda + \Pi}{1 - \rho_P} \leq \min_i \sigma_i. \tag{8.5}$$

**Proof** Assume that the condition (8.5) is satisfied but that there is a task that misses its deadline. Without loss of generality, assume that this happens for the first time for a task arriving at a time $t_0$ in an activity period that starts at $t = 0$. Since the policy is FIFO, the delay of a customer arriving at $t_0$ is precisely the smallest $\delta > 0$ such that

$$\int_{t_0}^{t_0+\delta} \pi(u) \, du = W(t_0^+).$$

This delay therefore satisfies $W(t_0^+) = \delta - \int_{t_0}^{t_0+\delta} [1 - \pi(u)] \, du$, or equivalently $\delta = W(t_0^+) + \int_{t_0}^{t_0+\delta} [1 - \pi(u)] \, du$. As no task arrived before $t_0$ missed its deadline, we have $W(t_0^+) \leq \Lambda$: otherwise, there would be at least two tasks of the same type in the queue. From the property (8.2), it follows that $\delta \leq \Lambda + \Pi + \rho_P \delta$, so that $\delta \leq (\Lambda + \Pi)/(1 - \rho_P)$. On the other hand, $\delta$ is larger than one of the $\sigma_i$. This is a contradiction. Thus the theorem is proved. □

When there is no preemption, the condition (8.5) is simply $\Lambda \leq \min \sigma_i$. This is actually a necessary and sufficient condition, because $W(t) = \Lambda$ if a task of each type arrives at time $t$ (synchronous arrivals).

### 8.3.3 Nonpreemptive priorities

This section is concerned with the study of nonpreemptive priorities. The main result (Theorem 8.6) gives a *feasibility test* for a given priority, using the beginning of particular trajectories. We also discuss the possibility of obtaining simpler conditions.

We consider a static, nonpreemptive priority **P** and assume, without loss of generality, that customer types are numbered according to decreasing priority, which we denote by $1 \succ 2 \succ .. \succ M$. Customers of type 1 therefore have the highest priority.

In the following we shall use the symbol "$0^+$" as a shorthand for "a time strictly positive as close as needed to 0". This symbol will always be used in places where the actual choice of the date is irrelevant. For type $i$, let

- $j(i) = \min\{l \prec i : L_l = \sup_{k \prec i} L_k\}$; in other words, $j(i)$ is a type with the longest task among those with a lower priority than $i$;
- $\Phi^i$ be the initial phase given by
  if $i = M$, $\Phi^M = (\mathbf{0}, .., \mathbf{0})$;
  if $i \succ M$, $\Phi^i_{j(i)} = 0$ and $\Phi^i_k = 0^+, k \neq j(i)$.

#### 8.3.3.1 Test by trajectories

Following Liu and Layland (1973), we seek a test based on the observation of particular trajectories, constructed with special initial phases. We first introduce some notation.

- Let $T_i$ be the first instant at which the queue contains only types with a lower priority than $i$, when the initial phase is $\Phi^i$. This instant is finite because of the stability of the queue. It is actually bounded by $S$.
- Let $\Delta_i = \sup_n D_i^n$ where $D_i^n$ is the delay of the $n$th customer of type $i$ to arrive in $[0, T_i]$.

**Theorem 8.6** *A task set $\mathcal{M}$ is feasible with the static nonpreemptive policy* **P** *if and only if for all $i$ the delay $\Delta_i$ with the initial phase $\Phi^i$ is less than the deadline $\sigma_i$:*

$$\mathcal{M} \text{ is feasible by } P \iff \Delta_i \leq \sigma_i \quad \forall i.$$

Before proceeding with the proof, let us make some comments on the difference between this theorem and Theorem 1 of Liu and Layland (1973). The argument used by Liu and Layland is that a given customer experiences the longest delay if it arrives at the same instant as customers of all types with higher priority. Actually, this fails to be true if the task set is not feasible. However, it is exact that if for all types this delay is less than its deadline then the task set is feasible. The nonpreemptive case differs in the fact that customers with lower priorities can not be ignored, and further by the fact that *even* if the task set is feasible, the first customer is not necessarily the one with the longest possible delay (see Remark 8.1). This accounts for the complexity of the following proof.

**Proof** Let us call "request" the arrival of a customer. If the service of this customer is completed before its deadline occurs, we say that the request is *satisfied*. Otherwise, the request is unsatisfied, and an *overflow* occurs when the deadline expires.

The direction ($\Leftarrow$) is easy to prove. If $\mathcal{M}$ is feasible then for all phase vectors and all types, the delay of any customer is less than its deadline. In particular, $\forall i, \forall n \geqslant 1, D_i^n \leqslant \sigma_i$, and, by extension, $\Delta_i \leqslant \sigma_i, \forall i$.

We prove ($\Rightarrow$) by contradiction. Assume there exists an initial phase vector where a request of type $i$ is not satisfied. We want to prove that a request of this type is not satisfied either in the interval $[0, T_i]$, with the phase $\Phi^i$. For this, we progressively change the phase while ensuring at each step that there is still an unsatisfied request. The steps are illustrated by an example in the appendix.

Assume therefore that in the system starting with an empty queue and with phase vector $\Phi$ (called "system 1") there is an overflow for a customer of type $i$ at instant $t$.

**Step 1** In this step we essentially set the origin of times at the beginning of the "busy period" in which the overflow occurred. The unsatisfied request never entered service during the interval $[t - \sigma_i, t - L_i]$. The instant $t - L_i$ belongs to a busy period relative to customers with priorities $k \succeq i$, which begins at an instant $d$. At $d$, two cases are possible:

- in case 1 there is no customer with priority $k \succeq i$ waiting in queue;
- in case 2 there is a customer with priority $k \succeq i$ in queue, and the server just completes the service of a customer with priority $j \prec i$.

In the first case let $t_0 = d$. In the second case, let $t_0$ be the date at which the customer of type $j$ enters service; that is, $d - L_j$.

In both cases the interval $[t_0, t - L_i]$ enjoys the following property $\mathscr{P}$: during the interval, the server is busy and it serves customers with higher priority than $i$ except possibly the first customer.

We move the origin of times at instant $t_0$ by defining the new phase vector $\Psi$:

- for $k \succeq i$, $\Psi_k = (\Phi_k - t_0)$ modulo $R_k$,
- for $l \prec i$, $\Psi_l = 0^+$, except for type $j$ in case 2, for which $\Psi_j = 0$.

Let us call "system 2" the system with phase $\Psi$ and initially empty. In system 2 the interval $[0, A - L_i]$ (with $A = t - t_0$) has property $\mathscr{P}$: indeed, during this interval, the server has the same sequence of services than in the system 1 during the interval $[t_0, t - L_i]$. This is due to the fact that there are no customers with higher priority than $i$ in queue at $t_0$. If it were not the case, the sequence of services might have changed in system 2. In particular, the request of type $i$ occurring now at $A - \sigma_i$ in system 2 is still not satisfied.

**Step 2** In the case $i \neq M$ we now make a request of type $j(i)$ enter service first. Intuitively, this is the worst that can be done to a customer of type $i$ with customers of lower priority than $i$. In addition, we make the requests of all other types but $i$ arrive as early as possible, while ensuring that the customer of type $j(i)$ will start service first. This is done by defining a second phase vector $\mathbf{Y}$:

- $\Upsilon_{j(i)} = 0$;
- $\Upsilon_i = \max\{\Psi_i, 0^+\}$, which will guarantee that $i$ does not start before $j(i)$;
- $\Upsilon_k = 0^+$, $k \neq i, j(i)$.

In the case $i = M$ we just set $\mathbf{Y} = (0, \ldots, 0, \Psi_M)$.

Let "system 3" be the system with phase vector $\mathbf{Y}$ starting with an empty queue. Property $\mathscr{P}$ for the interval $[0, A - L_i]$ in system 2 is equivalent to

$$A \leqslant \inf\{x > 0 : S_2(0, x) - x + \epsilon L_j \leqslant 0\}, \tag{8.6}$$

where $\epsilon = 1$ in case 2 (0 otherwise), and $S_2(0, x)$ is the load relative to customers of type $k \succeq i$ arrived during $[0, x[$ in system 2; that is (cf. Lemma 8.3)

$$S_2(0, x) = \sum_{k \succeq i} \left\lceil \frac{x - \Psi_k}{R_k} \right\rceil L_k.$$

For all customers of type $k$ with higher priority than $i$ note that $\Upsilon_k = 0$ or $0^+$, while $\Psi_k \geqslant 0$. If $S_3(0, x)$ is defined for system 3 in the same way as $S_2(0, x)$ for system 2 then $S_3(0, x) \geqslant S_2(0, x)$. Moreover, because of the choice of $j(i)$, we have $L_{j(i)} \geqslant \epsilon L_j$. It follows that (8.6) also holds for system 3. Thus the interval $[0, A - L_i]$ in system 3 satisfies property $\mathscr{P}$, and the request of type $i$ arriving at $A - \sigma_i$ is still unsatisfied.

**Step 3** Finally, let us consider the phase $\mathbf{\Phi}^i$ (system 4). There only remains to make the first request of type $i$ arrive at 0 if $i = M$, and at $0^+$ otherwise. We prove that this modification does not change the sequence of services during $[0, A - L_i]$.

First, note that the queue empties in neither systems during this interval. This is true in system 3 (property $\mathscr{P}$), and therefore in system 4 because $S_4(0, x) \geqslant S_3(0, x)$ for all $x$. One may show by simple induction that the dates of beginning of service are identical in both systems. This is true for the first service, which is of type $j(i)$ (or 1, if $i = M$) in both systems. For the following scheduling instants the only difference is that there may be one more customer of type $i$ (whose first request has been advanced) in system 4 than in system 3. As the queue is not empty in system 3, such a customer has the lowest priority among those present in queue. It follows that the choice of the next customer to serve is the same in both systems.

The interval $[0, A - L_i]$ therefore enjoys property $\mathscr{P}$ in system 4, and the request of type $i$ which arrived at $A - \sigma_i$ in system 3 is unsatisfied at date $A$ or before in system 4. This proves the theorem, because $T_i \geqslant A$, still by property $\mathscr{P}$. □

This test requires the observation of $M$ trajectories for a total duration less than max $\{\Lambda/(1-\rho), S\}$ (which is an upper bound for the lengths of busy periods, according to Theorem 8.3). This is reasonable, except if $\rho$ is very close to 1.

In the case of task sets with two types of tasks, it is easy to reduce the test to $L_1 + L_2 \leq \inf(\sigma_1, \sigma_2)$. In this case, the feasibility condition is identical for FIFO and for both fixed priorities $1 \succ 2$ and $2 \succ 1$.

**Remark 8.1** To conclude this section, we show a task set that is not feasible, and for which the first request of the faulty task is satisfied. The task set has size 6 and is described by $\mathbf{L} = (2, 2, 2, 2, 6, 2)$ and $\mathbf{R} = (17, 15, 13, 11, 16, 1000)$. Deadlines are equal to periods. The stability condition is satisfied: $\rho \approx 0.96$. On Figure 8.1 one can observe that with the initial phase $\Phi^5$, the first request of task 5 is served before its deadline ($D_5^1 = 16 \leq R_5$), while the second request does not meet its deadline ($D_5^2 = 18 > R_5$).

*8.3.3.2 Other tests*

From the analysis above, we deduce the following test on parameters for a task set to be feasible under policy **P**. This condition is necessary but not sufficient. It provides however a much simpler test than that of Theorem 8.6.

**Theorem 8.7** *A necessary condition for a task set $\mathcal{M}$ to be feasible with the nonpreemptive policy **P** is*

$$Y_i \leq \sigma_i, \quad \forall i,$$

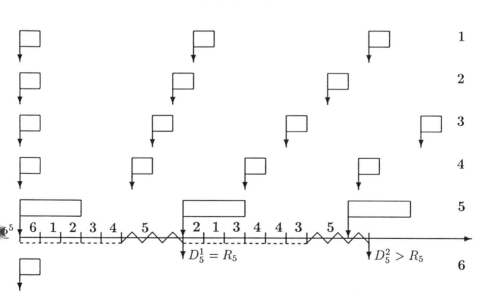

**Figure 8.1** The delay of the first request is not conclusive.

where

$$Y_i = \mathbf{1}_{\{i \neq M\}} L_{j(i)} + \sum_{k \succeq i} L_k.$$

**Proof** It is easily seen that $Y_i$ is the minimum delay experienced by a task of type $i$ for the initial phase $\Phi^i$. □

Following Liu and Layland, one may try to find feasibility tests based on the load of the task set, and look for optimal policies (see Liu and Layland 1973, Theorems 5 and 7). Unfortunately, neither of these approaches work for nonpreemptive policies.

There is no test using the load factor $\rho$: one may construct task sets that are not feasible with a load factor as small as required: for all $u \in [0, 1]$ and all $M > 1$ there exists a task set $\mathcal{M}$ of size $M$ such that $\rho \leqslant u$ and $\mathcal{M}$ is not feasible by any static, nonpreemptive policy. To see this, consider message sets of the form $(L_1, R_1) = (2n, 2n^2)$ and $(L_i, R_i) = (1, n)$ for 2, where $n = \lceil M/u \rceil$. Such message sets are never feasible by nonpreemptive policies, because $R_2 < L_1$.

Tests based on optimality principles also fail, since no optimal policy can be easily found. None of the following nonpreemptive static priorities is optimal:

- the one that assigns the highest priority to the smallest period (Liu and Layland 1973; Serlin 1972);
- the one that assigns the highest priority to the shortest task;
- the one that assigns the highest priority to the smallest critical delay; that is, the smallest value of $R_i - L_i$.

For each of these policies one may find a task set that is not feasible while it is feasible by another static nonpreemptive priority. We omit the details, which may be found in Lefebvre-Barbaroux (1992) and Lefebvre-Barbaroux et al. (1992).

## 8.4 A COMMUNICATION NETWORK WITH REAL-TIME CONSTRAINTS

We now turn to the study of a network of stations with periodic arrivals under a real-time constraint. Examples of such networks include manufacturing, signal processing and more general robotic architectures, where captors, processing units (signal processing, decision making etc.) and controllers are connected by some communication network and operate under time constraints. High-speed networks are another important application. For instance, the future integrated service digital networks (ISDN) will be used to carry data of various natures, some of which will be produced by periodic sources and will have strong constraints on response times (voice, video etc.).

In many applications it may be difficult (if possible at all) to cope with loss of data, which justifies the study of *hard* real-time constraints.

The study of real-time networks is complicated by the fact that, in addition to the contention at each station, one must take into account the contention for the communication medium.

The network motivating our study is a slotted, unidirectional ring. However, the topology and the protocol are not essential in our results, which may be adapted to other cases, for instance to store-and-forward networks.

We assume here that basic information units (slots) travel around the ring on a communication channel. Each station may occupy *free* slots to transmit its own messages. Slots are freed once they arrive at destination, and can be used by the emitting part of the station. An essential feature of this model is that slots occupied by message transiting "through" some station cannot be used by this station. Equivalently, one may consider that the bandwidth available to some station is limited by the communications between other stations.

In this networking context it is more natural to use the terms "messages" and "emission" instead of "task" and "service". We define

- $p$ as the section of the channel between station number $p$ and station number $p + 1$;
- $I(e)$ as the set of messages emitted by station $e$;
- $M(p)$ as the set of messages that go through section $p$;
- $\Lambda^e = \sum_{i \in I(e)} L_i$ as the largest quantity of work that can arrive simultaneously at the station $e$;
- $\rho^p = \sum_{i \in M(p)} L_i/R_i$ as the load factor of section $p$.

We shall assume that the following stability condition is satisfied for each section:

$$\rho^p < 1 \quad \forall p. \tag{8.7}$$

### 8.4.1 An optimal access policy

We describe here an optimal policy that ensures the feasibility of any task set which satisfies the network stability condition (8.7), in the case where deadlines and periods are equal. This policy can be seen as an extension of the ITS policy.

For any station $e$, one may decompose $I(e)$ into disjoint subsets $J_{e,1}, \ldots, J_{e,n(e)}$ such that every message type in $J_{e,k}$ has the same destination and the same period $R_{e,k}$. We call "elementary stations" the sets $J_{e,k}$ and note that $\Lambda_{e,k} = \sum_{i \in J_{e,k}} L_i$.

The principle of the parallel channel access (PCA) mechanism is to let each elementary station have access to the channel as if it were a real station of the network. At each instant, the server of station $e$ serves one customer in each set $J_{e,k}$ (if one is present) with speed $X_{e,k} = \Lambda_{e,k}/R_{e,k}$. Customers whose types belong to a particular set $J_{e,k}$ are served on a FIFO basis.

**Theorem 8.8** *The parallel channel access policy is optimal. Every message set for which the network is stable is feasible by PCA.*

**Proof** Under the parallel channel access policy, the network behaves like a network composed of elementary stations $J_{e,k}$, $1 \leqslant k \leqslant n(e)$. Each of these stations has access to a proportion $X_{e,k} = \Lambda_{e,k}/R_{e,k}$ of the channel. The stability condition (8.7) ensures that $\sum_{(e,k)} X_{e,k} < 1$. As far as feasibility is concerned, station $J_{e,k}$ behaves as a station with full access to the channel, but emitting messages whose length is multiplied by $1/X_{e,k}$. The quantity "$\Lambda$" corresponding to this system is $\sum_{i \in J_{e,k}} L_i/X_{e,k} = R_{e,k}$. According to Theorem 8.2, the task set is therefore feasible. $\square$

The implementation of the PCA policy on a slotted channel would be similar to that of the access control mechanism, at the extra cost (for the transmitter) to manage several message queues and poll among them in a cyclic way.

Finally, observe that under PCA, only a proportion $\rho^p$ of section $p$ is used, which may be useful if the network is also used by asynchronous traffic without real-time constraints.

### 8.4.2 Feasibility with trajectories

We illustrate in this subsection the difficulty in finding feasibility tests on trajectories for a network of stations. For a station in isolation, there is "monotonicity" of the notion of feasibility with respect to the message set, in that removing a task type from a feasible set gives another feasible set. This monotonicity does not exist in a network.

Consider a network of three stations in series, with three types of messages. Messages of type 1, 2 and 3 go from 1 to 3, 2 to 3 and 1 to 2 respectively. Let $I = \{1\}, J = \{2\}$ and $K = \{3\}$. Station 2 is preempted by messages going from 1 to 3. The following examples show that

(i) the feasibility of the whole task set $I \cup J \cup K$ does not imply the feasibility of $I \cup J$;
(ii) the feasibility of the task set reduced to $I \cup J$ does not imply the feasibility of the whole task set.

Figure 8.2 represents the workload processes for the three sets $I, J$ and $K$: $W_I$, $W_J$ and $W_K$. The parameters of the messages are $(L_1, R_1) = (4, 12)$, $(L_2, R_2) = (1, 4)$ and $(L_3, R_3) = (1, 2)$. Let $\sigma_i = R_i$, $i = 1, 2, 3$. The service discipline at station 1 is a preemptive priority of messages of type 3 on messages of type 1. The message set reduced to $I \cup J$ is not feasible, because $W_J$ exceeds $L_2 = 1$. However, with the complete task set, preemption periods at station 2 are split into pieces by messages of type 3, which allows the emission of messages of type 2 before their deadline.

In Figure 8.3 the messages are now described by $(L_1, R_1) = (1, 2)$, $(L_2, R_2) = (1, 4)$ and $(L_3, R_3) = (8, 10000)$. The service discipline is FIFO. The message set reduced to $I \cup J$ is clearly feasible: every message of type 2 will be

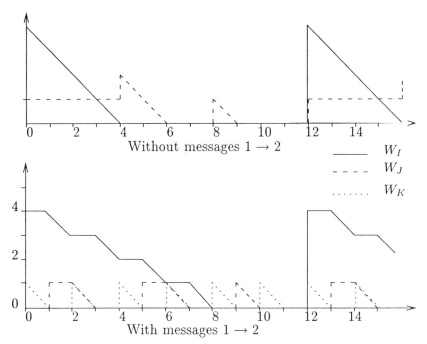

**Figure 8.2** Feasible with $K$ and not feasible without $K$.

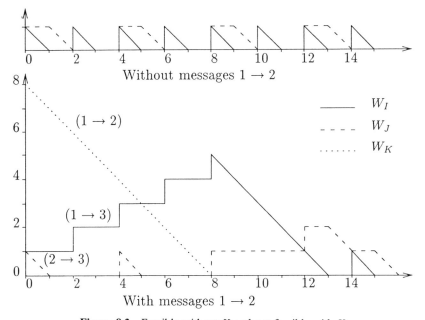

**Figure 8.3** Feasible without $K$ and not feasible with $K$.

emitted at worst two time units after its arrival. The whole task set is however not feasible: there are two messages of type 2 in queue at time 12.

Depending on the service disciplines, the presence of "local" messages going from one station to the next one may be either beneficial or detrimental to the satisfaction of real-time constraints of other messages.

The approach consisting in looking for a feasibility test based on a particular initial phase (typically, synchronous arrivals) also fails in general. Owing to the presence of local messages, it may happen that message sets satisfy the real-time constraint for synchronous arrivals, and not for some other phase. Details may be found in Chaouiya (1992).

### 8.4.3 Feasibility with bounds

Here we use an approach similar to that of Section 8.3 to derive sufficient conditions for feasibility based on bounds on the delay of messages.

In the remainder of this chapter we consider the simple case where stations are connected in series. Messages that have been emitted are assumed to preempt the stations through which they travel. It seems that our results could be adapted to other cases—for instance to feed forward acyclic networks.

In addition to the previous notation, let

- $\rho_I^e = \sum_{i \in I(e) \cap I} L_i / R_i$, be the load factor of messages of type belonging to some set $I$ at station $e$; in particular, $\rho_{M(e)}^e$ is the load of station $e$;
- $W^e$ be the workload process of station $e$,
- $f^e$ be defined by $f^e(t) = 0$ if station $e$ is not emitting at instant $t$ (either because $W^e(t) = 0$ or because of preemption) and $f^e(t) = i$ if the station is emitting a message of type $i$;
- $T^e$ be the least common multiple of the periods of all messages emitted by stations $p \leqslant e$.

Network delays are ignored: as far as periodicity and feasibility are concerned, a series network with communication delays is equivalent to one without delays in which the phases have been shifted.

The following theorem can be proved in the case where the service policy at the stations is FIFO or a static priority. It needs a proof of the periodicity of the *trajectory* of a station in isolation with preemption, operating under these service disciplines. The result then follows by induction. Details may be found in Chaouiya (1992).

**Theorem 8.9** *The system couples in finite time with a periodic regime: there exists a time $\tau$ such that for all station $e$, $W^e(t)$ is periodic for $t \geqslant \tau$. The period of station $e$ is $T^e$.*

**Theorem 8.10** *Define by recurrence the sequences*

$$\mathcal{B}^e = \sum_{p<e} B^p_{M(e)} + \Lambda^e \tag{8.8}$$

$$B^e_I = \Lambda_I + \rho_I \frac{\mathcal{B}^e - \Lambda_I}{1 - \rho^e + \rho^e_I}, \tag{8.9}$$

*for $2 \leqslant e \leqslant N$ and $I \subseteq I(e)$, with $B^1 = \Lambda^1$. Then, assuming that the system is stationary,*

$$\forall e, \ \forall t, \quad W^e(t) \leqslant B^e_{M(e)}.$$

**Proof** We actually prove by induction that for all $e$,

(i) $\qquad\qquad\qquad \forall t, \quad W^e(t) \leqslant B^e_{M(e)},$

(ii) $\qquad \forall I \subset I(e), \ \forall a < b, \quad \int_a^b \mathbf{1}_{\{f^e(u) \in I\}} \, du \leqslant B^e_I + \rho^e_I (b-a).$

For $e = 1$, (i) reduces to $W^1(t) B^1_{M(1)} = \Lambda^1$, which is true by Theorem 8.2. Likewise, (ii) is true by Lemma 8.4.

Assume the claims have been proved for all $p \leqslant e - 1$. The evolution equation for the workload of station $e$ can be expressed as:

$$W^e(b) = W^e(a) + S^e(a,b) - (b-a)$$
$$+ \int_a^b \mathbf{1}_{\{W^e(u)=0\}} \, du + \sum_{p<e} \int_a^b \mathbf{1}_{\{W^e(u)>0\}} \mathbf{1}_{\{f^p(u) \in M(e)\}} \, du.$$

This translates the fact that station $e$ does not emit if either it is empty or if some other station is currently emitting a message of type belonging to $M(e)$. Station $e$ can therefore be seen as a server in isolation preempted by a process characterized by

$$1 - \pi^e(u) = \sum_{p<e} \mathbf{1}_{\{f^p(u) \in M(e)\}}. \tag{8.11}$$

Using the induction hypothesis, we have for all $p < e$,

$$\int_a^b \mathbf{1}_{\{f^p(u) \in M(e)\}} \, du = \int_a^b \mathbf{1}_{\{f^p(u) \in M(e) \cap I(p)\}} \, du \leqslant B^p_{M(e)} + \rho^p_{M(e)}(b-a),$$

so that finally

$$\int_a^b [1 - \pi^e(u)] \, du \leqslant \sum_{p<e} [\rho^e_{M(e)}] = \mathcal{B}^e - \Lambda^e + (\rho^e - \rho^e_{M(e)})(b-a).$$

The preemption function satisfies (8.2) with $\Pi = \mathcal{B}^e - \Lambda^e$ and $\rho_P = \rho^e - \rho^e_{M(e)}$. Now let $I \subseteq I(e)$. Applying Lemma 8.4 yields (ii). (i) follows from Theorem 8.2. $\square$

We now give a bound on the lengths of the activity periods in any station. Based on equation (8.10), an argument similar to that of Theorem 8.3 yields the following.

**Theorem 8.11** *The length $\delta$ of any activity period of station e satisfies the bound*

$$\delta \leq \frac{\mathcal{B}^e}{1 - \rho^e}. \tag{8.12}$$

A general feasibility test similar to the condition (8.4) is derived from (8.12):

$$\forall e, \quad \frac{\mathcal{B}^e}{1 - \rho^e} \leq \min_{i \in I(e)} \sigma_i,$$

which is valid for any work-conserving service discipline. This condition can be improved if the service discipline is a fixed preemptive priority, as in Section 8.3.1. If messages are emitted in the FIFO order, we have the following.

**Corollary 8.1** *If all stations are FIFO, a sufficient condition for feasibility is*

$$\forall e, \quad \frac{\mathcal{B}^e}{1 - \rho^e + \rho^e_{M(e)}} \leq \min_{i \in I(e)} \sigma_i.$$

**Proof** This is a consequence of Theorem 8.5, when station $e$ is seen as a server preempted by the process $\pi^e$ given by (8.11). □

In conclusion, we briefly discuss the case of networks with more general topology. When there are loops, the analysis is more difficult, essentially because messages emitted by some station have an influence on the traffic of other stations, which influence in return this particular station. Even periodicity properties become difficult to establish, as shown in Chaouiya (1992). Bounds on workload process and on delays have been derived by Cruz (1991b), Chaouiya (1992) and Chang (1992), but they prove to be unaccurate when the load of the network is large. It is hoped that an approach based on evolution equations like (8.10) and careful bounding of the preemption processes will help improve them.

### APPENDIX: ILLUSTRATION OF THE PROOF OF THEOREM 8.6

We give an example to illustrate the three steps of the proof of Theorem 8.6. The message set is given by

- $M = 4, 1 \succ 2 \succ 3 \succ 4$;
- $\mathbf{L} = (3, 5, 5, 6)$;
- $\mathbf{R} = (9, 9, 1000, 1000)$, $\sigma_i = R_i \quad \forall i$

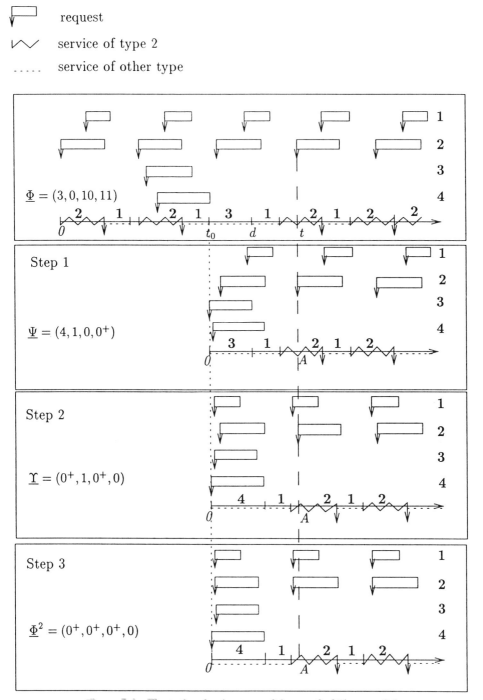

**Figure 8.4** Illustration for the steps of the proof of Theorem 8.6.

We first check that the stability condition is satisfied: $\sum_i i/R_i \approx 0.90$. Figure 8.4 shows the requests and the trajectory for the different steps (i.e. for the different initial phases). For the initial phase $\Phi = (3, 0, 10, 11)$, the third request of type 2 is not satisfied at $t = 27$.

*Step 1 (determination of $t_0$)* The beginning of the period of service of type 1 or 2 that contains $t$ is $d = 22$. At this instant, one message of each type 1 and 2 is waiting for service, and a message of type 3 completes service. We are in "case 1", and $t_0 = d - L_3 = 17$. The new initial phase is $\Psi(4, 1, 0, 0^+)$. The value of $A = t - t_0$ is 10.

*Step 2* The longest message with a strictly lowest priority than $i$ is of type $i_0 = 4$. Then the new initial phase is $Y = (0^+, 1, 0^+, 0)$.

*Step 3* Finally we consider the initial phase $\Phi^2 = (0^+, 0^+, 0^+, 0)$. The first request of type 2 is not satisfied at $t = 9$.

## REFERENCES

Chang C. S. (1992) Stability, queue length and delay. Part 1: deterministic queueing networks. IBM Research Report 17708, IBM Res. Division. In *Proceedings IEEE Conference on Decision and Control*, Vol. 1, pp 1005–1010.

Chaouiya C. (1992) Outils pour la validation de contraintes de synchronisation dans des systèmes distribués, PhD Thesis, University of Nice—Sophia Antipolis.

Cruz R. L. (1991a) A calculus for network delay. Part I: network elements in isolation *IEEE Trans. Inf. Theory*, **37**, 114–131.

Cruz R. L. (1991b) A calculus for network delay. Part II: network analysis *IEEE Trans. Inf. Theory*, **37**, 132–141.

Dhall S. K. (1977) Scheduling periodic-time critical jobs on single processor and multiprocessor computing systems. PhD Thesis, University of Illinois.

Kim K. H. and Naghibzadeh M. (1980) Prevention of task overruns in real-time non-preemptive multiprogramming systems *ACM Sigmetrics*, **9**, 267–276.

Labetoulle J. (1974) Some theorems on real time scheduling. In Gelenbe E. and Mahl R., editors, *Computer Architectures and Networks*, pp 285–297 North-Holland, Amsterdam.

Lawler E. L., Lenstra J. K., Rinnooy Kan A. H. G. and Shmooys B. B. (1989) Sequencing and scheduling: algorithms and complexity. Center for Mathematics and Computer Science, Amsterdam, Report BS-R8909.

Lefebvre-Barbaroux S. (1992) Files d'attente avec arrivées atypiques: environnement aléatoire et superposition de flux périodiques. PhD Thesis, University of Paris XI, Orsay.

Lefebvre-Barbaroux S., Jean-Marie A. and Chaouiya C. (1992) Problèmes d'ordonnancement d'une superposition de flux périodiques sous contrainte temps-réel. INRIA Research Report 1576.

Leung J. Y.-T. and Merril M. L. (1980) A note on preemptive scheduling of periodic, real-time tasks *Inf. Process. Lett.*, **11** 115–118.

Leung J. Y.-T. and Whitehead J. (1982) On the complexity of fixed priority scheduling of periodic real-time tasks *Perf. Eval.* **2**, 237–250.

Liu C. L. and Layland J. W. (1973) Scheduling algorithms for multiprogramming in hard-real-time environment *J. Assoc. Comput. Mach.*, **20**, 46–61.

Serlin O. (1972) Scheduling of time critical processes. In *AFIPS Conference Proceedings SJCC Atlantic City*, vol. 40.

Silly M. (1984) Contribution á l'ordonnancement de tâches temps-réel pour exécutif centralisé et réparti. PhD thesis, University of Nantes.

*INRIA Centre Sophia Antipolis, 2004 Route des Lucioles, BP 93, 06902 Sophia Antipolis Cedex, France*

CHAPTER 9

# Cyclic Scheduling on Parallel Processors: An Overview

C. Hanen and A. Munier
*LITP, Université Pierre et Marie Curie*

**Abstract**
Research effort has recently been devoted to cyclic scheduling problems that arise in the design of compilers for parallel architectures as well as in manufacturing systems. We focus here on the extensions of the basic cyclic scheduling problem (BCS), which seems to be one of the most suitable model for parallel processing applications. The properties of the earliest schedule of BCS are recalled and their most recent extensions are presented. Several generalizations of BCS that include resource constraints are then discussed. In particular, structural results and algorithms for periodic versions of job shop and $m$-machine problems are reported.

## 9.1 INTRODUCTION

Up to now, cyclic scheduling problems have been studied from several points of view, depending on the target application. Some theoretical studies have recently been devoted to these problems, in which basic results are often proved independently using different formalisms. We hope that this chapter, without pretending to be exhaustive, might contribute to the synthesis of this class of problems.

Cyclic scheduling deals with a set of "generic" tasks that have to be performed infinitely often. The usual scheduling constraints may be considered: precedences, resource sharing, deadlines, etc. However, these constraints must be formulated "generically". For example, cyclic versions of job shop scheduling problems can be defined. Usually, each job is a sequence of operations that outputs a manufactured product. In a mass production context, a great many identical products have to be produced, so one can assume that each job is to be performed infinitely often. Precedence as well as resource constraints are still expressed on one job. In this framework, the throughput is usually to be maximized, sometimes in combination with the minimization of the work-in-process.

*Scheduling Theory and its Applications* Edited by P. Chrétienne, E. G. Coffman, Jr., J. K. Lenstra and Z. Liu
© 1995 John Wiley & Sons Ltd

In other respects, cyclic scheduling also find applications in parallel computing. Loop scheduling is particularly important for the design of efficient compilers for parallel architectures. This application is detailed in Section 9.2, in order to present the main concepts of cyclic scheduling that we use in the rest of the chapter.

Scheduling periodic events, which arises for example in traffic-light scheduling, also induces related problems. However, this chapter will not tackle this area, except when the models have been designed for the two kinds of problems. We refer to Serafini and Ukovich (1989) and Korst (1992), which present many results and references on this subject.

Most of the problems that have been investigated up to now are related to the basic cyclic scheduling (BCS) problem, an instance of which is defined by a set of generic precedences called uniform constraints. It can be viewed as a cyclic version of the basic scheduling problem usually solved by the CPM (critical path method) or PERT algorithms. Section 9.2 presents the BCS problem and its simplest solution.

This chapter focuses on the different approaches to BCS and its extensions. Since many applications of cyclic scheduling are related to parallel processing, particular attention is paid to models that include resources like parallel processors. Two main approaches have been considered up to now.

The first aims at studying the asymptotic behavior of a dynamic scheduling policy. It has been applied to BCS and to some extensions. Since in BCS the tasks do not share any resource, the earliest schedule can be defined and has a maximum throughput. The steady state of this schedule has been characterized by studying timed event graphs (Carlier and Chrétienne 1988) or dioid algebra (Cohen et al. 1989; Gaubert 1990). Both theoretical tools have led to interesting extensions: deadlines, linear precedences and minimization of the work-in-process. Section 9.3 is devoted to these results.

The second approach deals with the construction of static schedules in the presence of resource constraints. Section 9.4 presents the main extensions of BCS that include shared resources. Most authors have tackled the construction of periodic schedules: the time interval between two successive executions of a generic task is constant. These schedules are not optimal in the general case (Munier 1991a; Hanen 1993), but they are easy to implement and their structure induces interesting dominating properties. Many heuristics have been developed for particular problems (Claver and Jackson 1987; Eisenbeis 1988; Graves et al. 1983; McCormick et al. 1989; Matsuo 1990) with good empirical behavior. We have chosen in this chapter to detail results that highlight the structure of the set of solutions, and seem to be among the most promising approaches to designing efficient algorithms.

In Section 9.5 we present some results on the structure of the set solutions of two cyclic versions of important classical scheduling problems: the periodic job shop and the periodic $m$-machine problem. We show how usual notions of scheduling theory can be extended to cyclic problems. We outline in Section 9.6 some exact algorithms and heuristics that have been designed for solving these problems.

Finally, two alternative models including resource constraints that have been proposed for cyclic scheduling are discussed in Section 9.7.

## 9.2 THE BASIC CYCLIC SCHEDULING PROBLEM

In order to introduce the basic cyclic scheduling problem (BCS), we present an application that arises in computer science, namely loop scheduling. We show that data transfers between the statements of the body's loop can be modeled by a uniform graph: each arc between two tasks corresponds to an infinite set of precedence constraints between their successive occurrences. We introduce the notion of critical circuit of a uniform graph. We show that they play the same role as critical paths for the basic (noncyclic) scheduling problem. The last subsection is devoted to the periodic solution of BCS, on which are based many results developed in the following sections.

### 9.2.1 Example

Loop scheduling plays an important role in the design of optimizing compilers for parallel architectures. Let us consider the loop shown in Figure 9.1.

On a parallel computer, the statements of the loop body may be performed simultaneously. Let us assume that the number of iterations $N$ is very large, and that the processing time of each statement $i$ is independent of the counter $I$ (see Figure 9.1). The loop can be considered as a set of generic tasks to be performed infinitely often. We denote by $\langle i, k \rangle$ the $k$th occurrence of the generic task $i$, which corresponds to the execution of statement $i$ in iteration $k$. A schedule assigns a starting time $t(i, k)$ to every occurrence $\langle i, k \rangle$.

The data dependences partially order the occurrences of the generic tasks. For example, the execution of $\langle 3, k \rangle$ must precede the execution of $\langle 4, k \rangle$ for any iteration $k$, since $A_3(k)$ is an input for the computation of $A_4(k)$. Note that some dependences may involve executions from different iterations: for any iteration $k$, $\langle 7, k \rangle$ must precede statement $\langle 3, k+1 \rangle$.

Thus a data dependence between two statements $i$ and $j$ induces, for every $k > 0$, a precedence constraint between the occurrences $\langle i, k \rangle$ and $\langle j, k + h_{ij} \rangle$. This defines a uniform constraint.

```
FOR I = 1 TO N DO
    A₁(I) = f₁(A₃(I-1))
    A₂(I) = f₂(A₁(I))
    A₃(I) = f₃(A₂(I), A₇(I-1))
    A₄(I) = f₄(A₃(I))
    A₅(I) = f₅(A₄(I))
    A₆(I) = f₆(A₅(I-1))
    A₇(I) = f₇(A₆(I))
```

| task i | $p_i$ |
|---|---|
| 1 | 1 |
| 2 | 3 |
| 3 | 2 |
| 4 | 5 |
| 5 | 1 |
| 6 | 3 |
| 7 | 2 |

**Figure 9.1** A recurrent loop and the processing times of the generic tasks.

Let us denote by $T$ the set of generic tasks. The set of uniform constraints can be modeled by a directed graph $G = (T, E)$. The arcs of $G$ are weighted by two functions $L : T \to \mathbb{N}^*$ and $H : T \to \mathbb{N}$, called length and height, such that if $e_{ij}$ denotes an arc from node $i$ to node $j$ then $L(e_{ij})$ is equal to the processing time of task $i$, and $e_{ij}$ corresponds to the following uniform constraint:

$$\forall k \geqslant 1, \quad t(i,k) + L(e_{ij}) \leqslant t(j, k + H(e_{ij}))$$

The triple $(G, L, H)$ is referred to in the following as a uniform graph. In our example the length of the arc $e_{73}$ is equal to the processing time of task 7, its height is equal to 1. The uniform graph associated with the recurrent loop is depicted in Figure 9.2.

It is usually assumed that two executions of a generic task $i$ cannot overlap. This "nonreentrant" constraint can be easily modeled by adding an arc from $i$ to $i$ with length $p_i$ and height 1. For a better understanding of figures, such arcs are omitted in the following.

If one assume that the number of processors is not limited, it is important to know how far the parallelism of the loop body can be exploited. The basic cyclic scheduling problem (BCS) is defined for this purpose. Given a uniform graph $(G, L, H)$, the problem is to find a schedule with a minimum average cycle time $w$ defined as follows:

$$w = \lim_{k \to \infty} \frac{\max_{i \in T}\{t(i,k) + p_i\}}{k}.$$

The inverse $\tau = 1/w$ is usually referred to as the throughput of the schedule.

### 9.2.2 Critical circuits

The circuits of $G$ are the key to both the problem of consistency and of the average cycle time minimization. For example, the circuit $c = (3, 4, 5, 6, 7)$ of $G$ induces the following constraints, for any integer $k$:

$$t(3, k + 2) \geqslant t(7, k + 1) + 2 \geqslant t(6, k + 1) + 5 \ldots \geqslant t(3, k) + 13.$$

Let us define the length $L(v)$ (respectively the height $H(v)$) of a path $v$ of $G$ to be the

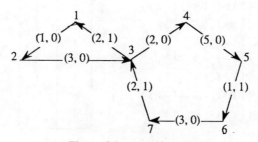

**Figure 9.2** A uniform graph.

sum of the lengths (respectively the heights). Then $c$ induces, for any integer $k$,

$$t(3, k + H(c)) \geq t(3,k) + L(c),$$

with $L(c) = 13$ and $H(c) = 2$. So, the average time interval between two successive occurrences of task 3 is at least $w(c) = \frac{13}{2} = L(c)/H(c)$.

In the general case, a uniform graph $(G, L, H)$ is consistent if and only if every circuit of $G$ has a positive height. Further, for any circuit $c$, the ratio $w(c) = L(c)/H(c)$ is a lower bound on the average cycle time. A circuit that maximizes this ratio is called a critical circuit: its value will be denoted by

$$w(G) = \max_{c \in C(G)} \frac{L(c)}{H(c)}.$$

Several algorithms have been proposed for the computation of the critical circuit of a uniform graph (Gondran and Minoux 1985). The most efficient is $O(n^3 \log n)$.

### 9.2.3 A periodic solution

The basic scheduling problem has been studied independently by many authors. To the best of our knowledge, the first solution of BCS was proposed by Ramchandani (1973) while studying the periodic behavior of a particular class of timed Petri nets, namely timed event graphs.

Indeed, a timed Petri net can be built from a uniform graph as follows: to each task $i$ there corresponds a transition $T_i$ with firing duration $p_i$; and to each arc $e_{ij}$ there corresponds a place $P_{ij}$ with input transition $T_i$ and output transition $T_j$, initially marked with the height of the arc. In such a net, each place has exactly one input and one output transition; thus this net is an event graph. The uniform constraints are just a particular expression of the marking equations of the net.

Let us define a periodic schedule as follows:

$$\forall i \in T, \quad \forall k \geq 1, \quad t(i,k) = t_i + w(k-1).$$

The period $w$ is the average cycle time of the schedule. The vector $t = (t_1, \ldots, t_n)$, referred to as the generic schedule, is the schedule of the first iteration that is repeated every $w$ time units. Thus a periodic schedule is entirely defined by the pair $\sigma = (t, w)$. Figure 9.3 illustrates a periodic schedule for our example.

Ramchandani proved that the bound $w(G)$ could be reached by a periodic schedule. Let us consider a period $w \geq w(G)$. If $\sigma = (t, w)$ is a feasible periodic schedule then the uniform constraint associated with any arc $e_{ij}$ of $G$ can be expressed as follows:

$$t_j - t_i \geq L(e_{ij}) - wH(e_{ij}).$$

So, if we denote by $a_w(e_{ij}) = L(e_{ij}) - wH(e_{ij})$ the amplitude of the arc $e_{ij}$ with respect to $w$, the generic schedule $t$ is a potential on the graph $(G, a_w)$. As $w \geq w(G)$, this graph has no circuit with positive amplitude, so one can build a periodic schedule $\sigma = (t, w)$ in polynomial time $O(n^3)$ (where $n$ is the number of

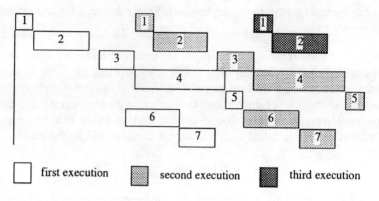

t = ( 0, 1, 4.5, 6.5, 11.5, 6, 9 )   w = 6.5

**Figure 9.3**  A feasible periodic schedule.

generic tasks) by computing the paths with maximal amplitude w.r.t. $w$ on $G$. Hence an optimal periodic schedule can be provided in $O(n^3 \log n)$.

Figure 9.4 shows the graph $(G, a_{6.5})$ of our example. A potential inducing the optimal schedule of Figure 9.3 is also reported.

These results are still valid if arbitrary length and height functions are considered (in particular, negative height values, and lengths different from the processing times).

## 9.3 DYNAMIC SCHEDULING

The solution of a scheduling problem can be either a static schedule, which assigns a starting time to each task, or a dynamic scheduling policy, which describes some simple rules in order to choose at each time $t$ which tasks are to be started. In a cyclic framework, the performances of such a policy can be analyzed through the study of the asymptotic behavior of the dynamic schedule.

Let us consider an instance of BCS. As the tasks do not share any resource (there are only precedence constraints), one can define the earliest schedule by a very

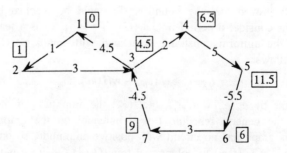

**Figure 9.4**  Graph $(G, a_{6.5})$ and a potential.

simple dynamic scheduling policy: execute a task as soon as possible. This schedule has obviously an optimal average cycle time.

There are two main studies devoted to the characterization of the asymptotic behavior of this schedule. Section 9.3.1 presents the first, developed by Chrétienne (1985), which uses graph theory arguments and longest-path computations. Section 9.3.2 is devoted to the second, developed by Cohen et al. (1985,1989), which uses algebraic results in the dioid ($\mathbb{R}$, max, +). Our purpose is to show the connections and the differences between these two approaches.

Both theoretical tools have been used to solve new problems described in Section 9.3.3 (existence of a "latest" schedule, generalization of uniform constraints, and work-in-process minimization).

### 9.3.1 Characterization of the earliest schedule from the longest paths of $g$

Let us consider a uniform graph $(G, L, H)$. The expanded graph $g$ associated with $(G, L, H)$ is an infinite graph whose vertices are the occurrences $\langle i, k \rangle, k > 0$ and whose arcs are the precedence constraints induced by $G$. Hence the earliest starting time of the occurrence $\langle i, k \rangle$ is the value of the longest path of $g$ ending with $\langle i, k \rangle$.

Now, if we consider two generic tasks $i$ and $j$, a $g$-path from $\langle i, k \rangle$ to $\langle j, k+h \rangle$ corresponds to a $G$-path from $i$ to $j$ with the same length and height $h$. Let us denote by $s_{ij}(h)$ the set of $G$-paths from $i$ to $j$ of height $h$, and by $S_{ij}(h)$ the set of $s_{ij}(h)$-paths of maximal length. The behavior of the schedule is clearly related to the study of the sets $S_{ij}(h)$.

Let $c_{ij} \in C(G)$ be the set of $G$-circuits $c$ that can be crossed by a path from $i$ to $j$, and let $C_{ij}$ be the set of $c_{ij}$-circuits of maximum value $L(c)/H(c)$.

Chrétienne (1985) proved that, for large enough $h \geqslant h_{ij}$, every path of $S_{ij}(h)$ passes through a circuit of $C_{ij}$ at least once. So, $\forall u \in S_{ij}(h)$, there are a circuit $c \in C_{ij}$ and two paths $v$ and $\mu$ from $i$ to $c$ and from $c$ to $j$ such that $u = vc\mu$ (see Figure 9.5). Moreover, he proved the existence of an integer $K_{ij}$ such that, for $h > h_{ij}$, any path of $S_{ij}(h + K_{ij})$ results from a path of $S_{ij}(h)$ by passing some more times through circuits of $C_{ij}$. He introduced the $K$-periodicity in order to characterize the length $L_{ij}(h)$ of $S_{ij}(h)$-paths: a sequence $u_n$ is $K$-periodic if there exist an integer $N_0$ and a pair $(K, a)$ such that $\forall N > N_0, u_{n+K} = u_n + aK$, where $K$ is the periodicity factor, $a$ the period and $K/a$ the frequency of the sequence.

Now, the main properties of the longest paths of $g$ can be summarized as follows.

- If there is only one circuit $c$ in $C_{ij}$, the sequences $u_h = L_{ij}(h)$ are $H(c)$-periodic with frequency $1/w(c)$.
- If $C_{ij}$ contains several elements, let $\Pi_{ij}$ be the product of their heights, $\prod_{c \in C_{ij}} H(c)$. The sequences $L_{ij}(h)$ are $\Pi_{ij}$-periodic with frequency $1/w(c), c \in C_{ij}$

The asymptotic behavior of the earliest schedule results from these properties, and is based on the analysis of the strongly connected components of $G$.

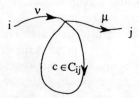

**Figure 9.5** A path $v c \mu$

- If $G$ is strongly connected then any circuit can be crossed by a path from $i$ to $j$. So, there exists $K$ depending as previously on the number of critical circuits of $G$, such that the earliest starting time sequence of every task $i, t(i, h) = \max_{j \in T} L_{ij}(h)$, is $K$-periodic with frequency $1/w(G)$. For example, the critical circuit of the uniform graph depicted in Figure 9.2 is $c = (3, 4, 5, 6, 7)$, and its value is $w(c) = 6.5$. Hence the earliest schedule is 2-periodic with an average cycle time equal to 6.5.
- In the general case let $Q_1, \ldots, Q_l$ be the strongly connected components of $G$. All tasks from a component $Q_k$ have the same frequency $\phi_k$. This value depends both on the critical circuits of $Q_k$ and its ancestor components. If we build the reduced graph $G_R$ by merging all the nodes from a same component of $G$, the frequencies $\phi_k$ are given as follows:

$$\phi_k = \min \left\{ \min_{q_r \in \Gamma^{-1}(Q_k)} \phi_r, \frac{1}{w(Q_k)} \right\},$$

where $\Gamma^{-1}(Q_k)$ denotes the set of predecessors of $Q_k$ in $G_R$. As $G_R$ is acyclic, the frequencies can be computed in polynomial time using a topological sort of the nodes. For example, the uniform graph depicted in Figure 9.6 has three strongly connected components $Q_1, Q_2$ and $Q_3$. Let us assume that the processing times of tasks 3 and 4 are 2 and 3. Now, $w(Q_1) = 2.5, w(Q_2) = 3$ and $w(Q_3) = 2$, so that $\phi_1 = \frac{2}{5}, \phi_2 = \frac{1}{3}$ and $\phi_3 = \frac{2}{5}$. The critical circuit of $Q_1$ slows down the executions of task 3. The average cycle time of the earliest schedule is thus 3.

Note that the average cycle time is always equal to the value of the critical circuit of $G$. However, when the frequencies of some tasks are different, the duration of one iteration may tend towards infinity, so that the work-in-process becomes infinite. The results recalled here can be used to detect such unsteadiness, which is useful in manufacturing systems.

All the above results can be extended to more general length and height functions: the heights of the arcs may have negative values, if the graph does not include circuits of nonpositive height, and the length of an arc may be different from the processing time of its origin (but still positive).

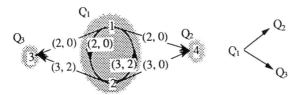

**Figure 9.6** A uniform graph $G$ and the reduced graph $G_R$.

### 9.3.2 Dioids

A dioid is a set $E$ with two associative laws $\oplus$ and $\otimes$ such that $\oplus$ is commutative and $\otimes$ is distributive over $\oplus$. The particular dioid considered here is $(\mathbb{R}, \max, +)$. It was introduced by Gondran and Minoux (1985) for the algebraic computation of the longest paths of valued graphs.

The first law $\oplus$ is associated with max, the second $\otimes$ with $+$. The neutral elements for the first and the second laws are $\epsilon = -\infty$ and $e = 0$ respectively.

The results recalled here are essentially due to Cohen et al. (1989) and addressed the previous problem: characterizing the asymptotic behavior of the earliest schedule of tasks submitted to uniform constraints.

#### 9.3.2.1 Linear algebra in $(\mathbb{R}, \max, +)$

The computation of the longest paths of a valued graph $(G, v)$ is strongly related to linear algebra results in the dioid $(\mathbb{R}, \max, +)$. Indeed, let us consider the vertex-incidence matrix $M$ of $G$: $M(i,j) = \epsilon = -\infty$ if there is no arc $(i,j)$, $M(i,j) = v(i,j)$ otherwise.

Let $k$ be an integer. Then

$$M^k = \underbrace{M \otimes M .. \otimes M}_{k \text{ times}}$$

is the incidence matrix of the longest paths of exactly $k$ arcs. We also define the star matrix

$$M^* = \bigoplus_{k=0}^{\infty} M^k;$$

$M^*(i,j)$ is the value (which may be infinite) of the longest path from $i$ to $j$ in $G$.

In other respects, the notion of eigenvalue is also related to longest paths. Indeed, if we write the usual definition of an eigenvalue $\lambda$, we get

$$MX = \lambda X \Leftrightarrow \lambda^{-1} MX = X$$

But in the dioid, $\lambda^{-1} M$ is the incidence matrix of the graph $(G, v - \lambda)$. Gondran and Minoux (1977) proved that if $G$ is strongly connected, the only eigenvalue of $M$ is

$$\lambda = \max_{c \in C(G)} \frac{v(c)}{n(c)},$$

where $v(c)$ denotes the sum of the values of the arcs of the circuit $c$, and $n(c)$ denotes the number of its arcs. Hence $\lambda$ is the minimum value that must be subtracted from all the arc valuations in order to avoid positive circuits, so that the longest paths of $(G, v - \lambda)$ become finitely valued.

### 9.3.2.2 Modeling the cyclic scheduling problem

Let us consider a uniform graph $(G, L, H)$ and denote by $\Gamma^{-1}(i)$ the set of predecessors of a node $i$ in $G$, and by $\bar{h}$ the maximum height of an arc of $G$.
The earliest schedule is defined by the following recurrence equations:

$$t(i,k) = \max_{j \in \Gamma^{-1}(i)} \{t(j, k - h_{ji}) + p_j\} \quad \forall k > \bar{h}, \quad \forall i \in T$$

These equalities can be expressed in the dioid as follows:

$$t(i,k) = \bigoplus_{j \in \Gamma^{-1}(i)} [p_j \otimes t(j, k - h_{ji})] \quad \forall k > \bar{h}, \quad i \in T.$$

Cohen et al. defined a sequence of square matrices of size $n = |T|, A_0, \ldots, A_{\bar{h}}$, as follows: $A_r[i,j] = p_i$ if there is an arc $e_{ij}$ of height $H(e_{ij} = r$. Setting $t(*, k) = (t(1, k) \ldots t(n, k))$, the earliest schedule is defined by a linear system of order $\bar{h}$:

$$t(*, k) = t(*, k)A_0 \oplus t(*, k-1)A_1 \oplus \oplus t(*, k-\bar{h})A_{\bar{h}} \quad \forall k > \bar{h}$$

Using computation in dioids (Gondran and Minoux 1985), the equation becomes

$$t(*, k) = t(*, k-1)A_1 A_0^* \oplus \cdots \oplus t(*, k-\bar{h})A_{\bar{h}} A_0^* \quad \forall k > \bar{h}$$

where $A_0^*[i,j]$ is the maximum length of a path of $G$ from $i$ to $j$ of null height. Setting $X^k = (t(*, k)t(*, k-1) \ldots t(*, k-\bar{h}+1))$, these equations can be rewritten as a linear recurrent system $X^k = X^{k-1}A$, where $A$ is the square matrix of size $\bar{h}n$ given by

$$A = \begin{pmatrix} A_1 A_0^* & e & \epsilon & \ldots & \ldots & \epsilon \\ A_2 A_0^* & \epsilon & e & \ddots & & \epsilon \\ \vdots & \vdots & \epsilon & \ddots & \ddots & \vdots \\ \vdots & \vdots & \vdots & \ddots & & \epsilon \\ A_{\bar{h}_1} A_0^* & \epsilon & \epsilon & & \ddots & \epsilon \\ A_{\bar{h}} A_0^* & \epsilon & \epsilon & \ldots & \ldots & \epsilon \end{pmatrix}.$$

The matrix $A$ of the uniform graph depicted in Figure 9.7 is

$$A = \begin{pmatrix} 3 & \epsilon & \epsilon & e & \epsilon & \epsilon \\ 6 & 2 & 4 & \epsilon & e & \epsilon \\ 4 & \epsilon & 2 & \epsilon & \epsilon & e \\ 7 & 3 & 5 & \epsilon & \epsilon & \epsilon \\ \epsilon & \epsilon & \epsilon & \epsilon & \epsilon & \epsilon \\ \epsilon & \epsilon & \epsilon & \epsilon & \epsilon & \epsilon \end{pmatrix}.$$

The recurrent system is solved by studying the properties of the powers of $A$ in the dioid $(\mathbb{R}, \max, +)$.

### 9.3.2.3 Powers of $A$ and the earliest schedule

Let $(G^A, v)$ be the simple-valued graph associated with $A$: if $A[i,j] \neq \epsilon$, $G^A$ has an arc $(i,j)$ valued by $A[i,j]$. We index the $n\hbar$ vertices of $G^A$ as follows: $(1,1),(2,1),\ldots(n,1),(1,2),\ldots,(n,\hbar)$ (see Figure 9.7).

An arc $e = ((i,\alpha_i),(j,1))$ of $G^A$ corresponds to a path of maximum length among the set of paths of $G$ from $i$ to $j$ that begins with an arc of height $\alpha_i$ and whose other arcs have a null height. Its valuation is equal to the length of the $G$-path. For example, the arc $((1,2),(3,1))$ of $G^A$ corresponds to the path $(1,2,3)$ of $G$.

If one assumes that $G$ is strongly connected then $G^A$ can be reduced to one strongly connected component by removing nodes without successors: in our example, we remove $(2,2)$ and $(3,2)$. The new associated matrix $A$ is

$$A = \begin{pmatrix} 3 & \epsilon & \epsilon & e \\ 6 & 2 & 4 & \epsilon \\ 4 & \epsilon & 2 & \epsilon \\ 7 & 3 & 5 & \epsilon \end{pmatrix}$$

Let us denote by $C(G^A)$ the set of circuits of $G^A$. As mentioned earlier, the eigenvalue of $A$ is $\lambda = \max_{c \in C(G^A)} [v(c)/n(c)]$. In the example, $\lambda = 3.5$.

$\lambda$ is also the value of the critical circuit of $G$ (i.e. $\lambda = w(G)$): indeed, a circuit $c$ of $G$ corresponds to a circuit $c^A$ of $G^A$. For example, if we consider the circuit

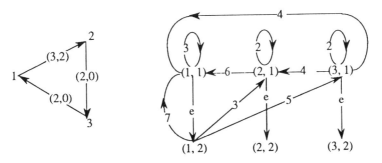

**Figure 9.7** A uniform graph and the associated graph $G^A$.

$c = (1, 2, 3, 1)$ of our example, the corresponding circuit of $G^A$ is $c^A = (1, 1), (1, 2), (1, 1))$. Now, by the properties of $G^A$, if $c$ is a critical circuit then $v(c_A) = L(c)$ and $n(c_A) = H(c)$. Conversely, any circuit $c^A$ of $G^A$ corresponds to a circuit $c$ of $G$ satisfying $v(c_A) = L(c)$ and $n(c_A) = H(c)$. For example, the circuit $((1,1),(1,2),(2,1)(1,1))$ of $G^A$ is associated with the circuit $(1,2,2,3,1)$ of $G$.

By definition of $\lambda$, the graph $G^{\lambda^{-1}A}$ associated with the matrix $\lambda^{-1}A$ has no positively valued circuit (see Figure 9.8).

Cohen et al. proved that there exists an integer $\rho$ depending on the height of the critical circuit of $G$ such that, for every $k > k_0$, $(\lambda^{-1}A)^{k+\rho} = (\lambda^{-1}A)^k$: the longest paths of $G^{\lambda^{-1}A}$ from $i$ to $j$ of $k + \rho$ arcs are obtained from paths of $k$ arcs by passing one more time through circuits with null valuation. So, the matrix $A$ is cyclic:

$$A^{k+\rho} = \lambda^\rho A^k \quad \forall k > k_0.$$

This property can be easily used to prove that the solution of the linear recurrence system $X^k = AX^{k-1}$ is $\rho$-periodic of frequency $1/\lambda$.

If $G$ is not strongly connected, Cohen et al. also used the same kind of arguments to get the behavior of the earliest schedule detailed in the previous section.

This linear algebra framework makes the proofs harder than those presented in the previous section. However, it highlights the underlying mathematical structures, and may lead to different kind of extensions, like the generalization of usual linear algebra results or the extensions to other dioids. Moreover, formal computational tools on dioids should be very useful when some features of the scheduling problem are unknown (Gaubert 1990).

### 9.3.3 Extensions

Up to now, three main extensions of the previous results have been developed. The first introduced deadlines and studied the properties of the latest schedule. The second defined a more general kind of precedence constraints, and studied the earliest schedule in this case. The third tackled the work-in-process minimization.

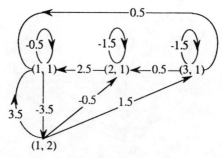

**Figure 9.8** Graph $G^{\lambda^{-1}A}$.

### 9.3.3.1 The basic cyclic scheduling problem with deadlines

Chrétienne has introduced a deadline function $\Delta$: for every occurrence $\langle i, k \rangle$ of $i \in T, \langle i, k \rangle$ must be achieved at the date $\Delta(i, k)$. If $\Delta(i, k), \forall (i, k) \in T \times N$, is finite and superior to the earliest starting time of $\langle i, k \rangle$, he proved (Chrétienne 1991) the existence of a latest schedule. He also showed that $\Delta$ needs to be defined only on a set of terminal tasks $\Pi$, such that $\forall \langle i, k \rangle \notin \Pi$, there is a path in the expanded graph $g$ from $\langle i, k \rangle$ to an element of $\Pi$.

Then he studied the asymptotic behavior of the latest schedule when the deadlines are defined as follows: for any task $i, \Delta(i, k) = a_k = \max_{\langle i,k \rangle} t(i, k) + p_i$, where $t$ denotes the earliest schedule. In other words, $a_k$ is the completion time of the $k$th first executions of the tasks in the earliest schedule. He proved that in this case the latest and the earliest schedules have the same behavior (periodicity and throughput).

### 9.3.3.2 The cyclic scheduling problem with linear precedence constraints

As shown in Section 9.2, uniform constraints model the data dependences of a particular kind of vector loop. Now, if we replace statement 6 of the recurrent loop depicted in Figure 9.1 by $A6(4I + 2) = f6(A5(3I - 1))$, the data dependences between statement 5 and statement 6 cannot be expressed by a uniform constraint anymore.

Thus Munier (1991b) has proposed the following generalization of uniform constraints: a linear precedence constraint between 2 generic tasks $i$ and $j$ is defined by four integers $\alpha_{ij} > 0, \alpha'_{ij} > 0, h_{ij}$ and $h'_{ij} \in \mathbb{Z}$. It corresponds to the following set of precedence constraints:

$$\forall k > 0, \quad t(i, \alpha'_{ij} k + h'_{ij}) + p_i \leq t(j, \alpha_{ij} k + h_{ij}).$$

If $\alpha_{ij} = \alpha'_{ij} = 1, h'_{ij} = 0, h_{ij} > 0$, it is a uniform constraint.

In our example the following linear precedence corresponds to the data dependence between tasks 5 and 6: $\alpha_{56} = 4, h_{56} = 2, \alpha'_{56} = 3$ and $h'_{56} = -1$. One can also model by linear precedences the firing rules of the transitions of a timed event graph with integer valuations on its arcs. Several authors have shown that this class of Petri nets is useful in modeling some assembly systems (Hillion 1989; Munier 1993).

Linear precedences are represented by a linear graph. Its vertices are the generic tasks, and its arcs are labeled with $(p_i, \alpha'_{ij}, h'_{ij}, \alpha_{ij}, h_{ij})$ (see Figure 9.9).

The consistency of a linear graph is still an open problem. However, a necessary (but not sufficient) condition has been established (Munier 1991b): if $G$ is consistent, then there is no circuit $c$, whose weight is lower than 1. The weight $W(c)$ of a circuit $c = i_1 e_1 i_2 e_2 \ldots i_r e_r i_1$ of $G$ is defined as

$$W(c) = \prod_{e_j \in c} \frac{\alpha_{e_j}}{\alpha'_{e_j}}.$$

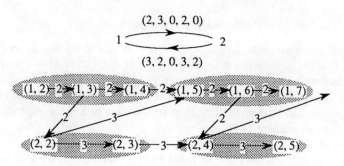

**Figure 9.9** A linear graph $G$ and the expanded graph $g$.

The basic idea for studying the asymptotic behavior of the earliest schedule in this context is to build an equivalent uniform graph by duplicating the generic tasks. For any task $i$, one can build $n_i$ representatives $i_1, \ldots, i_{n_i}$ of $i$, with processing time $p_i$ such that the $k$th execution of $i_l, l \in 1, \ldots, n_i$ is associated with the $[(k-1)n_i + l]$th execution of $i$.

Now, any linear constraint between $i$ and $j$ may be expressed by a set of uniform constraints on the representatives of $i$ and $j$ iff $n_i/\alpha'_{ij} = n_j/\alpha'_{ij} \in \mathbb{N}*$.

However, it is not always possible to find a solution $n_i$, $i \in T$ in which all the linear constraints are taken into account. Yet, if $G$ is a unitary graph, i.e. if $G$ is strongly connected and such that the weight of any circuit is equal to 1, it has been proved (Munier 1991b) that the system $\forall (i,j) \in E, n_i/\alpha'_{ij} = n_j/\alpha_{ij} \in \mathbb{N}*$ has infinitely many solutions. Furthermore, one can build a minimal solution, denoted by $N_i$, $i \in T$. $G^X$ denotes the uniform graph corresponding to this minimal solution.

For our example, $G$ is a unitary graph. The system is $\frac{1}{3}n_1 = \frac{1}{2}n_2 \in \mathbb{N}*$. The minimal solution is $N_1 = 3$ and $N_2 = 2$. Figure 9.10 shows the expanded graph $G^X$.

As $G^X$ is a uniform graph, the earliest schedule of its tasks is $K$-periodic of frequency $1/w(G^X)$. As $G$ and $G^X$ model the same precedence constraints, the earliest schedule of every task $i$ of $G$ is $K$-periodic of frequency $N_i/w(G^X)$. In our example $c = (1_1, 1_2, 2_1, 1_1)$ is the critical circuit of $G^X$ of value $w(c) = w(G^X) = 7$. So, the frequencies of the generic tasks 1 and 2 are respectively $\frac{3}{7}$ and $\frac{2}{7}$.

In the general case $G$ can be partitioned into unitary components $Q_1, \ldots, Q_l$. Using the values $w(Q_1^X), \ldots, w(Q_l^X)$, Munier proved that some arcs of $G$ between two components are not critical and can be removed without modifying the behavior of the earliest schedule. Moreover, once these arcs have been removed, one can compute the maximal frequencies of the generic tasks using a similar construction as for the basic cyclic scheduling problem. The frequencies corresponds to longest paths in a reduced graph $G_R$ whose nodes are the unitary components. Note that all the algorithms involved in this computation are polynomial in the size of the expanded components. But the number of representatives might be exponential in the size of the linear graph, and the overall complexity of the problem is still unknown.

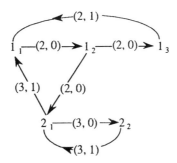

**Figure 9.10** Uniform graph $G^X$.

### 9.3.3.3 Optimization of the work-in-process

Several authors have tackled the problem of resource minimization in timed event graphs. In terms of uniform graphs, the problem can be expressed as follows. Given an average cycle time $w$ and a uniform graph $(G, L, H)$, one assumes that some of the heights of the arcs are unknown. The aim is then to minimize a linear function $f(h)$ of the heights $h = (\ldots, h_{ij}, \ldots)$ in order to achieve the cycle time $w$.

The interpretation is that the height of an arc $(i, j)$ may represent an initial availability of some resources, which is produced by each execution of task $i$ and consumed by each execution of task $j$ (like a queue of limited capacity). However, this notion must not be confused with what are referred to in this chapter as resource constraints. In this context, no decision has to be made about how to share the resources (unlike processor constraints, for which the assignment of tasks must be decided).

The applications proposed by the authors are associated with manufacture systems. Indeed, the height of a circuit of a uniform graph is the maximal number of tasks of this circuit that may be active simultaneously. So, the problem can be seen as minimizing a linear function of the work-in-process of a manufacture system.

This problem also applies to loop scheduling: suppose that some data transfers between statements corresponding to some arcs of $G$ do need intermediate storage. Assuming that a set of registers is assigned to each data transfer, it would be interesting to minimize the total amount of registers needed to achieve a given average cycle time.

Hillion and Proth (1989) modeled this by an integer linear program. The number of equations was proportional to the number of circuits of the graph. They proposed a heuristic for this problem. More recently, some authors have studied special cases and developed new exact algorithms.

- Gaubert (1990) modeled the uniform graph in the dioid ($\mathbb{R}$, min, +) with a linear system similar to the one presented in the previous subsection but where some values of the matrix are unknown. He proved that the size of the system may be

reduced by a formal computation of the eigenvalue of this matrix. This method is efficient if few heights are unknown.
- Lafitt et al. (1992) considered that all the heights are unknown and that $f$ is a linear function of the heights of one particular circuit. They proposed a heuristic for this problem based on a greedy construction of the heights. They also transformed the system into a mixed linear program with binary variables and a polynomial number of equations, and they proposed a convex optimization algorithm to solve it.

## 9.4 BCS WITH RESOURCE CONSTRAINTS

In many practical problems one has to take some resource constraints into account: for example, there may be a limited number of available processors. This section presents different models that have been proposed in order to introduce resource constraints in the BCS framework, and outlines the main results.

### 9.4.1 Reservation tables and timed Petri nets

The first model that was proposed, called reservation tables, arose from early pipelines applications (Davidson and Patel 1976; Kogge 1981; Hanen 1990). It assumes that a generic schedule of the first iteration is given, and that each iteration is computed according to this schedule. The problem is then to determine an optimal sequence of initiation times for the iterations with a minimum average cycle time. Uniform constraints as well as resource constraints (limited number of processors or registers) are modeled by a table that indicates the availability of resources during the generic schedule.

Some authors have studied the infinite schedules of a timed Petri net (Carlier and Chrétienne 1989). The problem was then to find a schedule of the transition firings that minimized the average cycle time. This rather general model can be used for most of the problems presented in this chapter. It has been applied in particular to a cyclic problem arising in pipelined architectures, involving uniform as well as resource constraints (Hanen 1989).

In both models it has been proved that the optimal solution is $K$-periodic and corresponds to the critical circuit of a state graph, whose size is exponential. Thus these results may not lead in general to efficient algorithms.

### 9.4.2 Doacross

Starting from the loop scheduling application presented in Section 9.2, Cytron (1984) studied the efficiency of the Doacross technique for vector loop parallelization, which aims at reordering the loop body to achieve maximum parallelism. This problem can be expressed as finding an optimal periodic schedule of a set of generic tasks subject to uniform constraints, assuming that the tasks belonging to the same iteration are performed by the same processor (Cytron 1984).

Moreover, Cytron assumed an infinite number of available processors. He proved the $\mathcal{NP}$-hardness of the subproblem in which tasks have unit processing times. He provided list heuristics for the general problem based on a level decomposition of the uniform graph inspired by Hu's (1961) algorithm for multiprocessor scheduling.

Munshi and Simons (1990) considered a special case in which tasks have unit processing times, heights are either 0 or 1, and the uniform graph is composed of $n$ disconnected chains. They proved that this subproblem can be solved with an $O(m(n \log n)^2)$ algorithm, where $m$ is the maximum length of a chain and $n$ the number of tasks. For the general case, these authors presented a worst case analysis of a "naive" algorithm, whose relative performance (i.e. the ratio between the period of their solution and the optimal one) was proved to be less than or equal to three. They also proposed a heuristic based on the polynomial solution of the latter subproblem that experimentally improved the solutions provided by Cytron's algorithm.

### 9.4.3 Cyclic versions of classical problems

The most usual way to introduce resources in the BCS model is to build cyclic versions of famous scheduling problems like job shop or $m$-processor problems. It is then interesting to generalize the well-known properties of the classical problems to their cyclic version.

*9.4.3.1 Definitions*

Let $\mathcal{P}$ be a classical makespan minimization problem, an instance of which is defined by an arbitrary precedence graph $\Gamma$ and a set of resource constraints $R$. The resource constraints may be characterized by a finite set of resources with an initial availability. Each task may consume and eventually produce a given amount of each resource during its execution. We define the cyclic version of $\mathcal{P}$, denoted by $\mathscr{C}(\mathcal{P})$, to be the cycle time minimization of a schedule, given an arbitrary uniform graph $(G, L, H)$ and the generic resource constraints $R$.

For example, an instance of the cyclic version of the $m$-processor problem is defined by a uniform graph and the additional constraint that no more than $m$ tasks can be processed concurrently. This problem will be referred to as cyclic scheduling on identical processors (CSIP).

Now, if $\mathcal{I}$ is an instance of $\mathcal{P}$, it can be reduced to an instance $\mathscr{C}(\mathcal{I})$ of the cyclic version $\mathscr{C}(\mathcal{P})$. The uniform graph associated with the precedence graph $\Gamma$ is built as follows: one adds to $\Gamma$ two dummy tasks *begin* and *end* that give a new precedence graph $\Gamma'$. Then the uniform graph $(G, L, H)$ is built by fixing the height of precedence arcs of $\Gamma'$ to 0, and by adding an arc (*end, begin*) with height 1. Figure 9.11 illustrates this transformation. The resource constraints of the classical scheduling problem are extended to the occurrences of the generic tasks.

As a simple outcome of the previous transformation, if $\mathcal{P}$ is $\mathcal{NP}$-hard then so is $\mathscr{C}(\mathcal{P})$. Unfortunately, this result does not highlight the relation between the cyclic

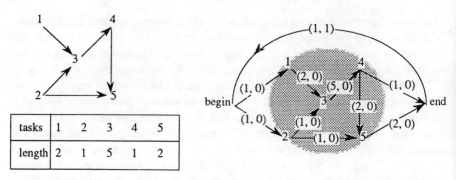

**Figure 9.11** A precedence graph $\Gamma$ and its associated uniform graph $(G, L, H)$.

feature of the problem and its complexity. So several authors studied special cases, as we shall see in the following subsections.

### 9.4.3.2 K-periodic schedules

In most cases, it can be proved using timed Petri nets (Carlier and Chr tienne 1989) or other techniques (Munier 1991a) that $K$-periodic schedules are dominant (i.e. there is at least an optimal schedule that is $K$-periodic).

Approximation algorithms producing $K$-periodic schedules have been developed for CSIP and related problems. In Aiken and Nicolau (1988) a list scheduling algorithm is performed on the expanded graph $g$ of $G$ until a $K$-periodic structure is detected.

In Parhi and Messerschmitt (1991) it is shown that, for every $p > 0$, one can build an unfolded uniform graph $G_p$ that expresses the same precedence constraints as $G$ by duplicating $p$ times every generic task. The authors proved that if $p$ is the least common multiple of the heights of the circuits of $G$ then the heights of the arcs of $G_p$ belong to $0, 1$. They showed that if the resource constraints are ignored, the periodic repetition of the optimal schedule of this new problem lead to an optimal solution of BCS. Moreover, they provided an upper bound on the number of processors that are necessary to compute this optimal schedule. If $m$ processors are available, an approximated solution can be computed by performing a list scheduling algorithm on $G_p$.

The complexity of such algorithms relies on two parameters: the number of duplicates of generic tasks that have to be scheduled until a periodic structure is detected, and the periodicity factor $K$ of the solution.

The number of duplicates in the previous algorithms may be great: the least common multiple of the heights of the circuits of $G$ can clearly be exponential. The computation of the minimum number of duplicates that are necessary to obtain a periodic structure is strongly related to the study of the length of the transitory part of the earliest schedule of BCS. Gao and Ning (1991) proved that this length is polynomial for a rather special case of timed event graphs.

The periodicity factor might also be exponential, so the implementation of $K$-periodic schedules can be difficult. The minimum value of $K$ such that there is an optimal schedule that is $K$-periodic is still an open problem in the presence of resource constraints (it is equal to 1 for BCS).

*9.4.3.3 Periodic scheduling*

Most authors have limited their studies to periodic solutions. Two main problems have been investigated up to now:

- the periodic job shop (PJS), an instance of which is defined by a uniform graph whose tasks are a priori mapped onto $m$ processors (Hanen 1993);
- the periodic scheduling problem on identical processors (PSIP), which has the same input as CSIP (a uniform graph and $m$ available processors) (Hanen and Munier 1992a).

In both cases, the aim is to find a periodic schedule with a minimal period.

Note that periodic schedules are not dominant with such resource constraints (Munier 1991a; Hanen 1993) (i.e. the schedule with minimum cycle time of an instance of CSIP may be $K$-periodic, with $K > 1$). However, they are easy to implement, and they have interesting properties that will be detailed in the next section.

The complexity of these two problems and of some of their special cases have been studied in several papers (Hanen 1993; Hanen and Munier 1992b; Roundy 1992; Munier 1990,1991a; Matsuo 1990).

If there are only two available processors in PSIP, the problem is weakly $\mathcal{NP}$-hard. However, Matsuo (1990) found some no-wait problems on two processors that could be solved using a polynomial subcase of the traveling salesman problem.

If $G$ is acyclic, it has been proved—first for PJS with unit processing times by Davidson and Patel (1976) and then in the general case for both problems by Munier (1991c)—that the induced cyclic problems can be solved in polynomial time. The optimal cycle time equals either the processing time of the bottleneck task or the loading of the bottleneck processor.

The second interesting subcase is when $G$ is reduced to a single circuit. In this case, Hanen (1993) proved, using the study of periodic scheduling of a single job in Roundy (1992), that the periodic job shop is still strongly $\mathcal{NP}$-hard. However, the problem with identical processors can be solved in polynomial time if $G$ is a circuit (Hanen and Munier 1992b).

From these two complexity results obtained with the same uniform graph, we can conclude that if we allow the mapping of tasks onto processors to vary in time then the expected cycle time is lower, and the problem might be easier to solve.

This approach of cyclic scheduling is particularly well suited for studying the structure of solutions, as we shall see in the following sections. In particular, the usual notion of "selection" in job shop problems, introduced to solve disjunctive resource conflicts, can be expressed in terms of uniform constraints in cyclic problems and generalized to identical-processor problems. Thus bounds on the optimum can be derived by solving the underlying basic cyclic scheduling problem.

## 9.5 PROPERTIES OF PERIODIC SCHEDULES

This section is devoted to the properties of the solutions of the periodic job shop problem and the periodic scheduling on identical processors problem. Both problems are cyclic versions of fundamental scheduling problems, and the way they have been studied provides a framework for transferring results on classical scheduling problems to their cyclic version.

We first tackle the periodic job shop problem. We show how the notion of selection, used to solve disjunctive resource conflicts, can be generalized in a cyclic context. A cyclic ordering of tasks on each processor induces a set of uniform constraints modeled by a selection graph. Then we show how these results have been extended to the problem with identical processors. Finally we present a dominating property in this case.

### 9.5.1 Periodic job shop: selection and uniform constraints

The structure of the set of solutions of PJS has been studied by several authors. We first mention Hillion and Proth (1989), who proved that if one is given a cyclic ordering of tasks on each processor, and if the number of initial waiting jobs in front of each machine is known, then the problem becomes a basic cyclic scheduling problem. Hence it can be solved in polynomial time. In other respects, Roundy (1992) defined the notion of selection for the problem of minimizing simultaneously the period and the work-in-process of one job (decomposed into a sequence of operations) to be repeated infinitely often.

We now present results (Hanen 1993; Hanen and Munier 1992b) that generalize these approaches.

Let us consider a disjunctive constraint between two tasks $i$ and $j$ performed by the same processor. For any executions of these two tasks $\langle i, k \rangle$ and $\langle j, l \rangle$, we get either

$$t_i + (k-1)w \geq t_j + p_j + (l-1)w$$

or

$$t_j + (l-1)w \geq t_i + p_i + (k-1)w.$$

This constraint can be rewritten as

$$\exists k_{ij} \in \mathbb{Z}, \quad t_j \geq t_i + p_i - k_{ij}w \quad \text{and} \quad k_{ij} + k_{ji} = 1$$

Note that in a classical job shop problem the same expression holds, except that the variables $k_{ij}$ belong to 0, 1 and indicate whether $i$ starts before or after $j$. In the cyclic case the $k_{ij}$ can also be interpreted in terms of ordering as follows: for any execution $\langle i, k \rangle$ of $i$, the next execution of $j$ on the processor is $\langle j, k + k_{ij} \rangle$.

We call *selection* of a disjunction between $i$ and $j$ the choice of a value $k_{ij} \in \mathbb{Z}$. A selection is said to be complete if all disjunctions have been selected.

In classical job shop problem, a complete selection is associated with a set of precedence constraints that defines an ordering of tasks on each processor. In a cyclic context, a complete selection induces a set of uniform constraints on the schedules with heights $k_{ij}$. So, once a complete selection is made, one can find the associated best cycle time by computing the critical circuit of a graph combining the arcs of $G$ and the uniform constraints induced by the selection.

Assume that in our example three processors are available. The first computes all the executions of tasks 1, 4 and 5, the second tasks 2 and 6, and the last tasks 3 and 7. A schedule is depicted in Figure 9.12. The associated complete selection is as follows:

$$k_{14} = -1, \quad k_{45} = 0, \quad k_{15} = -1, \quad k_{41} = 2, \quad k_{54} = 1, \quad k_{51} = 2,$$
$$k_{26} = 0, \quad k_{62} = 1, \quad k_{37} = 0, \quad k_{73} = 1.$$

Thus the problem can be reformulated as follows: find a complete selection minimizing the value of the critical circuit of the induced uniform graph.

The most efficient algorithms for job shop problems (Carlier and Pinson 1989) are designed to build the ordering of tasks on each processor. In a cyclic context, we can derive a similar feature by analyzing the relations between consecutive tasks in terms of uniform constraints.

Indeed, in the steady state of a given periodic schedule, if $\langle i, k \rangle$ is an execution of $i$ performed by processor $q$ then the next task performed by $q$ can be written $\langle j, k' \rangle$, where $j$ does not depend on $k$, and $k' = k + k_{ij}$. We then say that $j$ is the *next neighbor* of $i$.

The uniform constraints induced by a selection between tasks and their next neighbors define what we call a *selection graph*. Figure 9.13 shows the selection graph for our example. Note that this graph is composed of one circuit of height 1 per processor, each circuit involving of course all the tasks allocated to the corresponding processor.

Conversely, it can be proved that the uniform constraints defined by such a graph $X = (T, E_X)$ induce a complete selection. So, if we define the folded graph $GX = (T, E_X \cup E)$ by merging the edges of $G$ and $X$, we can reformulate the

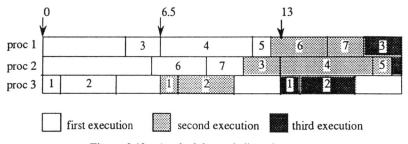

**Figure 9.12** A schedule on dedicated processors.

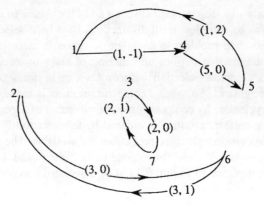

**Figure 9.13** A selection graph.

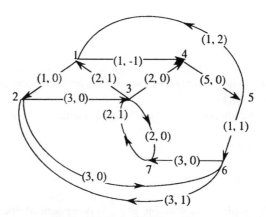

**Figure 9.14** A folded graph.

problem as follows: find a selection graph $X$ such that the value $w(GX)$ of the critical circuit of the folded graph is minimum.

The folded graph of our example is depicted in Figure 9.14. Its critical circuit $c = (2, 3, 4, 5, 6, 2)$ has a value of 7.5, corresponding to the optimal schedule shown in Figure 9.12.

These results can be extended to general length and height functions on the graph, i.e. arbitrary functions $L, H : E \to \mathbb{Z}$, assuming that the problem is consistent

### 9.5.2 Identical processors: selection and mapping structure

The approach developed previously for the periodic job shop has been extended for the identical-processor problem in Hanen and Munier (1992b). We now give an overview of the main results.

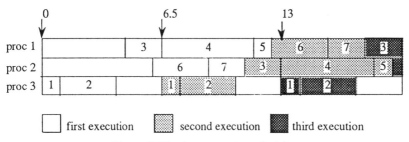

**Figure 9.15** A permutation schedule.

One of the important features of the identical-processor problem is that no assumption is made on the mapping of tasks onto processors, which may vary during the schedule. However, it can be proved that mappings built from a permutation induce a dominating subset of schedules (i.e. every feasible schedule with a given period may be associated with an element of this subset with a same period, so we can reduce our study to this subset). Such mappings are associated with a permutation $s$ of the set of processors $(1,\ldots,m$ such that if $r(i,k)$ denotes the processor that executes $\langle i,k \rangle$ then $r(i, k + 1) = s(r(i, k))$.

A schedule with such a mapping function is called a permutation schedule. Figure 9.15 shows a permutation schedule for our example on three identical processors. The permutation $s$ is defined by $s(1) = 2, s(2) = 1$ and $s(3) = 3$.

Now let $\langle i, k \rangle$ be an occurrence of $i$ performed by the processor $q$. The next execution of $i$ on the same processor is $\langle i, k + \delta(s) \rangle$, and it is performed $w\delta(s)$ time units later, where $\delta(s)$ denotes the degree of the permutation $s$. Hence the degree of $s$ determines the real periodicity of the schedule.

Now, if $\langle j, k' \rangle$ is the next task performed by processor $q$ then $j$ and $k' - k$ are independent of the iteration number $k$ and on the processor $q$. Thus one can say that $j$ is the next neighbor of $i$. Moreover, the schedule induces a uniform constraint of height $k' - k$ between $i$ and $j$.

If we build the graph of uniform constraints between tasks and their next neighbors, we get a graph also called a selection graph with the following structure: to each cycle (i.e. circular subpermutation) of the permutation $s$ there corresponds a circuit involving all tasks performed by the processors of the cycle. Note that the height of this circuit is equal to the cardinality of the cycle. Moreover, all circuits are disconnected, and the sum of their heights is less than or equal to the number of available processors, $m$.

In our example $s$ is composed of two cycles $s = (1, 2)(3)$. The corresponding selection graph is shown in Figure 9.16.

Conversely, let us consider a uniform graph $(X, L, H)$ of $n$ nodes composed of disconnected circuits whose heights are positive, such that the sum of these heights is less than or equal to $m$. It has been proved that the uniform constraints defined by $X$ are sufficient to meet the resource constraints. The main argument of the proof lies on the polynomial solution of the uniform circuit mentioned in Section 9.3.3.

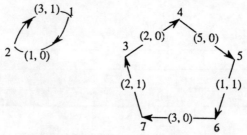

**Figure 9.16** Selection graph for identical processors.

Let us consider our example again and the circuit (3,4,5,6,7) with height 2. In terms of production scheduling, this corresponds to a limitation of the work-in-process to 2 in this part of the production process. Thus at most two of the tasks are active at the same time, and two processors are sufficient to perform any schedule of these tasks. So if $C$ is a circuit with a height $h$, and if there are at least $h$ processors available for the tasks of $C$, then any schedule will meet the resource constraints.

Moreover, it is always possible to build a permutation schedule by associating a circular subpermutation with each circuit of $X$. Thus the selection graph $X$ completely defines the periodicity of the associated schedules, since the degree of the permutation is just the least common multiple of the heights of the circuits of $X$.

Now, the problem can be reformulated as previously by defining the folded graph $GX$ (i.e. the graph obtained by merging $G$ and $X$). However, the selection graph is in this case more difficult to handle, since the partition of the nodes into circuits is not given.

Thus it is important to get dominating properties of particular structures of selection graphs. Hanen and Munier (1992b) have shown with local transformations of arbitrary selection graphs that the graphs composed of only one circuit of height $m$ are dominant. Such selection graphs define circular allocation functions associated with circular permutations, in which each task is processed successively by all the processors, following the order of the permutation.

Figures 9.17 and 9.18 show the optimal circular schedule and its selection circuit built from the previous schedule.

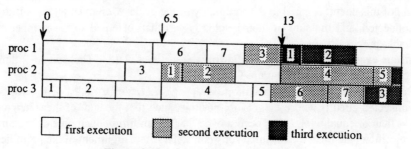

**Figure 9.17** An optimal circular schedule.

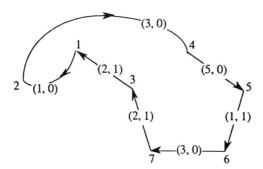

**Figure 9.18** Associated selection circuit.

## 9.6 ALGORITHMS FOR PERIODIC SCHEDULING ON PARALLEL PROCESSORS

In this section we present the main algorithms that have been designed up to now for the periodic job shop as well as for the periodic scheduling on identical processors problem. Many efficient heuristics for code generation have been proposed (Lam 1987; Parhi and Messerschmitt 1991; Eisenbeis and Windheiser 1992), which cannot be detailed here. We have chosen to mention the most promising approaches that consist either in designing algorithms based on the results presented in the previous section or in applying the powerful algorithms known in classical makespan minimization problems to cyclic problems.

### 9.6.1 Exact algorithms

To the best of our knowledge, the first exact algorithm developed for an $\mathcal{NP}$-hard cyclic problem (Roundy 1992) concerns the problem of minimizing simultaneously the period and the work-in-process in a cyclic job shop with only one job.

Roundy represents the set of solutions as a network. A vertex is associated with each feasible complete selection. An edge between two nodes $x$ and $y$ is built if a local transformation of the selection of $x$ called swap gives the selection associated with $y$. A swap corresponds to the exchange of two neighboring nodes in the selection graph. Roundy proved that a particular node is a root of this network. So, he developed an exact algorithm that constructs routes in the network, fathoming the nodes by evaluating the critical circuit of the associated folded graph.

A branch-and-bound algorithm based on the construction of partial selections has been proposed by Hanen (1993). Its basic features are as follows: a node of the search tree corresponds to bounds on each value $k_{ij}$ mentioned in Section 9.5.1, so that each $k_{ij}$ belongs to an interval. Such set of bounds is called a partial selection, and is computed using an upper bound $w^+$ on the optimal cycle time. $w^+$ is initially provided by a heuristic.

The branching rule consists in choosing a disjunction that is not yet selected and cutting its interval into two parts in order to create two new nodes.

A node is fathomed by means of an adjustment procedure that tightens up the bounds as far as possible using an approximation of the paths of maximal amplitude, and eventually detects a circuit with positive amplitude w.r.t. the best known cycle time $w^+$. In this case the node is removed from the search tree. If the node is a leaf, it corresponds to a complete selection. Thus the critical circuit of the folded graph is computed in order to update $w^+$.

This is the first branch-and-bound procedure that has been developed for the periodic job shop. Many features of this algorithm are like early branch-and-bound methods for classical job shops that directed disjunction edges one after the other. Thus it can surely be improved by using basic ideas developed in the most efficient algorithm for the job shop problem (Carlier and Pinson 1989).

### 9.6.2 Greedy construction of a selection graph by insertion

In this subsection we describe a heuristic that can be used to solve both problems (PJS and PSIP) and that is an extension of the one presented in Hanen and Munier (1992b).

The heuristic builds step by step a selection circuit with a given height $h$. At each step, the node $k+1$ is inserted in the current selection circuit composed of nodes $1, \ldots, k$. Figure 9.19 illustrates the principle of an insertion. The insertion parameters (position and height of the new arcs) are chosen so that the value of the critical circuit of the current folded graph, composed of the current selected arcs, remains as small as possible. It has been proved that for any numbering of nodes such a greedy process always provides a feasible solution.

As we have already seen in the previous section, the selection graph for a problem with identical processors can be restricted to a Hamiltonian circuit with height $m$. Thus such a greedy heuristic applied naturally in this case. In a periodic job shop the selection graph is composed of $m$ disconnected circuits of height 1 corresponding to the processors. Thus one can build the circuits one after the other using the heuristic. We propose to consider the circuits in decreasing order of the loading of the associated processors, in order to schedule first the bottleneck processor.

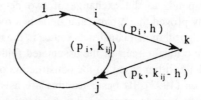

current selection circuit
$i, j \leq k - 1$

**Figure 9.19** Insertion.

The complexity of this heuristic is $O(n^3 \log n)$ in the worst case, but it can be experimentally improved if good data structures are used. Its application to our example led to the optimal schedules reported in Figures 4.12 and 4.15. This algorithm seems to provide good results for problems that have a really cyclic structure (with a work-in-process greater than 1). However, the worst case analysis of its performances is so far still an open problem.

### 9.6.3 Relaxation to noncyclic problems

In this subsection we present an original approach (Gasperoni and Schwiegelshohn 1992). Its principle is to associate a noncyclic problem instance $N(I)$ with a cyclic problem instance $I$, so that an approximate solution of $N(I)$ provides a feasible periodic schedule for the instance $I$. Moreover, it aims at performing a worst case analysis of the algorithm, using the performance guarantees of the heuristic used to solve $N(I)$. This approach has been investigated for PSIP and for one of its variants in which the processors are pipelined (i.e. a processor can start a new task at each time unit).

In order to get a noncyclic problem instance, a periodic schedule is represented by its pattern $\pi$, which is the schedule of the generic tasks during a period in the steady state:

$$\forall i \in T, \quad \pi_i = (t_i - t_1) \bmod w = t_i - t_1 - w \left\lfloor \frac{t_i - t_1}{w} \right\rfloor.$$

The occurrence vector $\eta$ is defined by

$$\eta_i = (t_i - t_1) \text{ div } w = \left\lfloor \frac{t_i - t_1}{w} \right\rfloor$$

Figure 9.20 illustrates the pattern of the periodic schedule depicted in Figure 9.3.

Now the basic idea is to remove arcs from the initial uniform graph in order to get an acyclic graph $H$. Thus $H$ defines a set of precedence constraints on the set of generic tasks. The tasks are then scheduled on $m$ processors according to these constraints. The resulting schedule defines a pattern for the cyclic problem.

In order to build the graph $H$, Gasperoni and Schwiegelshohn (1992) compute the pattern of the optimal periodic schedule without resource. The arcs of $G$ that are not

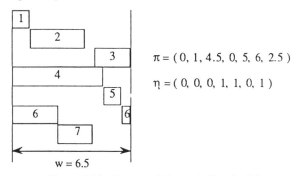

**Figure 9.20** Pattern of the periodic schedule.

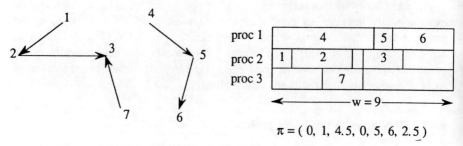

**Figure 9.21** Acyclic graph $H$ and the induced pattern.

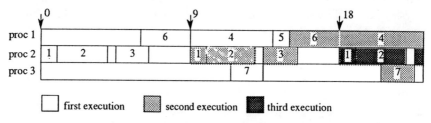

**Figure 9.22** The corresponding periodic schedule.

associated with a precedence in this pattern are removed. This in fact defines the occurrence vector (i.e. the relations between the occurrences of tasks in the pattern). Then they use any list-scheduling algorithm to get a schedule of tasks that meets the precedences defined by $H$ and some additional release times. It has been proved that this schedule provides a feasible pattern in polynomial time $O(n^3 \log n)$.

Figure 9.21 shows the graph $H$ associated with the optimal pattern of Figure 9.20 and the induced pattern. The corresponding periodic schedule is depicted in Figure 9.22.

This kind of utilization of list scheduling techniques has also been studied by Eisenbeis and Wang (1993), who defined a general unifying framework for such algorithms.

Gasperoni and Schwiegelsohn (1992) used the bound on the performances of any list scheduling algorithms on $m$ processors to perform a worst case analysis of their algorithm. If $w_{opt}$ denotes the optimal cycle time, $w_L$ the cycle time provided by the algorithm and $p_{max}$ the maximal processing time of a task, they proved that

$$w_L \leq \left(2 - \frac{1}{m}\right) w_{opt} + \frac{m-1}{m}(p_{max} - 1).$$

This approach has also been applied to the problem of scheduling a uniform graph on pipelined processors, leading to a performance ratio of roughly 2.

## 9.7 OTHER MODELS

The basic cyclic scheduling problem provides a framework for studying cyclic scheduling problems arising in parallel computing as well as in production systems. However, some other models have been proposed. This section is devoted to two that may be useful for modeling constraints other than precedences and that have led to interesting results.

The first model deals with the problem of scheduling periodic events within a given period, for which a branch-and-bound procedure has been developed. The second model is graph coloring, which has been used to derive complexity results on cyclic versions of the classical open shop problem.

### 9.7.1 Span constraints

Serafini and Ukovich (1989) introduced a general model for cyclic scheduling aimed at modeling various kinds of problems dealing with periodic events. Most of them are outside the scope of this chapter (e.g. traffic-light or vehicle scheduling). Note that Serafini and Ukovich (1989) give an important bibliography on this subject.

These authors were concerned with the existence of a periodic schedule with a given cycle time $w$. Uniform constraints can be seen as an extension of precedence constraints. The authors proposed another kind of generalization of these constraints: a span constraint between two generic tasks $i$ and $j$ is defined by a cyclic interval (a span) $\Delta_{ij}$ on the circle of length $w$ corresponding to the projection of a real interval $[d_{ij}^-, d_{ij}^+]$ on this circle by means of a Euclidean division. The constraint itself is given by

$$(t_i - t_j) \bmod w \in \Delta_{ij},$$

where $t$ denotes the generic schedule, i.e. $t_i$ denotes the starting time of the first occurrence of task $i$.

Such a constraint concerns all the occurrences of $i$ and $j$, say $\langle i, k \rangle$ and $\langle j, l \rangle$, that are performed within a period of length $w$. Thus it does not depend on the relation between the iteration numbers $k$ and $l$, but only on the closeness of the two occurrences. This is the main difference between a span constraint and a uniform constraint, for which this relation is fixed: there is a precedence between $\langle i, k \rangle$ and $\langle j, k + h \rangle$ for any iteration number $k$.

Serafini and Ukovich (1989) used a graphical representation of these constraints: generic tasks corresponded to the nodes, and each span constraint between two tasks $i$ and $j$ was modeled by an arc $(i, j)$ labeled by the span $\Delta_{ij}$. The span problem was then to find a periodic schedule of cycle time $w$ meeting the span constraints.

Note that the union of two disconnected spans may be obtained from the intersection of two spans. So, alternatives like $(t_i - t_j) \bmod w \in \Delta_{ij} \cup \Delta'_{ij}$ can easily be expressed in this model. In particular, one can model disjunctions with span constraints. If two generic tasks $i$ and $j$ are processed by the same processor, with processing times $p_i$ and $p_j$, then the disjunction can be formalized as follows:

$$w > (t_i - t_j) \bmod w \geqslant p_j, \quad w > (t_j - t_i) \bmod w \geqslant p_i.$$

However, this high modeling power of span constraints might also be a drawback, since it has been proved that the span problem is strongly $\mathcal{NP}$-complete.

Serafini and Ukovich established that if the graph of span constraints is a tree then the problem can be solved in polynomial time. They proposed a branch-and-bound algorithm for the general problem based on this relaxation. They also showed that the general problem is closely related to the traveling salesman problem.

Then they focused on the problem of scheduling a set of generic tasks subject to span constraints on $m$ identical processors assuming given setup times: if a processor performs $j$ just after $i$, there must be at least $s_{ij}$ time units between these two tasks. They proved that finding the minimum number of processors that realizes a given schedule is a $\mathcal{NP}$-hard problem, which can be seen in special cases as an assignment problem on a bipartite graph, a traveling salesman problem or a graph coloring problem.

They proved the dominance of allocation functions based on permutations, and showed that once this permutation is given, the resource constraints could be expressed as extended span constraints, so that the previous branch-and-bound algorithm could be used in this case. They suggested a two-step algorithm when the allocation function is not provided.

### 9.7.2 Graph coloring

Some authors (de Werra and Solot 1991; De Werra et al. 1992) have used graph coloring techniques to solve a cyclic version of the open shop scheduling problem, called cylindrical open shop: a job is split up into at most $m$ operations to be performed on $m$ special-purpose processors (so any operation is a priori allocated to a processor). A processor cannot process two operations at the same time, and two operations of the same job cannot overlap. Given $n$ different jobs that are to be performed infinitely often, and a period $w$, the problem is then to find a periodic schedule of period $w$.

As there are no precedence constraints, the disjunctive constraints surely could have been expressed as previously in a "span" formalism. However, their particular structure can be tackled by a much more powerful theoretical took, namely graph coloring. Indeed, with a problem instance, one associates a bipartite-valued graph, whose vertices are on one side the jobs and on the other side the processors. The weight $v(e)$ of an edge $e = (j, p)$ is defined to be the processing time of the operation of job $j$ performed by the processor $p$. Finding a periodic schedule of period $w$ is then equivalent to associating $v(e)$ colors that are cyclically consecutive in $1, \ldots, w$ with each edge $e$ so that no two adjacent edges have a color in common. This problem is known as weak ice $w$-coloring.

The study of this kind of coloring problem has led to some new results (de Werra and Solot 1991; de Werra et al. 1992). First, the weak ice $w$-coloring problem is $\mathcal{NP}$-complete. However, it becomes polynomial on three processors when the number of operations per job is less than or equal to two.

If no-wait constraints are added (i.e. every job is performed without waiting times, and every processor performs all its loading from a same period without interruption), the problem, referred to as compact ice coloring, has also been proved $\mathcal{NP}$-complete. Moreover, graph coloring techniques have been used to derive bounds on the optimal period in the general case. In particular, it has been proved that the absolute distance between the optimal solution and the lower bound given by the maximal loading and/or the maximal processing time of a job is lower than the maximal processing time of an operation.

## 9.8 CONCLUSIONS

The overview given in this chapter shows that the basic cyclic scheduling problem provides a good framework for many applications of cyclic scheduling. Other models (Serafini and Ukovich 1989; Korst 1992; de Werra and Solot 1991) should be preferred if there are no precedence constraints, as in scheduling periodic events with a given periodicity.

The study of the earliest schedule, by means of either graph theory or dioids, can be used to compute its asymptotic performances and its periodicity in polynomial time. Moreover, one can detect unsteadiness in a task system (i.e. tasks with different frequencies). Yet, the study of the transient rate of the earliest schedule is still an open area.

The extensions of this approach reported here have tackled basic problems in production systems, like work-in-process minimization and assembly systems problems, as well as in computer systems, like memory space minimization and linear recurrent loop parallelization. However, one can note that the complexity of these two problems is still open. Further, a necessary and sufficient consistency condition is still to be found for the cyclic problem with linear precedences.

In other respects, it would be very interesting to study the asymptotic behavior of the priority list scheduling heuristics that are very often used for noncyclic scheduling problems in the presence of resource constraints. The existence of a steady state and its periodicity properties should be investigated, so that several priority lists could be compared.

While studying the construction of periodic schedules with disjunctive or cumulative resources, we have shown that the notion of selection could be extended to cyclic problems by means of uniform constraints, inducing a powerful structuring of the set of solutions.

We have reported here some exact and approximation algorithms. However, they represent only the first steps in a wide open research area. In particular, no exact methods have been so far developed for the cyclic problem on identical processors. Furthermore, efficient algorithms for the recurrent job shop problem should be derived from the best algorithms for the usual job shop scheduling (Carlier and Pinson 1989).

The basic idea of Gasperoni and Schwiegelshohn (1992) that uses a makespan minimization underlying problem in order to build a pattern for the cyclic scheduling problem should certainly be developed further. One could, for example, determine the best acyclic subgraph of the uniform graph that provides a good pattern, or study problems for which such an approach provides the optimal solution. Moreover, all the heuristics presented in this chapter should be compared in terms of worst case performances or at least by experiments on the same problems.

Finally, many extensions of the basic cyclic scheduling problem could be fruitfully studied involving new resource constraints like registers in loop parallelization problems, storage areas in production systems, setup times and communication times in distributed systems.

## REFERENCES

Aiken A. and Nicolau A. (1988) Optimal loop parallelization. In *Proceedings of the SIGPLAN'88 Conference on PLDI*, pp 308–317 ACM.

Carlier J. and Chrétienne P. (1988) *Problèmes d'ordonnancement: modélisation/complexité/ algorithmes* Masson, Paris.

Carlier J. and Chrétienne P. (1988) Timed petri nets schedules. In *Advances in Petri Nets: Lecture Notes in Computer Science*, No. 340, pp 62–84, Springer, Berlin.

Carlier J. and Pinson E. (1989) An algorithm for solving the job-shop problem *Management Sci.*, **35**, 164–176.

Chrétienne P. (1985) Transient and limiting behavior of timed event graphs *RAIRO-TSI*, **4**, 127–192.

Chrétienne P. (1991) The basic cyclic scheduling problem problem with deadlines *Disc. Appl. Math.*, **30**, 109–123.

Claver J. F. and Jackson P. (1987) Lot sizing in cyclic scheduling. Technical Report 736, School of Operations Research and Industrial Engineering, Cornell University.

Cohen G., Moller P., Quadrat J.-P. and Viot M. (1985) A linear system theoretic view of discrete event processes and its use for performance evaluation in manufacturing *IEEE Trans. Autom. Control*, **30**, 210–220.

Cohen G., Moller P., Quadrat J.-P. and Viot M. (1989) Algebraic tools for the performance evaluation of discrete event systems. In *Proceedings on Discrete Event Dynamics Systems* IEEE.

Cytron R. (1984) Compile time scheduling and optimization for asynchronous machines. PhD Thesis, Computer Science Department, University of Illinois at Urbana Champaign.

Davidson E. S. and Patel J. H. (1976) Improving the throughput of a pipeline by insertion do delays. In *Proceedings on the 3rd Annual Symposium on Computer Architecture* IEEE.

de Werra D., Mahadev N. and Solot P. (1992) Cylindrical open-shop scheduling: some solvable cases. Technical Report, Ecole polytechnique de Lausanne.

de Werra D. and Solot P. (1991) Compact cylindrical chromatic scheduling *SIAM J. Discr. Math.*, 528–534.

Eisenbeis C. (1988) Optimization of horizontal microcode generation for loop structures. In *Proceedings of the International Conference on Super-computing*, pp 453–465. ACM.

Eisenbeis C. and Wang J. (1993) Decomposed software pipelining: a new approach to exploit instruction level parallelism for loop programs. In *IFIP-WG 10.3: Working conference on Architectures and Compilation Techniques for Fine and Medium Grain Parallelism.*
Eisenbeis C. and Windheiser D. (1992) A new class of algorithms for software pipelining with resource constraints. INRIA Technical Report.
Gao G. and Ning Q. (1991) A timed petri net model for fine-grain loop scheduling. In *Proceedings of SIGPLAN'91 Conference on PLDI.* ACM.
Gasperoni F. and Schwiegelshohn U. (1992) Scheduling loops on parallel processors: a simple algorithm with close to optimum performance. In *Proceedings of Joint Conference on Vector and Parallel Processing (CONPAR 92): Lecture Notes in Computer Science,* No. 634, pp 625–636, Springer, Berlin.
Gaubert S. (1990) An algebraic method for optimizing resources in timed event graphs. In *Proceedings of the 9th Conference on Analysis and Optimization of Systems: Lecture Notes in Control and Information,* No. 144, pp 957–966, Springer, Berlin.
Gondran M. and Minoux M. (1977) Valeurs propres et vecteurs propres dans les dioides et leur interprétation en théorie des graphes. *Bulletin de la Direction des Etudes et Recherches, Serie C, Mathématiques Informatique,* **2**, 35–41.
Gondran M. and Minoux M. (1985) *Graphes et algorithmes.* Eyrolles, Paris.
Graves S. C., Meal H. C., Stefek D. and Zeghmi A. H. (1983) Scheduling of re-entrant flow shops *J. Oper. Management,* August.
Hanen C. (1989) Optimizing microprograms for recurrent loops using timed petri nets. In *Advances in Petri Nets: Lecture Notes in Computer Science,* pp 97–122 Springer, Berlin.
Hanen C. (1990) Les tables de réservation numériques: un outil de résolution de certain problèmes d'ordonnancement cycliques *RAIRO Rech. Op r.,* **24**, 97–122.
Hanen C. (1994) Study of a np-hard scheduling problem: the recurrent job-shop *Eur. J. Oper. Res.,* **72**, 82–101.
Hanen C. and Munier A. (1992a) A study of the cyclic scheduling problem on parallel processors. Submitted to *Discr. Appl. Math.*
Hanen C. and Munier A. (1992b) A study of the cyclic scheduling problem on parallel processors. Rapport LRI 766, Université Paris-sud.
Hillion H. (1989) Timed petri nets and application to multi-stage production systems. In *Advances in Petri Nets: Lecture Notes in Computer Science* No. 424, pp 281–305, Springer, Berlin.
Hillion H. and Proth J.-M. (1989) Performance evaluation of a job-shop system using timed event graphs *IEEE Trans. Autom. Control,* **AC-1**.
Hu T. C. (1961) Parallel sequencing and assembly line problems *Oper. Res.,* **9**, 841–848.
Kogge P. M. (1981) *The Architecture of Pipelined Computers* McGraw-Hill, New York.
Korst J. (1992) Periodic multiprocessor scheduling. PhD Thesis, Technische Universiteit Eindhoven.
Laftit S., Proth J.-M. and Xie X. (1992) Optimization of invariant criteria for event graphs, *IEEE Trans. Autom. Control,* **37** (5).
Lam M. (1987) A systolic array optimizing compiler. PhD Thesis, Carnegie-Mellon University.
McCormick S. T., Pinedo M. L., Shenker S. and Wolf B. (1989) Sequencing in an assembly line with blocking to minimize cycle time *Oper. Res.,* **37** (6), 925–935.
Matsuo H. (1990) Cyclic sequencing problems in the two-machine permutation flow-shop: complexity, worst case and average case analysis *Nav. Res. Log. Quart.,* **37**, 679–694.
Munier A. (1990) The complexity of a cyclic scheduling problem with identical machines. Rapport MASI, Institut Blaise Pascal.

Munier A. (1991a) Contribution l'étude des problèmes d'ordonnancement cycliques. PhD Thesis, Université Pierre et Marie Curie.

Munier A. (1991b) The cyclic scheduling problem with linear precedence constraints. Rapport MASI 91/34, Institut Blaise Pascal.

Munier A. (1991c) Résolution d'un problème cyclique itérations indépendantes et contraintes de ressource *RAIRO Rech. Op r.*, **25**, 161–182.

Munier A. (1993) Régime asymptotique optimal d'un graphe d'événements généralisé. *RAIRO Automat.-Prod. Inform. Ind.*, to appear.

Munshi A. A. and Simons B. (1990) Scheduling sequential loops on parallel processors *SIAM J. Comput.*, **4**.

Parhi K. K. and Messerschmitt D. G. (1991) Static rate-optimal scheduling of iterative dataflow programs via optimum unfolding *IEEE Trans. Comput.* **40**, 178–195.

Ramchandani C. (1973) Analysis of asynchronous systems by timed Petri nets. PhD Thesis, MIT (MIT Technical Report 120).

Roundy R. (1992) Cyclic schedules for job-shops with identical jobs *Math. Oper. Res.* November.

Serafini P. and Ukovich W. (1989) A mathematical model for periodic scheduling problems *SIAM J. Disc. Math.*, **2**, 550–581.

---

*Laboratoire LITP, Case 168, Université Pierre et Marie Curie, 4 Place Jussieu, 75252 Paris Cedex 05, France*

CHAPTER 10

# Some Graph Coloring Models for Cyclic Scheduling

**D. de Werra**
*Swiss Federal Institute of Technology*

**P. Solot**
*Ciba-Geigy AG, Switzerland*

**Abstract**

We consider the open shop scheduling model and introduce some variations motivated by applications in production systems. They are based on edge coloring models in bipartite multigraphs and deal with the case of cyclic (periodic) scheduling with a fixed period length.

## 10.1 INTRODUCTION

Next to the well-known flow shop and job shop scheduling models, open shop scheduling models are often appropriate for capturing most of the constraints arising in simple automated production systems like some types of flexible manufacturing systems.

Our purpose is to formulate some types of cyclic (=periodic) open shop scheduling problems and to show how they can be approached by using some variations of edge coloring models in bipartite multigraphs.

This chapter is not primarily an exhaustive survey of cyclic scheduling or of variations on the open shop scheduling model. It is rather a contribution to the field of chromatic scheduling, and its aim is to show connections between variations of some classical scheduling models and graph coloring concepts.

The classical open shop scheduling model (OSS) is based on the following: a collection $P$ of processors $P_1, \ldots, P_m$ and a collection $J$ of jobs $J_1, \ldots, J_n$. Each job $J_j$ consists of tasks $T_{1j}, \ldots, T_{mj}$, where $T_{ij}$ has processing time $p_{ij} \geq 0$ and has to be processed on $P_i$. A processor cannot work on two tasks at a time, and no two tasks of

---

*Scheduling Theory and its Applications* Edited by P. Chrétienne, E. G. Coffman, Jr., J. K. Lenstra and Z. Liu
© 1995 John Wiley & Sons Ltd

the same job can be in process simultaneously. In the nonpreemptive OSS (NOSS) no interruption is allowed during processing of a task $T_{ij}$. In other words, this task is processed during $p_{ij}$ consecutive time units as soon as it starts on $P_i$. The preemptive OSS (POSS) allows interruptions at any time during the processing of any task $T_{ij}$. Other tasks of $J_j$ can then be (partially) processed, and $T_{ij}$ may be resumed later on $P_i$, possibly with additional interruptions. From now on, we shall assume that all processing times $p_{ij}$ are integral.

It will be convenient to associate with the above problems a bipartite graph $G = (P, J, E)$ with nodes $P_1, \ldots, P_m$ and $J_1, \ldots, J_n$. The set $E$ of edges corresponds to the tasks: $T_{ij}$ is represented by an edge $e = [P_i, J_j]$ with weight $w_e = p_{ij}$.

An *edge k-coloring* of $G$ is an assignment of $w_e$ colors in $1, \ldots, k$ to each edge $e$ in such a way that

(i) the $w_e$ colors of $e$ are all different;
(ii) no two adjacent edges have a color in common.

For POSS, a schedule in $k$ time units exists if and only if there exists an edge $k$-coloring of the associated graph $G$ (Blazewicz et al. 1986; Gonzalez and Sahni 1976).

Let

$$T(G, w) = \max \left\{ \max_i \sum_j p_{ij}, \max_j \sum_i p_{ij} \right\},$$

clearly no schedule can exist in less than $T(G, w)$ time units. It follows from the König theorem (Berge 1983, Chapter 12) that a preemptive schedule in $T(G, w)$ time units exists. Furthermore, for such a schedule, all preemptions occur at integral times. Finding a preemptive schedule with minimum completion time reduces to a sequence of network flow problems, and it can be done in polynomial time.

For NOSS, we consider *interval edge k-colorings* of $G$, these are edge $k$-colorings where (i) is replaced by

(i*) the $w_e$ colors of $e$ are consecutive in $1, \ldots, k$.

A nonpreemptive schedule in $k$ time units is represented by an interval edge $k$-coloring of the associated graph $G$. For an instance of NOSS with a given positive integer $k$, it is $\mathcal{NP}$-complete to decide whether there exists a nonpreemptive schedule in $k$ time units (Gonzalez and Sahni 1976).

The NOSS model applies to automated production systems: a job $J_j$ consists of the production of a part $j$; it has to be processed during $p_{ij}$ consecutive time units on each $P_i$ for which $p_{ij} > 0$, and no order of visits to the $P_i$ is specified.

For the POSS model, the situation is as follows: each job $J_j$ consists of the production of a batch of parts $j$, and on any processor $P_i$ with $p_{ij} > 0$ we can produce an integral number of parts $j$ of the lot in one time unit. The production of the batch of parts $j$ will occur on several processors $P_i$ (one after the other, with possible interruptions after an integral number of time units).

This chapter is structured as follows. Several cases of cyclic scheduling problems are considered in Section 10.2. In Section 10.3 a problem variation is dealt with in which both processor lead and job processing are balanced over time. Section 10.4 addresses a further specific cyclic scheduling problem, and open problems are mentioned in Section 10.5.

Graph-theoretical terms not defined here can be found in Berge (1983).

## 10.2 CYCLIC SCHEDULING

In many situations a production cycle is repeated periodically—such a procedure generally has the advantage of balancing the loads of the various processors.

Our model only needs to be slightly modified to handle this situation: we just consider that the set of $k$ time units we use for a schedule is cyclically ordered, i.e. period 1 follows period $k$. We shall call this problem NCOSS (nonpreemptive cyclic open shop scheduling).

The problem can be stated as follows.

**NCOSS** *Given a bipartite graph $G = (\mathcal{P}, \mathcal{J}, E)$ with weights $w_e = p_{ij}$ on the edges $e = [P_i, J_j]$ and given a positive integer $k$, does there exist a nonpreemptive cyclic schedule in $k$ time units?*

Such schedules are represented by edge $k$-colorings of $G$ satisfying

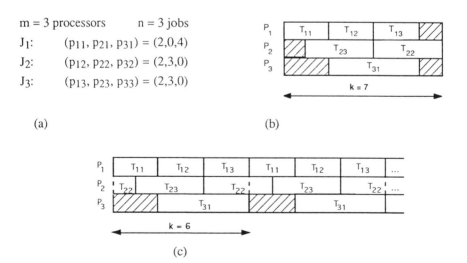

**Figure 10.1** An open shop cyclic scheduling problem: (a) an open shop scheduling problem; (b) a noncyclic schedule with $k = 7$; (c) a cyclic schedule with $k = 6$.

(i**) the $w_e$ colors of $e$ are cyclically consecutive in $\{1,\ldots,k\}$;
(ii) no two adjacent edges have a color in common.

These are called *weak interval cyclic edge k-colorings* (or weak ice *k*-colorings). An example of cyclic schedule is given in Figure 10.1. The schedule (b) is a normal nonpreemptive schedule in $k = 7$ time units; it could be considered as a particular cyclic schedule. The schedule in (c) is cyclic: for any choice of six consecutive time units, there is at least one task associated with a set of periods that are cyclically consecutive.

With a simple reduction from NOSS, the NCOSS problem was proved to be $\mathcal{NP}$-complete (de Werra and Solot 1993).

As far as special cases are concerned, partial results have been obtained concerning the class of graphs $G$ for which a weak ice $T(G,w)$-coloring exists for any choice of weights $w_e$ (Mahadev et al. 1992). These graphs are called *weakly ice-perfect*. In particular, it has been proven that this class contains, among others, all bipartite graphs $G = (\mathcal{J},\mathcal{J},E)$ satisfying the following two conditions:

(a) all nodes in $\mathcal{J}$ have degree 2 or less;
(b) all nodes in $\mathcal{P}$ belonging to the same 2-connected component of $G$ are on a single cycle.

A consequence of this result that has also been obtained by Mahadev et al. (1992) is that the NCOSS problem can be solved in polynomial time when there are three processors and there are less than three tasks per job.

Additional requirements on job scheduling in an NCOSS problem may occur. In some situations we may impose that, whenever a job starts being processed by the various machines, it visits the various machines without any waiting time until its entire processing is completed. Similarly, whenever a processor starts working in a cycle, it continues working without any interruption until its load of work for the cycle is completed. Cyclic schedules will be called *compact* when the no-wait requirements are satisfied for each processor and for each job.

They correspond in the associated graph $G$ with weights $w_e$ to a weak ice-$k$-coloring satisfying the additional requirement

(iii) the cyclic intervals $I_e$ of colors associated with the edges $e$ in a bundle $B(v)$ can be combined into a single cyclic interval of $\{1,\ldots,k\}$.

An example is shown in Figure 10.2; the corresponding colorings will be called *ice k-colorings*.

Results holding in the case of flow shop scheduling with no-wait constraints can be found in Reddi and Ramamoorthy (1972).

We have proved (de Werra and Solot 1993) that the nonpreemptive compact cyclic open shop scheduling problem (NCCOSS) is $\mathcal{NP}$-complete.

The problem of deciding whether, for an instance represented by a graph $G = (\mathcal{P},\mathcal{J},E)$ with weights $w_e$ and an integer $k$, there exists a nonpreemptive cyclic

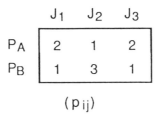

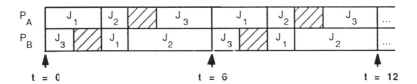

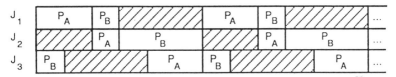

**Figure 10.2** A cyclic schedule (compact for all jobs and for all processors) with period $k = 6$.

schedule with compactness requirements only for jobs, is called NSCCOSS (nonpreemptive semicompact cyclic open shop scheduling). It turns out that this somewhat-relaxed version of NCCOSS is also $\mathcal{NP}$-complete (De Werra and Solot 1993). Since the graph model shows the symmetry of the problem with respect to jobs and processors, the same complexity result holds when compactness requirements hold only for processors.

In terms of the associated graph $G = (\mathcal{P}, \mathcal{J}, E)$ (with weights $w_e$), we can define an *ice-perfect* graph as a graph for which there is an ice $T(G, w)$-coloring for any choice of weights $w_e$. Similarly, $G$ is *Nice-perfect* if, for any choice of weights $w_e$ in $N = \{0, 1, 2, 3, 4\}$, there is an ice $T(G, w)$-coloring. Note that any ice-perfect graph is also weakly ice-perfect. With these definitions, we can state the following.

**Proposition 10.1** *For a bipartite graph G, the following statements are equivalent:*

(a) *G is ice-perfect;*
(b) *G is Nice-perfect;*
(c) *G is a bipartite outerplanar graph.*

This result provides an upper bound on the minimum completion time for an NOSS problem.

Denoting by $q_{int}(G, w)$ the *interval chromatic index*, namely the smallest $k$ for which a bipartite graph $G$ with weights $w_e$ has an interval edge $k$-coloring, and denoting by $w_{max}$ the largest $w_e$ occurring on the edges of $G$, one can show (de Werra and Solot 1993) that, for a weakly ice-perfect graph $G$ with weights $w_e$,

$$q_{int}(G) \leq T(G, w) + w_{max} - 1.$$

In fact, the bipartite outerplanar graphs have a stronger property, which will be explained in Section 10.4. We first need some preliminaries.

## 10.3 EQUITABLE EDGE COLORINGS AND SCHEDULING

In order to introduce the concepts that will be needed later, we have to consider an extension of edge $k$-colorings. Let $G$ be a graph and consider an assignment $F$ of $w_e$ colors in $1, \ldots, k$ to each edge $e$ of $G$. Denote by $f_i(x)$ the number of occurrences of color $i$ on edges adjacent to node $x$. Note that a color may occur several times on an edge $e$ with $w_e > 1$.

We shall say that $F$ is an *equitable edge $k$-coloring* of $G$ if for each node $x$ and for any two colors $i$ and $j$, we have

$$-1 \leq f_i(x) - f_j(x) \leq +1.$$

In other words, for any node $x$, all colors occur almost the same number of times on edges adjacent to $x$. An example of equitable $k$-coloring is shown in Figure 10.3(b) for the graph associated with the instance of the open shop problem given in Figure 10.3(a).

If $k \geq T(G, w)$, an equitable edge $k$-coloring reduces to an edge $k$-coloring (at each node $x$, every color occurs on one or on no edge). It is well known that, for any $k$, a bipartite multigraph $G$ has an equitable $k$-coloring (de Werra 1971). A simple way to see that for a bipartite graph $G$ an equitable edge $k$-coloring exists for any $k$ is the following: we assume $k < T(G, w)$, since otherwise any edge $T(G, w)$-coloring is an equitable edge $k$-coloring and it exists from the König theorem.

The degree $d(x)$ of a node $x$ is the sum of the weights of the adjacent edges. Each node $x$ with degree $d(x)$ is split into $\lceil d(x)/k \rceil$ nodes $x', x'', \ldots$ (here $\lceil q \rceil$ is the upper integer part of $q$). We split the weights of the edges initially adjacent to node $x$ in such a way that as many copies of $x$ as possible have degree $k$ (at most one node $x', x'', \ldots$ will have degree smaller than $k$). The resulting graph $G'$ is the *transformed graph*. Since $T(G', w') = k$, it has an edge $k$-coloring. An example of $G'$ is given in Figure 10.3(c) with an edge 4-coloring of $G'$. Clearly, such a coloring of $G'$ gives (by merging for each node $x$ the nodes $x', x'', \ldots$ corresponding to $x$) an equitable edge $k$-coloring of $G$.

Let us now consider an instance of POSS associated with a bipartite multigraph $G$ where each edge $e = [P_i, J_j]$ has a weight $w_e = p_{ij}$. A positive integer $k$ is given; it is the length of a cycle in a cyclic schedule. We impose that in each cycle jobs $J_1, \ldots, J_n$ are entirely completed. Note that this is possible if $k \geq T(G, w)$; here we may have $k < T(G, w)$. This means that a job $J_j$ for which $\sum_i p_{ij} > k$ must be

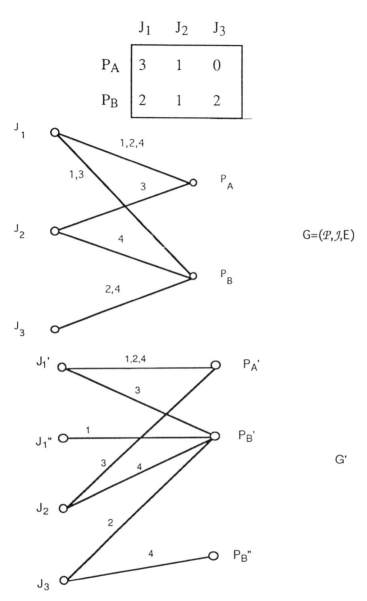

**Figure 10.3** (a) The data. (b) An equitable edge 4-coloring of the associated graph $G$. (c) An edge 4-coloring of the transformed graph $G'$.

started in a cycle previous to the one in which it is to be completed.

This also means that in each cycle we shall have several jobs $J_j$ that are simultaneously in process, so we shall talk of jobs of type $j$ rather than of job $J_j$.

Also, if a processor $P_i$ has a load $\sum_j p_{ij} > k$, it will not be able to process its entire work within a cycle of length $k$. Hence it should be duplicated. So we shall talk of processors of type $i$, since we may need more than one in the shop.

The preemptive cyclic open shop scheduling problem (or PCOSS) consists in constructing a cyclic schedule with cycles of given length $k$, where in each cycle one job of each type $j$ is completed. This clearly may require multiplications of processors and of jobs, as mentioned above.

Does such a schedule exist for any $k$?

We shall now show that a cyclic schedule with cycle length $k$ can be associated with any equitable $k$-coloring of the graph $G$ corresponding to the problem. This can be seen as follows. Suppose we have an equitable edge $k$-coloring of $G$; it corresponds to an edge $k$-coloring of $G'$. In order to obtain $G'$, we have simply replaced each processor $P_i$ by $r_i = \lceil \sum_j p_{ij}/k \rceil$ processors $P'_i, P''_i, \ldots$ of type $i$, and each job $J_j$ by $q_j = \lceil \sum_i p_{ij}/k \rceil$ jobs $J'_j, J''_j, \ldots$ of type $j$; furthermore, the edges (with their weights) have been split as described above.

The coloring of $G'$ corresponds to a cyclic schedule with cycle length $k$: assume that edge $[P_i^{(r)}, J_j^{(s)}]$ has received color $t$ (among others); this means that at the $t$th time unit of a cycle, job $J_j^{(s)}$ is in process on processor $P_i^{(r)}$.

Figure 10.4 shows a cyclic schedule (for two consecutive cycles) corresponding to

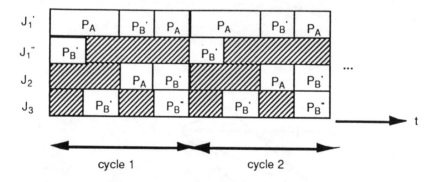

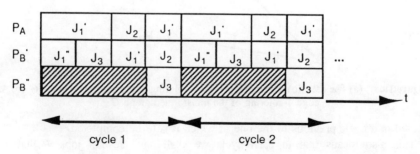

**Figure 10.4** A cyclic schedule for the PCOSS in Figure 10.3.

the coloring of Figure 3(c): the schedule for jobs and the schedule for processors are given.

For processors, the interpretation is immediate: the workload of $P_i$ has forced us to introduce several processors $P'_i, P''_i, \ldots$ identical to $P_i$.

For jobs $J'_j, J''_j, \ldots$ representing $J_j$, we have to observe that we must have several copies of job $J_j$ simultaneously present in the shop (with a time shift of one cycle length) to make sure that within each cycle one copy exactly of each job in $J_1, \ldots, J_n$ is completed. So the interpretation is as follows: a first copy of a job, say $J_j$, follows the schedule of $J'_j$ in the first cycle, the schedule of $J''_j$ in the second cycle, and so on. It is completed in the $q_j$th cycle. Another copy of $J_j$ starts in cycle 2 with the schedule of $J'_j$, continues in cycle 3 with the schedule of $J''_j$, and so on. Finally, the $q_j$th copy of $J_j$ starts in cycle $q_j$ with the schedule of $J'_j$, continues in cycle $q_j + 1$ with the schedule of $J''_j$, and so on. So each copy of $J_j$ follows consecutively the schedules of $J'_j, J''_j, J_j^{(3)}, \ldots, J_j^{(q_j)}$.

As an example, consider the schedule in Figure 10.4: the first copy of $J_1$ follows the schedule of $J'_1$ in cycle 1 (it goes consecutively to processors $P_A, P'_B$ and $P_A$) and the schedule of $J''_j$ in cycle 2 (then it goes to $P'_B$, and it is ready); a second copy follows the same program with a time shift of 4 time units.

As a consequence of the above observation, we can state the following.

**Proposition 10.2** *Given an instance of PCOSS represented by a weighted bipartite graph $G = (\mathcal{P}, \mathcal{J}, E)$ and given a positive integer $k$, there exists a cyclic schedule with cycle length $k$ that needs exactly $r_i$ processors of type $i$ and in which exactly $q_j$ copies of job $J_j$ are simultaneously in process.*

Note that no cyclic schedule with cycle length $k$ can use less than $r_i$ processors of type $i$ or have less than $q_j$ copies of $J_j$ simultaneously in process.

Additional properties of the schedules based on equitable edge $k$-colorings can be expressed: let $\Pi(i, t)$ be the number of processors of type $i$ that are active in the $t$th time unit of the cycle. Then it follows that for any $t, u \leqslant k$ and for any $i$, the cyclic schedule satisfies

$$-1 \leqslant \Pi(i, t) - \Pi(i, u) \leqslant 1 \qquad (10.1)$$

In other words, for any given processor type $i$, the number of processors of type $i$ that are simultaneously active will take at most two values (which differ by one).

Similarly, let $\gamma(j, t)$ be the number of copies of $J_j$ that are in process during the $t$th time unit of the cycle. We have for any $t, u \leqslant k$ and for any $j$,

$$-1 \leqslant \gamma(j, t) - \gamma(j, u) \leqslant 1 \qquad (10.2)$$

This means that the number of copies of $J_j$ that are simultaneously in process will take at most two values (which differ by one).

It is, furthermore, an easy matter to show by simple alternating chain arguments that an equitable edge $k$-coloring can be constructed where all colors occur almost the same number of times. In terms of scheduling, if we call $n(t)$ the total number of processors (or equivalently of copies of jobs) that are active (or in process) during the $t$th time unit of the cycle, we have for any $t, u \leq k$,

$$-1 \leq n(t) - n(u) \leq 1 \qquad (10.3)$$

The schedule is such that the total number of processors simultaneously active can take at most two values (which differ by one).

As an illustration, the schedule in Figure 10.4 has these properties: at any time, the total number of processors that are active is two or three.

Furthermore, there are one or two copies of $J_1$ that are simultaneously in process at any time, and finally one or two processors of type B are active at any time.

In summary, the concept of equitable $k$-colorings has allowed us to study a type of cyclic schedule that satisfies some balancing requirements. In the next section we shall also be concerned with cyclic schedules with a fixed cycle length, but there will be another type of requirement (compactness), which will need another type of edge coloring.

## 10.4  CYCLIC COMPACT SCHEDULES

Before stating the scheduling problem that will be studied, let us first introduce some additional definitions related to edge colorings.

Let $K = \{1, \ldots, k\}$ be a cyclically ordered set of colors. A c.p.i. (cyclic pseudointerval) of length $q$ is a sequence of $q$ consecutive integers in $K$; if $q > k$ the same integer may occur several times in a c.p.i.

Given a graph $G$ with integral weights $w_e$ on each edge $e$, a *c.p.i. $k$-coloring* of $G$ is an assignment of a c.p.i. of length $w_e$ to each edge $e$ in such a way that in each

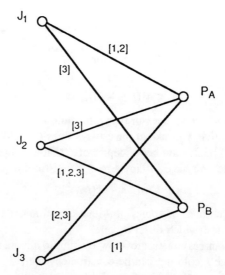

**Figure 10.5**  A c.p.i. 3-coloring of $G$.

bundle $B(x)$ of edges adjacent to the same node $x$ the c.p.i.s can be combined into a c.p.i.

Notice that a c.p.i. $k$-coloring is always an equitable $k$-coloring, but the converse is not true.

For the weighted bipartite graph corresponding to the problem in Figure 10.2, a c.p.i. 3-coloring is given in Figure 10.5.

Such a type of coloring may be interpreted as a schedule in which some processors may be duplicated and some jobs may be also duplicated in order to meet the production requirements in a cycle of $k$ time units. Figure 10.6 shows a schedule corresponding to the c.p.i. 3-coloring of Figure 10.5.

It is not true that a bipartite graph $G$ has a c.p.i. $k$-coloring for each $k$. However, the following result holds.

**Proposition 10.3** (de Werra and Solot 1991) *For a graph $G$ the following statements are equivalent:*

(a) *for any $k$ and for any choice of integral weights $w_e$ for all edges $e$, $G$ has a c.p.i. $k$-coloring;*
(b) *$G$ is bipartite outerplanar.*

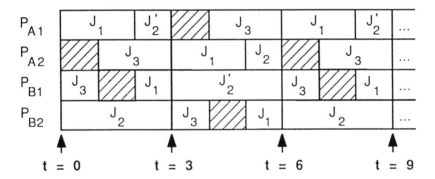

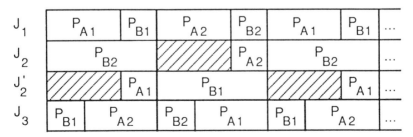

**Figure 10.6** A periodic schedule with period $k = 3$ (for the same data as in Figure 10.5).

## 10.5 CONCLUSIONS AND ADDITIONAL QUESTIONS

Several questions are still unanswered in the area of cyclic compact scheduling. For instance, it is not known whether the same balancing properties hold as in the case of CPOSS: can one find a cyclic compact schedule such that (10.3) also holds (the total number of processors simultaneously active takes at most two values (which differ by one))? Clearly, (10.1) and (10.2) hold, but the alternating chain techniques used to obtain (10.3) in PCOSS do not seem to apply for cyclic compact scheduling.

One may explore other types of chromatic scheduling problems. For example, it would be interesting to characterize the graphs $G$ for which, given any choice of integral weights $w_e$, there exists a compact ice $k$-coloring for some $k \leq T(G, w) + 1$.

Also, in many circumstances we may accept preemptions in the open shop model: a task may be interrupted (say after an integral number of time units), and resumed later on the same or on another machine. The problem can be solved in polynomial time if preemptions are allowed on all tasks, as discussed in Section 10.1. Trivially, one may see that it is still the case if preemptions are forbidden just for (all tasks to be processed on) one specific machine. Furthermore, if NOSS has at most two processors, it is solvable in polynomial time.

What can be said about the case where among the $m$ processors there are exactly two for which preemptions are forbidden and for all remaining processors preemptions are allowed?

Also, resources (renewable or nonrenewable) may have to be taken into account. Partial results (complexity and solvable cases) are given by de Werra and Blazewicz (1992) for the case where one type of resource is present. What can be said in the situation where two types of resources are needed?

In some circumstances a scheduling problem may be modeled as a POSS with some additional constraints of simultaneity: some pairs of tasks $T_{ij}, T_{kl}$ are given, and it is required that for each such pair, there is some period where both $T_{ij}$ and $T_{kl}$ are in process (this means that in such a pair of edges it is required that both have the same color). Partial results are available for handling special cases of such requirements, but more research is needed.

The cases studied here are extremely special. Open shop models, and in particular their cyclic versions, may provide tools for dealing with simplified scheduling problems occurring in automated production systems. A further step would be to develop heuristic procedures for the general case of cyclic compact scheduling. The coloring techniques used for the special case of bipartite outerplanar graphs may provide a starting point for this task.

## REFERENCES

Berge C. (1983) *Graphes* Gauthier-Villars, Paris.
Blazewicz J., Cellary W., Slowinski R. and Weglarz J. (1986) *Scheduling under Resource Constraints: Deterministic Models* Baltzer, Basel.
Burkard R. E. (1986) Optimal schedules of periodically recurring events *Discr. Appl. Math.*, **15**, 167–180.
de Werra D. (1971) Equitable colorations of graphs *RAIRO*, **R-3**, 3–8.
de Werra D. and Blazewicz J. (1992) Some preemptive open shop scheduling problems with a renewable or a non-renewable resource *Discr. Appl. Math.*, **35**, 205–219.
de Werra D. and Solot P. (1991) Compact cylindrical chromatic scheduling *SIAM J. Discr. Math.*, **4**, 528–534.
de Werra D. and Solot P. (1992) Some graph-theoretical models for scheduling in automated production systems *Networks*, **23**, 651–660.
Gonzalez T. and Sahni S. (1976) Open shop scheduling to minimize finish time, *J. Assoc. Comput. Mach.*, **23**, 665–679.
Mahadev N. V. R., Solot P. and de Werra D. (1993) The cyclic compact open shop scheduling problem *Discr. Math.*, **111**, 361–366.
Mahadev N. V. R., Solot P. and de Werra D. (1992) Cylindrical open shop scheduling: some solvable cases, *Vishwa Int. J. Graph Theory*, **1**, 29–52.
Reddi S. S. and Ramamoorthy C. V. (1972) On the flowshop sequencing problem with no wait in process *Oper. Res. Quart.*, **23**, 323–331.

*Swiss Federal Institute of Technology, CH-1015 Lausanne, Switzerland*

*Ciba-Geigy AG, IS7.2—Operations Research, R-1008.72.27, CH-4002 Basel, Switzerland*

CHAPTER 11

# Transforming Cyclic Scheduling Problems into Acyclic Ones

**Franco Gasperoni**
*Telecom Paris, ENST*

**Uwe Schwiegelshohn**
*Institute for Information Technology Systems, University of Dortmund*

## Abstract

A cyclic task models the behavior of a system which must continuously execute a fixed set of interdependent operations. A typical example of a cyclic task is a loop in a computer program. In this work we provide a general framework for transforming a cyclic task $T_c$ into an acyclic one, say $T$, such that any scheduling algorithm $A$ for $T$ can be transformed into a scheduling algorithm for $T_c$ which preserves all of $A$'s performance guarantees.

## 11.1 INTRODUCTION

A cyclic task $T_c$ models the behavior of a system that must continuously execute a fixed set of interdependent operations. A typical example of a cyclic task is a loop in a computer program. Given a target machine $M$, a cyclic scheduling problem consists in finding a periodic schedule $\sigma$ which respects $T_c$'s dependences and can be executed on $M$. Like its acyclic counterpart, a cyclic scheduling problem is in general $\mathcal{NP}$-hard (Serafini and Ukovich 1989).

A significant amount of research has been devoted to the problem of finding an approximate solution to acyclic scheduling problems (Coffman 1976; Graham *et al.* 1979; O'kEigeartaigh *et al.* 1985). This is in contrast with the situation for the corresponding cyclic problems. It is therefore highly desirable to transform a given cyclic task $T_c$ into an acyclic one, say $T$, such that the approximation results and algorithms obtained for $T$ on a machine $M$ could be carried over to $T_c$.

By providing a general framework for performing such transformation, this chapter goes one step closer to achieving this goal. More precisely, given a dependence graph $G$ for $T_c$ we suggest doing the following.

(1) Cut as many edges from $G$ as allowed by a simple condition (Theorem 11.4) to "maximize the parallelism" in the derived acyclic graph $G'$. $G'$ now embodies the dependences of some acyclic task $T$.
(2) Generate an acyclic schedule $\mu$ that respects $T$'s dependences, is acceptable for the target machine $M$ and is as close as possible to the optimum.
(3) Find the minimum interval $\Delta$ for which an infinite repetition of $\Delta$-spaced copies of $\mu$ yields a periodic schedule $\sigma$ that is acceptable for the original cyclic task $T_c$ and the machine $M$ (Theorem 11.4 and Corollary 11.1).

In Theorem 11.5 it is established that every periodic schedule that is admissible for $M$ and that preserves $T_c$'s dependences can be constructed in this fashion.

The general question of cutting edges is somewhat linked to the underlying machine model. Our framework gives the general rule that every edge cutting algorithm must satisfy, but, because our machine model is extremely general, it does not provide an actual algorithm for cutting edges.

We have applied the technique described in this chapter to a particular instance of our general machine model and have devised a polynomial-time algorithm that is able to guarantee a worst case factor from optimality of roughly 2 (Gasperoni and Schwiegelshohn 1994).

## 11.2 MACHINE MODEL

The processing framework $M$ considered in this chapter is very general and subsumes most of the machine models that have appeared in the scheduling literature. $M$ provides a set of operations that it is capable of executing. Each operation requires a certain amount of processing in order to be completed. The result of a given operation can be transmitted to another operation in a predetermined amount of time depending on the operations in question. Time is considered to be continuous and is modeled by a real number. The amount of machine resources required to process each stage of an operation are embodied in a set of valid machine states listing all the activities that $M$ is allowed to concurrently undertake. We have assumed that the validity of a machine state is independent of the history of computation.

After formally defining our machine framework $M$, we show how a less abstract machine model is represented in our framework.

**Definition 11.1 (Machine)** *A machine* $M = (\mathcal{O}, \delta, \Phi, \mathcal{S})$ *is a quadruple, where the components are as follows.*

(1) $O$ is the set of operations that the machine is capable of performing. An element of $O$ is called a machine operation.
(2) $\delta$ is the duration function of $M$, a function mapping $\mathcal{O}$ into the positive reals. For $op \in \mathcal{O}$, $\delta(op)$ is the overall amount of time it takes for $M$ to process $op$.
(3) $\Phi$ is the flow function of $M$, a function mapping $\mathcal{O} \times \mathcal{O}$ into the nonnegative reals. For any two machine operations $op$ and $op'$, $\Phi(op, op')$ is the time it takes in $M$ for information produced by $op$ to flow or be transported to $op'$.
(4) $\mathscr{S}$ is the set of valid machine states. A machine state is a subset of $\{(op, t) : op \in \mathcal{O}, t \in [0, \delta(op)]\}$. A machine state is said to be valid iff it belongs to $\mathscr{S}$. Intuitively $\mathscr{S}$ portrays the machine configurations that $M$ is allowed to take.

**Example 11.1** The above machine framework is very general. As an example, we show how to model a machine with two processors, $P_1$ and $P_2$, and two identical resources such that an operation executing on this machine needs from zero to two resources and takes 1 or 2 units of time depending on whether it executes on $P_1$ or $P_2$. Thus

$$\mathcal{O} = \{\underbrace{(P_1,0)}_{op_1}, \underbrace{(P_1,1)}_{op_2}, \underbrace{(P_1,2)}_{op_3}, \underbrace{(P_2,0)}_{op_4}, \underbrace{(P_2,1)}_{op_5}, \underbrace{(P_2,2)}_{op_6}, \},$$

$$\delta(op_1) = \delta(op_2) = \delta(op_3) = 1, \quad \delta(op_4) = \delta(op_5) = \delta(op_6) = 2,$$

where $(P_i, j)$ models an operation executing on processor $P_i$ and using $j$ resources. In this machine it takes one unit of time to forward the results of one operation executing on $P_1$ to another operation executing on $P_2$, whereas it takes no time to forward results in all other cases. Thus

$$\Phi(op_i, op_j) = \begin{cases} 1 & \text{if } i \in \{1,2,3\} \text{ and } j \in \{4,5,6\}, \\ 0 & \text{otherwise.} \end{cases}$$

Two operations can simultaneously execute on the machine iff they can execute on different processors and they need all together two resources at most. Thus the set of valid machine states is

$$\mathscr{S} = \left( \bigcup_{i=1}^{3} \bigcup_{j=4}^{7-i} \{(op_i, t_1), (op_j, t_2) : t_1 \in [0, 1[, t_2 \in [0, 2[\} \right)$$

$$\cup \left( \bigcup_{i=1}^{3} \{(op_i, t_1) : t_1 \in [0, 1[\} \right) \cup \left( \bigcup_{j=4}^{6} \{(op_j, t_2) : t \in [0, 2[\} \right).$$

After introducing the processing framework, it is important to define the notion of computation or schedule.

**Definition 11.2 (Schedule)** A schedule $\sigma$ for a machine $M = (\mathcal{O}, \delta, \Phi, \mathscr{S})$ is a function mapping each time instant, i.e. a real, into a set of machine operations, such that the following conditions are satisfied.

(1) *For any machine operation op, either op does not execute in $\sigma$, or it executes in $\sigma$ for $\delta(op)$ time, that is,*

$$\forall op \in \mathcal{O}, \quad \text{either} \quad \{t : op \in \sigma(t)\} = \emptyset \quad \text{or} \quad \|\{t : op \in \sigma(t)\}\| = \delta(op),$$

*where $\|\cdot\|$ denotes the Lebegue measure of a set of reals. The starting and completing times of an operation op in $\sigma$ are denoted respectively by $\sigma_s(op) = \min\{t : op \in \sigma(t)\}$ and $\sigma_c(op) = \max\{t : op \in \sigma(t)\}$.*

(2) *For each time instant t, the computation state of schedule $\sigma$ at time t, denoted by $S(\sigma, t)$, is a valid machine state, i.e. $S(\sigma, t) \in \mathcal{S}$. If for each operation op we define $\sigma_t(op) = \|\{t' \leqslant t : op \in \sigma(t')\}\|$ to be the total execution time of op in schedule $\sigma$ up to time t, then*

$$S(\sigma, t) = \{(op, \sigma_t(op)) : op \in \sigma(t)\}.$$

*If the time instants $\sigma_s = \min_{op} \sigma_s(op)$ and $\sigma_c = \max_{op} \sigma_c(op)$ are defined, the schedule is said to be bounded and its length $|\sigma|$ is defined to be $|\sigma| = \sigma_c - \sigma_s$.*

**Example 11.2** The following is a schedule for the machine model given in Example 11.1:

for $t < 0$ $\quad \sigma(t) = \emptyset,$ $\quad\quad\quad\quad S(\sigma, t) = \emptyset$
$t \in [0, 1[, \quad \sigma(t) = \{op_3\}, \quad\quad\quad S(\sigma, t) = \{(op_3, t)\}$
$t \in [1, 2[, \quad \sigma(t) = \{op_1, op_5\}, \quad S(\sigma, t) = \{(op_1, t-1), (op_5, t-1)\}$
$t \in [2, 3[, \quad \sigma(t) = \{op_5\}, \quad\quad\quad S(\sigma, t) = \{(op_5, t-2)\},$
$t \geqslant 3, \quad\quad \sigma(t) = \emptyset, \quad\quad\quad\quad S(\sigma, t) = \emptyset.$

This schedule can be portrayed graphically as follows:

| Processors | | Resources | |
|---|---|---|---|
| $P_1$ | $P_2$ | $R$ | $R$ |
| $op_3$ | | $op_3$ | |
| $op_1$ | $op_5$ | $op_5$ | $op_3$ |
| | $op_5$ | $op_5$ | |
| | | $op_5$ | |

$\sigma_s(op_3) = 0$
$\sigma_c(op_3) = \sigma_s(op_1) = \sigma_s(op_5) = 1$
$\sigma_c(op_1) = 2$
$\sigma_c(op_5) = 3$

The schedule given in this example is bounded, since $\sigma_s = 0$ and $\sigma_c = 3$. Its length is of 3 units of time.

## 11.3 TASKS AND CYCLIC TASKS

Informally, a task is a collection of several interdependent activities, called conceptual operations, all of which must be executed for the task to complete.

**Definition 11.3 (Task)** *A task $T$ is a pair $T = (O, \prec)$, where the components are as follows.*

(1) *$O$ is the conceptual operation set of $T$. A conceptual operation $\overline{op} \in O$ is a subset of machine operations. Intuitively, every $op \in \overline{op}$ is a machine operation by which the conceptual operation $\overline{op}$ can be implemented in $M$.*
(2) *$\prec$ is the dependence relation of $T$, a partial order on $O$. Informally, $\overline{op} \prec \overline{op}'$ means that operation $\overline{op}'$ requires the information produced by $\overline{op}$ to start executing: $\overline{op}'$ is said to depend on $\overline{op}$.*

The partial order $\prec$ can be represented by a directed graph $G$, called the dependence graph of $T$. The vertex set of $G$ is $O$ and there exists a path in $G$ from a conceptual operation $\overline{op}$ to another conceptual operation $\overline{op}'$ iff $\overline{op} \prec \overline{op}'$. Note that under this definition a task does not necessarily have a unique dependence graph.

We now explain the conditions under which a machine schedule $\sigma$ is valid for a given task $T$.

**Definition 11.4 (Valid schedule)** *Let $T = (O, \prec)$ be a task and $\sigma$ a schedule. $\sigma$ is said to respect $T$'s dependences or to be valid for $T$ iff:*

(1) *every conceptual operation $\overline{op}$ of $T$ is implemented in $\sigma$ by a unique machine operation $op \in \overline{op}$; that is,*

$$\forall \overline{op} \in O, \quad \exists! op \in \overline{op} \text{ s.t. } op \text{ executes in } \sigma$$

*(to simplify notation, the machine operation implementing a conceptual operation $\overline{op}$ will always be denoted by $op$);*
(2) *the conceptual operation dependences of $T$ are respected in $\sigma$; that is, for all $\overline{op}, \overline{op}' \in O$,*

$$\overline{op} \prec \overline{op}' \Rightarrow \sigma_c(op) + \Phi(op, op') \leq \sigma_s(op').$$

Note that the restriction that a conceptual operation be implemented by a unique machine operation could be lifted. However, for the sake of simplicity and conciseness, we do not consider schedules where a conceptual operation can be implemented by multiple copies of a machine operation. The results stated in this chapter can be generalized to allow for this possibility.

**Example 1.3** Consider the machine described in Example 11.1 and the following task $T$. The conceptual operation set of $T$ is $O = \{\overline{op}_a, \overline{op}_b, \overline{op}_c\}$, where

$\overline{op}_a = \{op_3, op_4\}$ (that is, we can implement $op_a$ by either executing it on $P_1$ in 1 time unit but using both of the machine resources $R$, or by executing it on P2 in 2 time steps but using no resources),

$\overline{op}_b = \{op_2, op_5\}$

$\overline{op}_c = \{op_1, op_6\}$

The dependences of $T$ are $\overline{op}_a \prec \overline{op}_c$. The schedule given in Example 11.2 is valid for $T$. In fact,

$\overline{op}_a$ is implemented by $op_3$

$\overline{op}_b$ is implemented by $op_5$

$\overline{op}_c$ is implemented by $op_1$

Furthermore, the dependence $\overline{op}_a \prec \overline{op}_c$ is satisfied, since

$$\underbrace{\sigma_c(op_3)}_{=1} + \underbrace{\Phi(op_3, op_1)}_{=0} \leq \underbrace{\sigma_s(op_1)}_{=1}.$$

We can now introduce the notion of cyclicity. Informally, a cyclic task $T_c$ can be seen as an infinite sequence of tasks, called iterations. Iterations share the same conceptual operation set $O$, but may have different dependence relations. Operations in one iteration may depend on operations in preceding iterations.

**Definition 11.5 (Cyclic task)** *A cyclic task $T_c = (O \times \mathbb{Z}, \prec)$ is a task satisfying the following conditions.*

*(1) The operation set of $T_c$ is the cartesian product of a finite set of conceptual operations $O$ and the integers. For $\overline{op} \in O$ and $i \in \mathbb{Z}$, $\overline{op}[i]$ denotes the instance of conceptual operation $\overline{op}$ in the ith iteration and for $op \in \overline{op}$, $op[i]$ the instance of machine operation $op$ implementing $\overline{op}[i]$. The set $O$ is called the core operation set of $T_c$. Note that for $i \neq j$, $op[i]$ and $op[j]$ are considered two distinct machine operations, although they are identical from the standpoint of machine characteristics.*
*(2) For all $\overline{op}, \overline{op}' \in O$ and $i, j \in \mathbb{Z}$, $\overline{op}[i] \prec \overline{op}'[j]$ implies $i \leq j$.*

**Example 11.4** Consider the machine described in Example 11.1 and the following cyclic task $T_c$. The core operation set of $T_c$ is the same as that given in Example 11.3: $O = \{\overline{op}_a, \overline{op}_b, \overline{op}_c\}$. The dependence relation of $T_c$ is

$\forall i, \quad \overline{op}_a[2i] \prec \overline{op}_c[2i]$

$\forall j > 0, \quad \overline{op}_c[2j+1] \prec \overline{op}_a[2j+3]$

Let $T_c$ be some cyclic task with core operation set $O$. Given any two operations $\overline{op}, \overline{op}' \in O$, it will become important, for the purpose of generating regular schedules, to identify the most stringent dependence between $\overline{op}$ and $\overline{op}'$. To this end, we introduce the notion of dependence distance between $\overline{op}$ and $\overline{op}'$.

**Definition 11.6 (Dependence distance)** *Let $T_c$ be some cyclic task with core operation set $O$. For all $\overline{op}, \overline{op}' \in O$, the dependence distance from $\overline{op}$ to $\overline{op}'$, denoted by $d(\overline{op}, \overline{op}')$, is defined as*

$$d(\overline{op}, \overline{op}') = \min \{d : \exists k \in \mathbb{Z}, \overline{op}[k] \prec \overline{op}'[k+d]\},$$

*where the minimum of the empty set is equal to $\infty$ by definition.*

The dependence distances of a cyclic task $T_c = (O \times \mathbb{Z}, \prec)$ can be portrayed by a weighted graph $G = (O, E, d)$, called the *cyclic dependence graph*. $G$'s vertex set is $O$, the core operation set of $T_c$. $G$'s edge set $E$ must satisfy the following two requirements.

(1) If $e = (\overline{op}, \overline{op}') \in E$ then $d(\overline{op}, \overline{op}') < \infty$. The weight of $e$ is set to $d(\overline{op}, \overline{op}')$.
(2) $\forall \overline{op}, \overline{op}' \in O, d(\overline{op}, \overline{op}') < \infty \Rightarrow \exists$ path $P$ from $\overline{op}$ to $\overline{op}'$ s.t. $d(P) \leq d(\overline{op}, \overline{op}')$, where $d(P)$ denotes the sum of the weights of the edges of $P$.

Note that $E$ is not necessarily unique. The edge set

$$E = \{(\overline{op}, \overline{op}') : d(\overline{op}, \overline{op}') < \infty\}$$

has the biggest cardinality, whereas the edge set

$$E = \{(\overline{op}, \overline{op}') : d(\overline{op}, \overline{op}') < \infty \text{ and } \not\exists \overline{op}'' d(\overline{op}, \overline{op}'') + d(\overline{op}'', \overline{op}') \leq d(\overline{op}, \overline{op}')\}$$

is the edge set with the smallest cardinality. By computing all pairs of shortest paths of $G$, it is easy to transform the original edge set into the one with the fewest edges. This step can be implemented in $O(nm + n^2 \log n)$ time, where $n$ and $m$ denote the cardinalities of $G$'s vertex and edge sets respectively (Cormen *et al.* 1990).

**Example 11.5** The dependence distances of the cyclic task given in Example 11.4 are all infinity except for

$$d(\overline{op}_a, \overline{op}_c) = 0, \quad (\overline{op}_c, \overline{op}_a) = 2.$$

There is a unique dependence graph for this cyclic task. It contains three vertices, $\overline{op}_a$, $\overline{op}_b$ and $\overline{op}_c$, and two edges, $\overline{op}_a \xrightarrow{0} \overline{op}_c$ and $\overline{op}_a \xleftarrow{2} \overline{op}_c$ with weights 0 and 2 respectively.

## 11.4 PERIODIC SCHEDULES

For cyclic tasks one is interested in generating regular schedules that can be finitely encoded. To this end we introduce the notion of a periodic schedule.

**Definition 11.7 (Periodic schedule)** *A schedule $\sigma$ is said to be periodic iff there exists a time delay $\Delta > 0$, called the initiation interval of $\sigma$, such that*

$$\forall t \in \mathbb{R}, \quad op[i] \in \sigma(t) \Leftrightarrow op[i+1] \in \sigma(t+\Delta)$$

Intuitively, $\Delta$ represents the asymptotic performance of $\sigma$, and will be regarded as its performance estimator. More specifically, $1/\Delta$ denotes $\sigma$'s throughput, that is, the number of iterations performed by $\sigma$ per unit of time.

The results stated in this chapter can be extended to account for a more general definition of periodicity, namely a period greater than one. For the sake of conciseness, we have omitted such generalization.

**Example 11.6** The following is a valid periodic schedule for the cyclic task of Example 11.4:

$$\forall k, \quad \text{for } t \in [2k, 2k+1[, \quad \sigma(t) = \{op_1[k-1], op_4[k]\},$$
$$\text{for } t \in [2k+1, 2k+2[ \quad \sigma(t) = \{op_2[k], op_4[k]\},$$

where $op_4$ implements $\overline{op}_a$, $op_1$ implements $\overline{op}_c$ and $op_2$ implements $\overline{op}_b$. Note that the dependences of $T_c$ are satisfied, since

$$\forall i, \quad \overline{op}_a[2i] \prec \overline{op}_c[2i] \text{ and } \underbrace{\sigma_c(op_4[2i])}_{=4i+2} + \underbrace{\Phi(op_4, op_1)}_{=0} \leq \underbrace{\sigma_s(op_1[2i])}_{=4i+2}$$

$$\forall j > 0 \quad \overline{op}_c[2j+1] \prec \overline{op}_a[2j+3] \text{ and } \underbrace{\sigma_c(op_1[2j+1])}_{=4j+5} + \underbrace{\Phi(op_1, op_4)}_{=1}$$

$$\leq \underbrace{\sigma_s(op_4[2j+3])}_{=4j+6}.$$

This schedule can be portrayed graphically as follows:

| Processors | | Resources | |
|---|---|---|---|
| $P_1$ | $P_2$ | $R$ | $R$ |
| ... | | ... | |
| $op_1[-2]$ | $op_4[-1]$ | | |
| $op_2[-1]$ | $op_4[-1]$ | $op_2[-1]$ | |
| $op_1[-1]$ | $op_4[0]$ | | |
| $op_2[0]$ | $op_4[0]$ | $op_2[0]$ | |
| $op_1[0]$ | $op_4[1]$ | | |
| $op_2[1]$ | $op_4[1]$ | $op_2[1]$ | |
| ... | | ... | |

We now look at two properties of periodic schedules. The first pinpoints the repetitive nature of their computational states. This theorem allows us to characterize the $M$-admissibility of a periodic schedule $\sigma$ in terms of a restricted number of computational states.

**Theorem 11.1** *Let $\sigma$ be a periodic schedule with initiation interval $\Delta$. If we disregard iteration indices, we have*

$$\forall t \in \mathbb{R}, \quad S(\sigma,t) = S(\sigma,t+\Delta).$$

**Proof** This is trivial. □

A periodic schedule $\sigma$ that respects the dependences of a cyclic task $T_c$ is completely determined by its initiation interval $\Delta$ and the starting cycles of operations of the form $op[0]$. It should therefore be possible to characterize the validity of $\sigma$ for $T_c$ solely in terms of $\Delta$, $\sigma_s(op[0])$ and $\sigma_c(op[0])$ for each machine operation $op$ that appears in $\sigma$. Indeed one can show that:

**Theorem 11.2** *Let $T_c = (O \times \mathbb{Z}, \prec)$ be a cyclic task and $\sigma$ a periodic schedule with initiation interval $\Delta$. $\sigma$ respects $T_c$'s dependences iff*

$$\forall \overline{op}, \overline{op}' \in O, \quad \sigma_s(op'[0]) \geq \Phi(op,op') - d(\overline{op},\overline{op}')\Delta.$$

*Note that for $d(\overline{op},\overline{op}') = \infty$ the constraint is true by definition.*

**Proof** Suppose that $\sigma$ respects $T_c$'s dependences but there exist two conceptual operations $\overline{op}$ and $\overline{op}'$ in $T_c$ such that

$$\sigma_s(op'[0]) - \sigma_c(op[0]) < \Phi(op,op') - d(\overline{op},\overline{op}')\Delta.$$

Because of the definition of $d(\overline{op},\overline{op}')$, we have

$$\exists k \in \mathbb{Z}, \quad \overline{op}[k] \prec \overline{op}'[k + d(\overline{op},\overline{op}')].$$

This implies that

$$\sigma_c(op[k]) + \Phi(op,op') \leq \sigma_s(op'[k + d(\overline{op},\overline{op}')]).$$

Because $\sigma$ is periodic, we can write

$$\sigma_c(op[0]) + \Phi(op,op') + k \cdot \Delta \leq \sigma_s(op'[0]) + (k + d(\overline{op},\overline{op}'))\Delta$$

which contradicts the assumption stated at the beginning of the proof.

Conversely let $\sigma$ be a periodic schedule with initiation interval $\Delta$ such that for all $\overline{op}, \overline{op}' \in O$,

$$\sigma_c(op[0]) + \Phi(op, op') \leq \sigma_s(op'[0]) + d(\overline{op}, \overline{op}')\Delta.$$

Then, because of the periodicity of $\sigma$, we have, for all integers $k$ and $k' \geq k$,

$$\sigma_c(op[k]) + \Phi(op, op') \leq \sigma_s(op'[k' + d(\overline{op}, \overline{op}')]).$$

Because of the definition of $d(\overline{op}, \overline{op}')$, for all $q, q' \in \mathbb{Z}$,

$$\overline{op}[q] \prec \overline{op}[q'] \;\Rightarrow\; q + d(\overline{op}, \overline{op}') \leq q'.$$

By letting $k = q$ and $k' = q' - d(\overline{op}, \overline{op}')$ in the previous inequality, we can see that $\sigma$ preserves the dependence from $\overline{op}[q]$ to $\overline{op}[q']$ for any $\overline{op}, \overline{op}' \in O$ and any $q, q' \in \mathbb{Z}$. □

## 11.5 CORE SCHEDULES

In this section we introduce the central notion of core schedule. This notion plays a fundamental role in the transformation of a cyclic scheduling problem into an acyclic one. Informally, the core schedule $\mu$ of a periodic schedule $\sigma$ with initiation interval $\Delta$ is a pattern of operations whose infinite repetition, spaced by a delay of $\Delta$ cycles, yields $\sigma$.

**Definition 11.8 (Core schedule)** *Let $\mu$ be a schedule, $\Delta > 0$ a time delay and $I$ an indexing function, that is, a function mapping the operations executing in $\mu$ into the integers. Then $\mu^*_{\Delta,I}$, called the $\Delta$-repetition of $\mu$ indexed by $I$, is defined by*

$$\forall t, \; \forall i, \quad op \in \mu(t) \iff op[i] \in \mu^*_{\Delta,I}(t + [i - I(op)]\Delta).$$

*Let $\sigma$ be a periodic schedule with initiation interval $\Delta$. Then $\mu$ is said to be a core schedule of $\sigma$ iff there exists an indexing function $I$ such that*

$$\sigma = \mu^*_{\Delta,I}.$$

*If $\mu_s + \Delta \geq \max_{op} \mu_s(op)$, we say that $\mu$ is a dense core schedule of $\sigma$.*

**Example 11.7** We give two core schedules for the periodic schedule $\sigma$ given in Example 11.6. The first core schedule $\mu$ is

$$\begin{aligned} \text{for } t < 0, & \quad \mu(t) = \emptyset, \\ t \in [0, 1[, & \quad \mu(t) = \{op_1, op_4\}, \\ t \in [1, 2[, & \quad \mu(t) = \{op_2, op_4\}, \\ t \geq 2, & \quad \mu(t) = \emptyset. \end{aligned}$$

This schedule can be portrayed graphically as follows:

| Processors | | Resources | |
|---|---|---|---|
| $P_1$ | $P_2$ | $R$ | $R$ |
| $op_1$ | $op_4$ | | |
| $op_2$ | $op_4$ | $op_2$ | |

If we set $I(op_4) = I(op_2) = 0$ and $I(op_1) = -1$ then $\sigma = \mu^*_{2,I}$. Schedule $\mu$ is dense, since $\Delta = 2 > \max_{op} \mu_s(op) = 1$.

The second core schedule $\mu'$ for $\sigma$ is

$$\begin{aligned} \text{for } t < 0, & \quad \mu'(t) = \emptyset, \\ t \in [0, 2[, & \quad \mu'(t) = \{op_4\}, \\ t \in [2, 3[, & \quad \mu'(t) = \{op_1\}, \\ t \in [3, 4[, & \quad \mu'(t) = \{op_2\}, \\ t \geq 4, & \quad \mu'(t) = \emptyset. \end{aligned}$$

This schedule can be portrayed graphically as follows:

| Processors | | Resources | |
|---|---|---|---|
| $P_1$ | $P_2$ | $R$ | $R$ |
| | $op_4$ | | |
| | $op_4$ | | |
| $op_1$ | | | |
| $op_2$ | | $op_2$ | |

If we set $I(op_4) = I(op_1) = 0$ and $I(op_2) = 1$ then $\sigma = \mu' *_{2,I}$. Schedule $\mu'$ is not dense, since $\Delta = 2 \leq \max_{op} \mu'_s(op) = 3$.

Let $t_0 \in R$ be any time instant and $op$ any core operation of a periodic schedule $\sigma$. Because of the definition of periodicity, there must exist an iteration index $i$ such that $op[i]$ is initiated in the interval $[t_0, t_0 + \Delta[$. Thus every periodic schedule contains at least one dense core schedule. A periodic schedule $\sigma$ contains several core schedules. It is important to note that all dense core schedules of $\sigma$ have the same length, which is the smallest among $\sigma$'s core schedules. Also, all dense core schedules can be obtained from each other by "circular rotation". Informally, a dense core schedule $\mu$ is the shortest-length schedule such that $\mu *_{\Delta,I} = \sigma$. Even more informally, it is the shortest pattern such that its infinite $\Delta$-spaced repetition yields $\sigma$. The proofs and formalizations of these last remarks is left as an exercise.

The computational states of a periodic schedule can be expressed in terms of the computational states of any of its core schedules. Before stating the exact relationship, let us extend the modulo notion to the reals. Let $x$ and $y > 0$ be two reals. We define $x \bmod y$ to be $x - \lfloor x/y \rfloor y$.

**Theorem 11.3** *Let $\sigma$ be a periodic schedule with initiation interval $\Delta$ and $\mu$ any core schedule of $\sigma$. Then, if we disregard iteration indices*

$$\forall t \in \mathbb{R}, \quad S(\sigma, t) = \bigcup_{k=0}^{\lfloor |\mu|/\Delta \rfloor} S(\mu, \mu_s + (t - \mu_s) \bmod \Delta + k\Delta).$$

**Proof** Because of Theorem 11.1, it suffices to show that the theorem applies only for $t \in [\mu_s, \mu_s + \Delta[$. Thus it suffices to show that

$$\forall t \in [\mu_s, \mu_s + \Delta[, \quad S(\sigma, t) = \bigcup_{k=0}^{\lfloor |\mu|/\Delta \rfloor} S(\mu, t + k\Delta).$$

For all $t \in R$ we have

$$op[i] \in \sigma(t) \Leftrightarrow op[I(op)] \in \sigma(t + [I(op) - i]\Delta) \Leftrightarrow op \in \mu(t + [I(op) - i]\Delta).$$

and therefore $(op[i], x) \in S(\sigma, t) \Leftrightarrow (op, x) \in S(\mu, t + [I(op) - i]\Delta)$. Let $k = I(op) - i$. Then if $t \in [\mu_s, \mu_s + \Delta[$, $k$ can span all the integers between 0 and $|\mu|/\Delta$. $\square$

**Corollary 11.1** *Let $\sigma$ be a periodic schedule with initiation interval $\Delta$ and $\mu$ any core schedule of $\sigma$. Then $\sigma$ is valid for machine M iff*

(1) *for each op executing in $\mu$, $\|\{t : op \in \mu(t)\}\| = \delta(op)$;*
(2) *$\forall t \in [\mu_s, \mu_s + \Delta[, \bigcup_{k=0}^{\lfloor |\mu|/\Delta \rfloor} S(\mu, t + k\Delta) \in \mathscr{S}$ (i.e. it is a valid machine state).*

## 11.6 MAIN RESULT

Let $T_c = (O \times \mathbb{Z}, \prec)$ be a cyclic task and $\mu$ some schedule such that for every conceptual operation $\overline{op} \in O$, there exists a machine operation $op \in \overline{op}$ executing in $\mu$. The following two theorems give a necessary and sufficient condition for $\mu$ to be the core schedule of a periodic schedule respecting $T_c$'s dependences and having initiation interval $\Delta$. To this end, we introduce the notion of *feasibility graph*. A feasibility graph is basically a cyclic dependence graph incorporating the additional constraints on execution imposed by $\mu$. For any two conceptual operations $\overline{op}, \overline{op}' \in O$, we define $f(\overline{op}, \overline{op}')$ as follows:

$$f(\overline{op}, \overline{op'}) = \begin{cases} 0 & \text{if } \mu_s(op') \geqslant \mu_c(op) + \Phi(op, op') \\ \mu_c(op) + \Phi(op, op') - \mu_s(op') & \text{otherwise} \end{cases}$$

Let $\Delta > 0$ be some real. The feasibility graph $F_\Delta = (O, E, d\Delta - f)$ of $\mu$ with respect to $T_c$ is defined as a weighted parametric graph whose vertex set is the set of conceptual operations of $T_c$, such that every two vertices $\overline{op}$ and $\overline{op'}$ are connected by an edge $e$ iff $d(\overline{op}, \overline{op'}) < \infty$. The weight of $e$ is defined to be $\Delta d(\overline{op}, \overline{op'}) - f(\overline{op}, \overline{op'})$.

We can now state a sufficient condition for the existence of an indexing function $I$ and an initiation interval $\Delta$, such that $\mu^*_{\Delta, I}$ preserves $T_c$'s dependences.

**Theorem 11.4** (**Constructing periodic schedules**) *Let $T_c = (O \times \mathbb{Z}, \prec)$ be a cyclic task and $G$ any cyclic dependence graph for $T_c$. Then the following procedure yields a valid periodic schedule for $T_c$.*

(1) *Let $G'$ be any graph obtained from $G$ by deleting between 1 and $d(C)$ edges from every cycle $C$ of $G$, where $d(C)$ denotes the sum of the $d(e)$ for each edge $e$ in $C$. The graph $G'$ induces a unique dependence relation $\prec'$ on $O$.*
(2) *Let $T = (O, \prec')$ and $\mu$ any schedule preserving $T$'s dependences.*
(3) *Let $\Delta_{\min}$ be the smallest value of $\Delta$ such that the feasibility graph $F_\Delta$ of $\mu$ with respect to $T_c$ has no negative cycle. $\Delta_{\min}$ can be computed in polynomial time (Lawler 1967; Young et al. 1991).*

*Then there exists an indexing function $I$ computable in polynomial time in the size of $F_\Delta$ such that*

$$\forall \Delta \geqslant \Delta_{\min}, \quad \mu^*_{\Delta, I} \text{ preserves } T_c\text{'s dependences}$$

*Furthermore, $\Delta_{\min} \leqslant \max_{op, op'} \mu_c(op) + \Phi(op, op') - \mu_s(op')$.*

**Proof** Let us first show that $\Delta_{\min}$ exists and that

$$\Delta_{\min} \leqslant \max_{op, op'} \mu_c(op) + \Phi(op, op') - \mu_s(op').$$

Clearly if the feasibility graph $F_\Delta$ has a cycle then, for a sufficiently small positive value of $\Delta$, $F_\Delta$ will have a negative cycle. Thus, to show the existence of $\Delta_{\min}$, the smallest positive value of $\Delta$ such that $F_\Delta$ has no negative cycle, it suffices to exhibit a positive real $\Delta_0$ such that $F_{\Delta_0}$ has no negative cycle. Let $\Delta_0 = \max_{op, op'} \mu_c(op) + \Phi(op, op') - \mu_s(op')$; then $0 \leqslant f(\overline{op}, \overline{op'})/\Delta_0 \leqslant 1$ for any two conceptual operations $\overline{op}$ and $\overline{op'}$ of $T_c$.

Let $(\overline{op}_1, \overline{op}_2)$ be an edge in $F_{\Delta_0}$. Then $d(\overline{op}_1, \overline{op}_2) < \infty$, and, by the definition of cyclic dependence graphs, there must exist a path $P$ in $G$ from $\overline{op}_1$ to $\overline{op}_2$ such that $d(\overline{op}_1, \overline{op}_2) \geqslant d(P)$. If $f(\overline{op}_1, \overline{op}_2) > 0$ then the schedule $\mu$ does not respect the dependence from $\overline{op}_1$ to $\overline{op}_2$, and consequently at least one edge must have been deleted from path $P$. Let $C$ be a cycle in $F_{\Delta_0}$ comprising the edges $(\overline{op}_1, \overline{op}_2), \ldots, (\overline{op}_c, \overline{op}_1)$. Then, by the previous remark, for each $1 \leqslant i \leqslant c$, there must exist a path $P_i$ in $G$ from $\overline{op}_i$ to $\overline{op}_{1+i \bmod c}$ such that $d(\overline{op}_i, \overline{op}_{1+i \bmod c}) \geq d(P_i)$. Let $Q$ be the cycle in $G$ comprising all of the paths $P_i$. Then

$$\sum_{i=1}^{c} d(\overline{op}_i, \overline{op}_{1+i \bmod c}) \geq \sum_{(\overline{op},\overline{op}') \in Q} d(\overline{op}, \overline{op}') = d(Q).$$

Because there are at most $d(Q)$ edges deleted from cycle $Q$ and because every time $0 < f(\overline{op}_i, \overline{op}_{1+i \bmod c})/\Delta_0 \leq 1$ at least one edge has been deleted from $P_i$, we can write

$$\sum_{i=1}^{c} d(\overline{op}_i, \overline{op}_{1+i \bmod c}) \geq d(Q) \geq \sum_{i=1}^{c} f(\overline{op}_i, \overline{op}_{1+i \bmod c})/\Delta_0$$

As a result, every cycle $C$ in $F_{\Delta_0}$ has nonnegative weight, and $0 < \Delta_{\min} \leq \Delta_0$ is well defined.

Let $\Delta \geq \Delta_{\min}$. The existence of an indexing function $I$ such that $\mu_{\Delta,I}^*$ preserves $T_c$'s dependences is shown by a constructive argument.

Let $\overline{op}_0$ be some vertex in the feasibility graph $F_\Delta$. Since for $\Delta \geq \Delta_{\min}$, $F_\Delta$ contains no negative cycles, the shortest path from $\overline{op}_0$ to any other vertex $\overline{op}$ in $F_\Delta$ is well defined. Let us denote this shortest path by $SP(\overline{op})$. $SP$ can be computed in $O(nm)$ time, where $n$ and $m$ are the cardinalities of $F_\Delta$'s vertex and edge sets respectively, by employing the Bellman–Ford algorithm (Cormen et al. 1990). Let the index function $I = \lceil SP/\Delta \rceil$. Then for all conceptual operations $\overline{op}$ and $\overline{op}'$ the following indexing constraint is satisfied:

$$\Delta(I(op) - I(op')) \geq f(\overline{op}, \overline{op}') - \Delta d(\overline{op}, \overline{op}').$$

In fact, by virtue of the shortest-path inequality, we have

$$SP(\overline{op}) + \Delta d(\overline{op}, \overline{op}') - f(\overline{op}, \overline{op}') \geq SP(\overline{op}').$$

Since $f(\overline{op}, \overline{op}') \geq 0$ and for $y \geq 0$, $\lceil x+y \rceil \geq \lceil x \rceil + y$, we have

$$I(op) + d(op, op') \geq I(op') + f(op, op')/\Delta,$$

from which the above mentioned inequality is easily derived.

Using the previously established indexing inequality, we can show that $\mu_{\Delta,I}^*$ preserves $T_c$'s dependences. Let $\sigma = \mu_{\Delta,I}^*$. Because $\sigma$ is periodic for any two machine operations $op$ and $op'$ executing in $\mu$, we can write

$$\sigma_c(op[I(op)]) = \sigma_c(op[0]) + I(op)\Delta,$$
$$\sigma_s(op'[I(op')]) = \sigma_s(op'[0]) + I(op')\Delta.$$

If $\mu_c(op) + \Phi(op, op') \leq \mu_s(op')$ then $f_\Delta(op, op') = 0$ by definition, and

$$\sigma_s(op'[0]) - \sigma_c(op[0]) = \mu_s(op') - \mu_c(op) + \Delta(I(op) - I(op'))$$
$$\geq \Phi(op, op') - \Delta d(\overline{op}, \overline{op}')$$

If $\mu_c(op) + \Phi(op, op') > \mu_s(op')$ then

$$\sigma_s(op'[0]) - \sigma_c(op[0]) \geq \mu_s(op') - \mu_c(op) + f(\overline{op}, \overline{op'}) - \Delta d(\overline{op}, \overline{op'})$$
$$\geq \Phi(op, op') - \Delta d(\overline{op}, \overline{op'}).$$

Thus, by Theorem 11.2, the periodic schedule $\mu_{\Delta,I}^*$ preserves $T_c$'s dependence constraints. □

Note that the feasibility graph $F_\Delta$ may have a size quadratic in the number of conceptual operations, whereas $T_c$ could have a cyclic dependence graph with a much smaller edge set. If the flow function $\Phi$ verifies the triangle inequality, we can improve on this state of affairs. In fact, any dependence graph $G$ for $T_c$ can serve as the feasibility graph. More specifically, if for all machine operations $op, op'$ and $op''$ we have

$$\Phi(op, op') \leq \Phi(op, op'') + \Phi(op'', op')$$

then, by changing the weight of every edge $(\overline{op}, \overline{op'})$ in $G$ from $d(\overline{op}, \overline{op'})$ to $\Delta d(\overline{op}, \overline{op'}) - f(\overline{op}, \overline{op'})$, we can use $G$ instead of $F_\Delta$ as the feasibility graph for any schedule $\mu$. For conciseness, the proof of this property is left as an exercise.

We now show that the procedure stated in the previous theorem is not only sufficient but also necessary. More specifically, we have the following.

**Theorem 11.5 (All periodic schedules can be constructed using Theorem 11.4)** *Let $T_c = (O \times \mathbb{Z}, \prec)$ be a cyclic task, $G = (O, E)$ any cyclic dependence graph for $T_c$ and $\sigma$ a periodic schedule with initiation interval $\Delta$ respecting $T_c$'s dependences. Then*

(1) *for each dense core schedule $\mu$ of $\sigma$ there exists a graph $G'$ obtained from $G$ by deleting between 1 and $d(C)$ edges from every cycle $C$ of $G$ such that $\mu$ preserves the dependences embodied by $G'$;*
(2) *the feasibility graph $F_\Delta$ of $\mu$ with respect to $T_c$ has no negative cycle.*

**Proof** Let $\mu$ be any dense core schedule of $\sigma$. $G'$ is constructed from $G$ as follows. Let $(\overline{op}, \overline{op'})$ be an edge of $G$; if $\mu_c(op) + \Phi(op, op') > \mu_s(op')$ then we delete $(\overline{op}, \overline{op'})$ from $G$, otherwise we leave the edge untouched. By construction, $\mu$ respects the dependences embodied by $G'$; furthermore, at least one edge has been deleted in $G'$, or else for every operation $\overline{op}$ traversed by a cycle $Q$ in $G'$ we should have $\mu_c(op) \leq \mu_s(op)$, which implies that $\delta(op) = 0$, an impossibility.

Because of Theorem 11.2, we can write, for any two operations $op$ and $op'$ executing in $\sigma$,

$$\sigma_s(op'[0]) - \sigma_c(op[0]) = \mu_s(op') - \mu_c(op) + \Delta(I(op) - I(op'))$$
$$\geq \Phi(op, op') - \Delta d(\overline{op}, \overline{op'}).$$

Thus

$$\Delta(I(op) - I(op')) \geq \mu_c(op) + \Phi(op, op') - \mu_s(op') - \Delta d(\overline{op}, \overline{op'}).$$

If $\mu_c(op) + \Phi(op, op') - \mu_s(op') \geq 0$, we have
$$\Delta[I(op) - I(op')] \geq f(\overline{op}, \overline{op'}) - \Delta d(\overline{op}, \overline{op'});$$
otherwise, if $\mu_c(op) + \Phi(op, op') - \mu_s(op') < 0$, since $\mu$ is dense in $\sigma$, we have $\Delta \geq \max_{op} \mu_s(op) - \mu_s > \mu_s(op') - \mu_c(op) - \Phi(op, op')$, and therefore

$$\underbrace{I(op) - I(op')}_{\text{integer}} \geq \underbrace{\frac{\mu_c(op) + \Phi(op, op') - \mu_s(op')}{\Delta}}_{\in [0,-1[, \text{ since } \mu \text{ is dense in } \sigma} - \underbrace{d(\overline{op}, \overline{op'})}_{\text{integer}}$$

$$\geq \underbrace{0}_{=f(\overline{op}, \overline{op'})} - d(\overline{op}, \overline{op'}).$$

Thus in all cases $\Delta(I(op) - I(op')) \geq f(\overline{op}, \overline{op'}) - \Delta d(\overline{op}, \overline{op'})$. By summing this inequality over every cycle $Q$ of the feasibility graph $F_\Delta$ of $\mu$, we obtain $0 \geq f(Q) - \Delta d(Q)$, and therefore $F_\Delta$ contains no negative cycle.

To show that at most $d(C)$ edges have been deleted from any cycle $C$ of $G$, it suffices to remark that if the edge $(\overline{op}, \overline{op'})$ has been cut in $G$ then we have $\mu_c(op) + \Phi(op, op') - \mu_s(op') > 0$, and therefore

$$\underbrace{I(op) - I(op')}_{\text{integer}} \geq \frac{\mu_c(op) + \Phi(op, op') - \mu_s(op')}{\Delta} - \underbrace{d(\overline{op}, \overline{op'})}_{\text{integer}}$$

$$\geq 1 - d(\overline{op}, \overline{op'}),$$

and if the edge has not been cut then
$$I(op) - I(op') \geq -d(\overline{op}, \overline{op'}).$$

Again by summing these inequalities over the cycle $C$, we obtain the desired result. □

## 11.7 APPLICATIONS AND CONCLUSIONS

By combining the results in Sections 11.5 and 11.6, we can derive a powerful methodology to convert a cyclic scheduling problem into an acyclic one. More precisely, assume that we are given a machine model $M$ as described in Definition 11.1 and a cyclic task $T_c$. The goal is to generate an $M$-admissible schedule $\sigma$ which respects $T_c$'s dependences. As stated in Section 11.1, it would be highly desirable to transform the cyclic task $T_c$ into an acyclic one, say $T$, such that the approximation results obtained for $T$ could be transposed to $T_c$. The previous section shows how to carry out such a transformation. More precisely, given any cyclic dependence graph $G$ for $T_c$, we suggest doing the following.

(1) Cut as many edges from $G$ as allowed by Theorem 11.4 so as to "maximize the parallelism" in the derived graph $G'$.

(2) Generate an $M$-admissible schedule $\mu$ that is as close as possible to the optimum $M$-admissible schedule for the acyclic task system $T$ whose dependences are embodied by $G'$. Suppose that $\mu$ is within a factor $k_a$ from the optimum.
(3) Find the smallest initiation interval $\Delta$ that satisfies both Theorem 11.4 and Corollary 11.1. Note that $\Delta = \max_{\text{op,op}'} \mu_c(\text{op}) + \Phi(\text{op},\text{op}') - \mu_s(\text{op}')$ satisfies both.
(4) Compute the index function $I$ as suggested in the proof of Theorem 11.4. The periodic schedule $\mu^*_{\Delta,I}$ is both $M$-admissible and preserves $T_c$'s dependences.

Note that because of Theorem 11.5, every $M$-admissible periodic schedule that preserves $T_c$'s dependences can be constructed in this fashion.

Let $\Phi_{\max} = \max_{\text{op,op}'} \Phi(\text{op},\text{op}')$, $\delta_{\max} = \max_{\text{op}} \delta(\text{op})$ and $k_c$ be the ratio between the length of the best possible acyclic schedule that can be generated from $G'$ and the length of any dense core schedule of the optimum periodic schedule for our cyclic problem. Alternatively, let $k_c = 1$ if the cutting phase guarantees that the minimality of the dependence criterion employed in the proof for the approximate factor $k_a$ has been preserved by the cutting edge algorithm. If the desired periodic schedules are restricted to be nonpreemptive, we can guarantee a factor of $k_c k_a + (\delta_{\max} + \Phi_{\max})/\Delta_{\text{opt}}$, where $\Delta_{\text{opt}}$ is the optimum initiation interval for our case. If we are allowed to have a preemptive schedule then the bound becomes $k_c k_a + [(\mu_c - \max_{\text{op}} \mu_s(\text{op})] + \Phi_{\max})/\Delta_{\text{opt}}$.

The general question of cutting edges is of course entangled with operation selection. However, this problem does not arise if the choice of the machine operation with which a conceptual operation should be implemented can be made locally, i.e. it does not depend on the operation selection for other conceptual operations.

We have applied the technique described in points (1)–(4) above to a particular instance of the machine model described in Definition 11.1. The actual machine model comprises $p$ identical pipelined or unpipelined processors, and incurs no communication costs; that is, for all machine operations op and op', $\Phi(op,op') = 0$. By being careful in the edge cutting phase, we have been able to guarantee a worst case factor from optimality of roughly 2 (Gasperoni and Schwiegelshohn 1994).

## ACKNOWLEDGMENTS

The work of Franco Gasperoni was partially performed at New York University and was supported by Air Force Contract F08630-92-C0019.

## REFERENCES

Coffman E. G. (1976) *Computer and Job-shop Scheduling*, Wiley, New York.
Cormen T. H., Leiserson C. E. and Rivest R. L. (1990) *Introduction to Algorithms*, MIT Press, Cambridge, MA, and McGraw-Hill New York.
Gasperoni F. and Schwiegelshohn U. (1994) Generating close to optimum loop schedules on parallel processors, *Parallel Processing Lett.*, **4** (4).

Graham R. L., Lawler E. L., Lenstra J. K. and Rinnooy Kan A. H. G. (1979) Optimization and approximation in deterministic sequencing and scheduling: a survey, *Ann. Discr. Math.* 287–326.

O'kEigearthaich M. , Lenstra J. K. and Rinnooy Kan A. H. G. *Combinatorial Optimization— Annotated Bibliographies*, pp. 165–189, Wiley, New York (1985).

Lawler E. L. (1967) Opatimal cycles in doubly weighted directed lineargraphs. In Rosenstiehl P., editor, *Theory of Graphs: An International Symposium*, pp. 209–213 Gordon and Breach, New York.

Serafini, P. and Ukovich W. (1989) A mathematic model for periodic scheduling problems *SIAM J. Discr. Math.* **2**, 550–581.

Young, N. E. Tarjan R. E. and Orlin J. B. (1991) Faster parametric shortest path and minimum-balance algorithms, *Networks*, **21**, 205–221.

---

*Telecom Paris, ENST, 46 Rue Barrault, 75634 Paris Cedex 13, France*
gasperon@inf.enst.fr

---

*Institute for Information Technology Systems, University of Dortmund, 44221 Dortmund, Germany*
uwe@carla.e-technik.umi-dortmund.de

CHAPTER 12

# Recent Advances in Branch-and-Bound Procedures for Resource-Constrained Project Scheduling Problems

**Willy S. Herroelen and Erik L. Demeulemeester**
*Department of Applied Economic Sciences, Katholieke Universiteit Leuven*

**Abstract**

We discuss some recent advances in branch-and-bound algorithms for optimally solving three types of resource-constrained project scheduling problems: the classical resource-constrained project scheduling problem (RCPSP), the generalized resource-constrained project scheduling problem (GRCPSP) and the preemptive resource-constrained project scheduling problem (PRCPSP). The RCPSP involves the scheduling of a project to minimize its duration subject to zero-lag finish–start precedence constraints of PERT/CPM type and constant availability constraints on the required set of renewable resources. We discuss recent advances in dealing with this problem using a new depth-first branch-and-bound procedure. We then discuss how the procedure can be extended for solving the GRCPSP, which extends the RCPSP to the case of the precedence diagramming type of precedence constraints, activity ready times and due dates and variable resource availabilities. Finally, we elaborate on the extension of the procedure to the PRCPSP, which relaxes the nonpreemption condition of the RCPSP, thus allowing activities to be interrupted at integer points in time and resumed later without any additional penalty cost.

## 12.1 INTRODUCTION

The objective of this chapter is to discuss some recent advances in branch-and-bound algorithms for optimally solving three types of resource-constrained project

*Scheduling Theory and its Applications* Edited by P. Chrétienne, E. G. Coffman, Jr., J. K. Lenstra and Z. Liu
© 1995 John Wiley & Sons Ltd

scheduling problems: the classical resource-constrained project scheduling problem (RCPSP), the generalized resource-constrained project scheduling problem (GRCPSP) and the preemptive resource-constrained project scheduling problem (PRCPSP).

The classical *resource-constrained project scheduling problem (RCPSP)* involves the scheduling of a project to minimize its total duration subject to zero-lag finish–start precedence constraints of the PERT/CPM type and constant availability constraints on the required set of renewable resources. The RCPSP can be formulated as follows (Christofides *et al.* 1987; Demeulemeester 1992; Demeulemeester and Herroelen 1992*a*):

minimize $f_n$ (12.1)

subject to

$$f_1 = 0 \quad (12.2a)$$

$$f_i \leq f_j - d_j, \quad (i,j) \in H, \quad (12.2b)$$

$$\sum_{i \in S_t} r_{ik} \leq a_k, \quad t = 1, 2, \ldots, f_n, \quad k = 1, 2, \ldots, K. \quad (12.3)$$

The *problem parameters* are as follows: $H$ is the set of pairs of activities indicating precedence constraints, $d_i$ is the processing time of activity $i$, $r_{ik}$ is the amount of resource type $k$ required by activity $i$, and $a_k$ is the total availability of resource type $k$. The *problem output* consists of the following: $f_i$ is the finish time of activity $i$ ($i = 1, 2, \ldots, n$) and $S_t = \{i : f_i - d_i < t \leq f_i\}$ is the set of activities in progress in the time interval $]t - 1, t]$.

It is assumed that activity $i$ has a fixed processing time $d_i$ (setup times are negligible or are included in the processing time). We further assume activity-on-the-node networks where activities 1 and $n$ are dummy activities indicating the single start and end nodes of a project respectively. The resource requirements $r_{ik}$ are known constants over the processing interval of the activity. The availability of resource type $k$, $a_k$, is also a known constant during the project duration interval.

Equation (12.2a) assigns a completion time of 0 to dummy start activity 1. The precedence constraints given by (12.2b) indicate that an activity $j$ can only be started if all predecessor activities $i$ are completed. Once started, activities run to completion (nonpreemption condition). The resource constraints given in (12.3) indicate that for each time period $]t - 1, t]$ and for each resource type $k$, the resource amounts required by the activities in progress cannot exceed the resource availability. The objective function is given by (12.1). The project duration is minimized by minimizing the finishing time of the unique dummy ending activity $n$.

The RCPSP is a generalization of the well-known job-shop scheduling problem and as such is $\mathcal{NP}$-hard (Blazewicz *et al.* 1983). Reviews of the RCPSP have been given by Davis (1966, 1973), Herroelen (1972) and Patterson (1984).

In the *generalized resource-constrained project scheduling problem (GRCPSP)* three of the basic assumptions of the RCPSP are relaxed. First, the GRCPSP allows for precedence relations of the precedence diagramming type, namely start–start, finish–finish, finish–start and start–finish constraints, with the only restriction being that activities are not allowed to start before one of their predecessors has started. Secondly, ready times and due dates may be specified for each activity. An activity cannot be started earlier than its ready time and must be completed by its due date. Last but not least, the resource availabilities may be variable over the project horizon.

Conceptually, the GRCPSP problem may be formulated as follows:

minimize $f_n$ (12.4)

subject to

$$f_i - d_i + SS_{ij} \leq f_j - d_j \quad \text{for all } (i, j) \in H_1, \quad (12.5)$$
$$f_i - d_i + SF_{ij} \leq f_j \quad \text{for all } (i, j) \in H_2, \quad (12.6)$$
$$f_i + FS_{ij} \leq f_j - d_j \quad \text{for all } (i, j) \in H_3, \quad (12.7)$$
$$f_i + FF_{ij} \leq f_j \quad \text{for all } (i, j) \in H_4, \quad (12.8)$$
$$f_1 = 0 \quad (12.9)$$
$$f_i - d_i \geq g_i, \quad i = 1, 2, \ldots, n, \quad (12.10)$$
$$f_i \leq h_i, \quad i = 1, 2, \ldots, n, \quad (12.11)$$
$$\sum_{i \in S_t} r_{kt} \leq a_{kt}, \quad k = 1, 2, \ldots, K, \quad t = 1, 2, \ldots, f_n, \quad (12.12)$$

where $H_1$ is the set of pairs of activities indicating start–start relations with a time lag of $SS_{ij}$, $H_2$ is the set of pairs of activities indicating start–finish relations with a time lag of $SF_{ij}$, $H_3$ is the set of pairs of activities indicating finish–start relations with a lag of $FS_{ij}$, $H_4$ is the set of pairs of activities indicating finish–finish relations with a lag of $FF_{ij}$, $g_i$ is the ready time of activity $i$, $h_i$ is the due date of activity $i$ and $a_{kt}$ is the availability of resource type $k$ during the period $]t - 1, t]$.

The objective function (12.4) is to minimize the project duration by minimizing the finish time of the unique dummy end activity $n$. Equations (12.5)–(12.8) ensure that the various types of precedence constraints are satisfied. Equation (12.9) assigns the dummy start activity 1 a completion time of 0. Equations (12.10) guarantee that the ready times are respected, while (12.11) guarantee that no due dates are violated. Equation (12.12) specify that the resource utilization during any time interval $]t - 1, t]$ does not exceed the resource availability levels during that time interval for any of the resource types.

There seems to be no extensive literature on the GRCPSP. Bartush et al. (1988) report on computational results with a branch-and-bound procedure that is applicable to the case with arbitrary precedence constraints involving minimal and maximal time lags between starting times and completion times of any two activities, different resource types whose availability may change in discrete jumps over time, and resource requirements per activity involving several types and amounts that may vary in discrete jumps with the processing of each activity. They mention that due dates and release times can be represented by the model. The case of variable resource availabilities over time has been dealt with by Simpson (1991).

The *preemptive resource-constrained project scheduling problem (PRCPSP)* allows activities to be preempted at integer points in time; i.e. the fixed duration $d_i$ of an activity may be split into $j = 1, 2, \ldots, d_i$ duration units. Each duration unit $j$ of activity $i$ is then assigned a finish time $f_{i,j}$. The variable $f_{i,0}$ denotes the earliest time that an activity $i$ can be started. As only finish–start relations with a time lag of zero are allowed, $f_{i,0}$ equals the latest finish time of all the predecessors of activity $i$. An activity $i$ belongs to the set $S_t$ of activities in progress at time $t$ if one of its duration units $j = 1, 2, \ldots, d_i$ finishes at time $t$ (i.e. if $f_{i,j} = t$). The PRCPSP can now be conceptually formulated as follows (Demeulemeester 1992):

minimize $\quad f_{n,0}$ \hfill (12.13)

subject to

$$f_{i,d_i} \leq f_{j,0} \quad \text{for all } (i, j) \in H, \tag{12.14}$$

$$f_{i,j-1} + 1 \leq f_{i,j} \quad i = 1, \ldots, n, \quad j = 1, \ldots, d_i, \tag{12.15}$$

$$f_{1,0} = 0, \tag{12.16}$$

$$\sum_{i \in S_t} r_{ik} \leq a_k, \quad k = 1, \ldots, f_{n,0}. \tag{12.17}$$

The objective function (12.13) minimizes the project length by minimizing the earliest start time of the dummy end activity, which by assumption has a duration of 0. Equations (12.14) ensure that all precedence relations are satisfied; the earliest start time of an activity $j$ cannot be smaller than the finish time of the last unit of duration of its predecessor $i$. Equations (12.15) specify that the finish time for every unit of duration of an activity has to be at least one time unit larger than the finish time for the previous unit of duration. Activity 1 is assigned an earliest start time of 0 through (12.16), while (12.17) stipulate the resource constraints. The size of the formulation given by (12.13)–(12.17) is pseudopolynomial in problem size.

Slowinski (1980) and Weglarz (1981) have presented optimal solution procedures for the case of continuous processing times for the different activities. Davis and Heidorn (1971) developed an implicit enumeration scheme based on the splitting of activities in unit-duration tasks. Kaplan (1988, 1991) presented a dynamic program formulation and suggested a solution by a reaching procedure.

In the next section we start by giving a survey of the optimal procedures for the RCPSP, and continue with a description of the basic principles of a recently developed depth-first branch-and-bound scheme (Demeulemeester 1992; Demeulemeester and Herroelen 1992a). In Section 12.3 we discuss extensions of the basic scheme to the GRCPSP. Extensions to the PRCPSP are reserved for Section 12.4. Section 12.5 deals with overall conclusions.

## 12.2 BRANCH-AND-BOUND FOR THE RCPSP

### 12.2.1 Optimal procedures for the RCPSP

The RCPSP has been extensively studied. Previous research on optimal procedures basically involved the use of integer linear programming procedures and implicit enumeration (dynamic programming and branch-and-bound).

*Integer linear programming procedures* have been presented by Bowman (1959), Brand *et al.* (1964), Elmaghraby (1967), Moodie and Mandeville (1966), Patterson and Huber (1974), Patterson and Roth (1976), Pritsker *et al.* (1969) and Wiest (1964). However, the general conclusion can be drawn that a frontal attack on the RCPSP by applying integer linear programming is not efficient.

Numerous *implicit enumeration procedures* have been developed for solving the RCPSP. Carruthers and Battersby (1966) and Petrovic (1968), among others, tried to use *dynamic programming*, but the "dimensionality curse" precludes the use of dynamic programming as a practical, efficient solution method. Other implicit enumeration procedures have been developed by Balas (1971), Bell and Park (1990), Carlier and Latapie (1991), Christofides *et al.* (1987), Davis and Heidorn (1971), Demeulemeester and Herroelen (1992a), Johnson (1967), Radermacher (1985), Schrage (1970), Stinson *et al.* (1978) and Talbot and Patterson (1978).

Johnson (1967) was among the first to develop an implicit enumeration algorithm for the RCPSP involving one single resource. Limited computational experience indicated that his procedure could solve 50-activity projects within about 5 min of CPU-time on an IBM 360 Model 65.

Davis (1968) presented a different scheme applicable to the RCPSP involving multiple resources and variable levels of activity resource requirements. The approach required job durations to be represented as a series of unit-duration tasks, making the computational effort dependent upon the sum of activity durations of the project. Davis and Heidorn (1971) developed an IBM 360-programmed version of the procedure capable of handling projects with a total duration sum of 220 units and up to 5 resource types per activity and project.

Talbot and Patterson (1978) developed an implicit enumeration procedure involving a systematic enumeration of all possible finish times for the activities of a project. Most noteworthy in their algorithm is the concept of a *cut*, a strong fathoming rule used to eliminate from explicit consideration possible inferior activity completion times earlier in the enumeration phase of the algorithm. From their computational experience, the authors conclude that the likelihood of obtaining an optimal solution for projects containing more than 50 activities within a reasonable amount of computation time is low.

Stinson *et al.* (1978) developed a best-first branch-and-bound procedure based on strong bounding criteria and dominance rules. The nodes of the branch-and-bound tree correspond to partial schedules based on semi-active timetabling; i.e. activities are started as early as possible, satisfying both precedence and resource constraints. The bounding arguments involve the computation of a precedence-based lower bound, a resource-based bound and a critical sequence bound. The precedence-based lower bound is computed as the remaining critical path length. The resource-based bound is computed as the maximum ratio over all resource types of the sum of the resource duration products over all activities and the resource availability. The critical sequence bound takes into account both precedence and resource constraints. The computational results (Stinson et al. 1978) indicate that the program can solve most (56 out of 60) 43-activity problems with 4 resource types within an average CPU-time of 5.84 min on an IBM 370-155. The constraining factor in solving the RCPSP with the Stinson procedure is the memory requirement, which seems to be an exponential function of the number of activities.

Patterson (1984) conducted a computational experiment in which he compared the solution methods of Davis and Heidorn (1971), Talbot and Patterson (1978) and Stinson et al. (1978). The three algorithms were coded in Fortran V for use on an Amdahl 470/V8 computer. A total of 110 test problems were gathered from the literature, involving from 7 to 50 activities and from 1 to 3 resource types. The computational results show that while the implicit enumeration procedure of Talbot and Patterson required very little primary computer storage, only the branch-and-bound procedure of Stinson et al. was able to solve all 110 test problems within the time limit of 5 minutes per problem (as compared to 96 (Davis and Heidorn) and 97 (Talbot and Patterson)). Moreover, the average CPU time required to solve a problem to optimality was much smaller (0.82 s compared with 14.02 s (Davis and Heidorn) and 14.98 s (Talbot and Patterson)).

Radermacher (1985) developed a structural approach to the RCPSP based on an order-theoretic approach (Möhring and Radermacher 1989) that views project analysis and scheduling as combinatorial optimization problems over partially ordered sets, and is related to the disjunctive graph concept introduced by Balas (1971). The basic idea of the model is to replace schedules, which depend on the activity durations, by partial orders, which are independent of the durations. The procedure starts with the original activity-on-the-node network and iteratively introduces new relations between suitable pairs of jobs, aiming at the iterative destruction of all *forbidden sets*. Forbidden sets are sets of activities that cannot be scheduled together because of the resource constraints. The different choices yield the branches of the branch-and-bound tree, with all nodes corresponding to feasible partial ordered sets. The objective value associated with the occurring partial ordered sets is obtained by an early-start schedule and is monotonically increasing with respect to the introduction of additional precedence constraints. It forms a lower bound in the sense that the shortest project length of any early-start schedule over all feasible partial orders constitutes the optimal project length. Important speed-ups are obtained by reduction techniques that reduce the dimension of the problem to the number of essential jobs that actually require scarce resources (are in some forbidden set) or whose completion time is essential in the objective function. Further reductions are obtained by time windows that restrict the choice of precedence constraints on a forbidden set (Möhring and Radermacher 1989).

Most computational experience has been obtained on different constellations of a bridge construction project. The average CPU time for a total of 192 instances of this problem involving 16 nondummy activities and 4 resource types was 8.952 s on a CYBER 175. The maximum computation time occurred on an instance with resource availabilities 2/2/2/1, requiring a CPU time of 327.187 s. We solved the same instance using a PC version of the Stinson *et al.* code on an IBM PS/2 Model 70 in 0.50 s. This seems to confirm that the Radermacher procedure works very well when the number of reduced forbidden sets is not too large and these sets are small. This is usually the case when there are many temporal constraints. When the problem has a large degree of parallelism, i.e. if there are many (large) forbidden sets or if there are not many temporal constraints, establishing optimality of a solution will require too much time (see also Bartusch *et al.* 1988).

Christofides *et al.* (1987) (*CAT*) have developed a depth-first branch-and-bound procedure that generates a branch-and-bound tree, whose nodes correspond to semi-active partial schedules that are feasible, satisfying both the precedence and resource constraints. Partial schedule decisions are made at time instants $m$ corresponding to the completion time of one or more project activities. The procedure only branches in order to resolve a resource conflict. At every time instant $m$, eligible (candidate) activities are defined as those activities that are not yet in the partial schedule and whose predecessor activities have finished. These activities can start at time $m$ if the resource constraints are not violated. The candidates are sorted in decreasing order of $L(i)$, the remaining critical path length from activity $i$ to the end of the project, and considered in that order. The candidate activity is either put in progress or, if it cannot be scheduled within the resource constraints, a *resource conflict* occurs. The candidate is entered into a conflict set. For a conflict activity $i$ the set of *delaying alternatives* $A(i)$ consists of the possible minimal combinations of activities, the delay of which would allow the conflict activity to be scheduled. The combinations are *minimal* in the sense that no alternative in the set $A(i)$ is a subset of another alternative in that set. $S^*$ is then constructed as the set of all activities that belong to at least one alternative in the set $A(i)$. They determine how to delay conflict activity $i$ using the earliest finishing activity $a^*$ in the set $S^*$ as a predecessor. This leads to a corresponding new node in the solutions tree. The remaining branching alternatives are generated by taking for each alternative $A$ in the set $A(i)$ precisely that activity $a'$ that has the smallest finishing time among the activities in $S^* - A + \{i\}$.

The branching alternative is always to delay the candidate (depth-first). The other delaying alternatives are to be considered during *backtracking*, which occurs when a schedule is completed or fathomed by the lower bound calculation. At the backtrack the added arcs, corresponding to the last alternative studied, are removed and new arcs are added for the next one at the same level. If there is no alternative left at that level, the procedure backtracks to the previous level. The alternative $A$ with most elements is branched on first (ties are broken arbitrarily). The process is finished when level zero is reached. In order to calculate the *precedence-based lower bound LB0*, each arc added at level $p$ in the solutions tree is considered. One arc is added for each delayed activity. The path length through each added arc is calculated, and $LB0$ is the longest of these paths. If $LB0$ fails to fathom the partial schedule, an attempt is made to strengthen the bound. The disjunctive arcs added to levels earlier than the current level $p$ are searched. The idea is to find an earlier added arc whose source node is the same as the destination node of the arc(s) just added. If one exists, the path through these linked disjunctive arcs may be longer than the $L(i)$ originally used in computing the $LB0$. The path length through the linked disjunctive arcs is calculated and the partial schedule fathomed if $LB0$ is greater than or equal to the current best feasible schedule length $T$.

Computational experience obtained with the CAT algorithm by Demeulemeester (1992) and Demeulemeester *et al.* (1994) on the 110 Patterson problems indicates that the computation times do not compared favorably with those obtained with the Stinson *et al.* procedure. The program solved 109 problems in an average time of 54.635 s (with a standard deviation of 193.565 s) on an IBM 3090. Remember that the Stinson *et al.* code requires an average of 0.82 s to solve all 110 test problems on an Amdahl 470/V8. This indicates that the CAT claim that their algorithm is competitive with the Stinson *et al.* procedure cannot be confirmed. In addition, it is shown in Demeulemeester *et al.* (1994) that the CAT algorithm may occasionally fail to generate the optimal solution.

Bell and Park (1990) have developed an $A^*$ *algorithm* that is a best-first search algorithm, where the first feasible solution found is guaranteed to be optimal. Each *state* in the search tree is a set of precedence constraints expressed in activity-on-the-node network form. From such a state, one can compute the early-start time schedule. The *initial state* network contains only those precedence constraints expressed in the original problem description. A *goal state* is the minimal makespan network that satisfies resource constraints as well as precedence constraints. Imposing new disjunctive constraints allows one to generate the children of a state (one child for each disjunct). A parent state is transformed into a child state by the addition of exactly one arc to the network. At each step of the $A^*$ search process, the state with the shortest makespan is selected. If the selected state has no resource conflict then an optimal schedule is found. Otherwise, the selected state is expanded by discovering and using a disjunctive constraint to generate the state's successors.

In order to do so, *minimum resource violating sets (MRVS)* of activities are identified such that:

(1) all activities in a MRVS are executed concurrently under the state's schedule;
(2) the use of some resource exceeds its availability;
(3) an MRVS has no proper resource-violating subsets.

The algorithm looks for the smallest time $t$ such that a resource-violating set of activities are in progress in $]t, t+1]$. This value of $t$ is called the *RVST (resource-violating set time)*. Appropriate disjunctive arcs are introduced to break up the MRVS. Two dominance rules are used: the already-mentioned left-shift dominance rule and a second rule that is used to dominate state $S'$ by state $S$ if

(1) $S$ and $S'$ have the same scheduled set of activities;
(2) the precedence networks of $S$ and $S'$ have the same subsets of arcs connecting pairs of unscheduled activities;

(3) the schedule of $S$ has start times for unscheduled activities that are less than or equal to the corresponding start time in the schedule of $S'$.

The algorithm has been coded for an Apple Macintosh Plus using a version of Common Lisp. Among the 110 Patterson test problems, 75 problems were solved in les than 300 s and 96 problems were solved in less than 600 s. However, 8 problems in the set could not be solved optimally within 1000 s.

Carlier and Latapie (1991) have presented a novel and very ingenious branch-and-bound procedure that extends earlier work on the interval method (Carlier 1987) for the job shop scheduling problem. An execution interval is associated with each activity based on its computed ready time and due date. Branching consists in choosing a task and splitting its interval of execution into two intervals, whose union gives the original interval again. For each node in the search tree, an upper bound is obtained by computing a single-machine Jackson preemptive schedule (Carlier and Pinson 1988), which gives priority to the available task with maximal remaining critical path length. In addition, a lower bound is computed by solving $m$-machine problems generated from the initial problem formulation by relaxing some of the constraints. The authors have implemented both depth-first and breadth-first branching strategies. The depth-first strategy seems to outperform the breadth-first procedure on the larger problem instances. During the course of the algorithm, no extra arcs are added: only the execution interval is replaced by two intervals by changing only two parameters during branching.

The program has been coded in Pascal for the VAX 8530 computer. It appears that on some of the Patterson test problems the procedure does not perform very well. For one of the Patterson problems, optimality could not be established after 2 h of CPU time on a VAX 8530. However, the procedure is very promising for flow shop problems and RCPSP instances characterized by high degrees of resource constrainedness.

Recently, Demeulemeester and Herroelen (1992a) have extended the CAT procedure. Their depth-first procedure seems to be the fastest exact solution method for solving the RCPSP.

### 12.2.2 The Demeulemeester–Herroelen (DH) procedure

The *DH-procedure* (Demeulemeester 1992; Demeulemeester and Herroelen 1992a) generates a search tree whose nodes correspond to partial schedules in which finish times have been temporarily assigned to a subset of the activities of the project. The partial schedules are feasible, satisfying both the precedence and resource constraints. *Partial schedules $PS_m$* are only considered at those time instants $m$ corresponding to the completion time of one or more project activities. The partial schedules are constructed by semi-active timetabling. In other words, each activity is started as soon as it can within the precedence and resource constraints. A partial schedule $PS_m$ at time $m$ thus consists of the set of temporarily scheduled activities. Scheduling decisions are temporary in the sense that temporarily scheduled activities may be delayed as a result of decisions made at later stages in the search process. Partial schedules are built up starting at time 0 and proceed systematically throughout the search process by adding at each decision point subsets of activities, including the empty set, until a complete feasible schedule is obtained. In this sense, a complete schedule is a *continuation of a partial schedule*.

At every time instant $m$ we define the *eligible set* $E_m$ as the set of activities that are not in the partial schedule and whose predecessor activities have finished. These eligible activities can start at time $m$ if the resource constraints are not violated. Demeulemeester and Herroelen (1992a) have proved two theorems that allow the procedure, at decision point $m$, to decide on which eligible activities must be scheduled by themselves, and which pair of eligible activities must be scheduled concurrently.

**Theorem 12.1** *If at time $m$ the partial schedule $PS_m$ has no activity in progress and an eligible activity $i$ cannot be scheduled together with any other unscheduled activity at any time instant $m' \geq m$ without violating the precedence and resource constraints then there exists an optimal continuation of the partial schedule with the eligible activity $i$ put in progress (started) at time $m$.*

**Theorem 12.2** *If at time $m$ the partial schedule $PS_m$ has no activity in progress, if there is an eligible activity $i$ that can be scheduled concurrently with only one other unscheduled activity $j$ at any time instant $m' \geq m$ without violating precedence or resource constraints, and if activity $j$ is both eligible and not longer in duration than activity $i$, then there exists an optimal continuation of the partial schedule in which both activities $i$ and $j$ are put in progress at time $m$.*

If it is impossible to schedule all activities at time $m$, a *resource conflict* occurs, which will produce a new branching in the branch-and-bound tree. The branches describe ways to resolve the resource conflict by deciding on which combinations of activities are to be delayed. A *delaying set* $D(p)$ consists of all subsets of activities $D_q$, either in progress or eligible, the delay of which would resolve the current resource conflict at level $p$ of the search tree. A *delaying alternative* $D_q$ is minimal if it does not contain other delaying alternatives $D_v \in D(p)$ as a subset. Demeulemeester and Herroelen (1992a) have proved that in order to resolve a resource conflict, it is sufficient to consider only minimal delaying alternatives.

One of the minimal delaying alternatives (nodes in the search tree) is arbitrarily chosen for branching. The delay of a delaying alternative $D_q$ is accomplished by adding a *temporal constraint* causing the corresponding activities to be delayed up to the *delaying point*, which is defined as the earliest completion of an activity in the set of activities in progress that does not belong to the delaying alternative. The delayed activities are removed from the partial schedule and the set of activities in progress, and the algorithm continues by computing a new decision point.

The search process continues until the dummy end activity has been scheduled. Every time such a complete schedule has been found, *backtracking* occurs: a new delaying alternative is arbitrarily chosen from the set of delaying alternatives $D(p)$ at the highest level $p$ of the search tree that still has some unexplored delaying alternatives left, and branching continues from that node. When level zero is reached in the search tree, the search process is completed.

Two *dominance rules* are used to prune the search tree. The first is a variation of the well-known left-shift dominance rule, and can be stated as follows.

**Theorem 12.3** *If the delay of the delaying alternative at the previous level of the branch-and-bound tree forced an activity i to become eligible at time m, if the current decision is to start activity i at time m, and if activity i can be left-shifted without violating the precedence or resource constraints (because activities in progress were delayed), then the corresponding partial schedule is dominated.*

The second dominance rule is based on the concept of a cutset. At every time instant $m$ a cutset $C_m$ is defined as the set of unscheduled activities for which all predecessor activities belong to the partial schedule $PS_m$. The proof of the following theorem can be found in Demeulemeester (1992) and Demeulemeester and Herroelen (1992a):

**Theorem 12.4** *Consider a cutset $C_m$ at time m that contains the same activities as a cutset $C_k$ that was previously saved during the search of another path in the search tree. If time k was not greater than time m and if all activities in progress at time k did not finish later than the maximum of m and the finish time of the corresponding activities in $PS_m$ then the current partial schedule $PS_m$ is dominated.*

The procedure has been tested with three *lower bounding rules*. The well-known remaining critical path length bound and critical sequence lower bound (Stinson et al. 1978) are supplemented by an extended critical sequence lower bound that is computed by repetitively looking at a path of unscheduled, noncritical activities in combination with a critical path. The extended critical sequence lower bound calculation starts by calculating the Stinson critical sequence lower bound. This allows us to determine which activities cannot be scheduled within their slack time. Subsequently, all paths consisting of at least two unscheduled, noncritical activities, which start and finish with an activity that cannot be scheduled within its slack time, are constructed. A simple type of dynamic programming then allows us to calculate the extended critical sequence bound for every noncritical path.

The branch-and-bound procedure has been programmed in Turbo C Version 2.0 for a personal computer IBM PS/2 Model 70 A21 (or compatibles) running under the DOS operating system. The procedure solves all the 110 Patterson test problems in an average CPU time of 0.204 s and a standard deviation of 0.450 s. Demeulemeester (1992) directly translated the Stinson et al. procedure from Fortran to C. The resulting code required an average of 2.494 s to solve all 110 problems, with a standard deviation of 3.762 s. This indicates that the DH procedure is on the average more than 12 times faster than that of Stinson et al. making it the fastest exact procedure on this problem set so far available. Looking at the 10 hardest problems, one observes that the DH procedure outperforms the Stinson *et al.* code in terms of CPU-time by factor of more than 13.

## 12.3 EXTENSION OF THE DH PROCEDURE TO THE GRCPSP

In order to extend the DH procedure described in the previous section to the GRCPSP, all precedence constraints are converted to finish–start precedence relations using the following conversion formula:

$$FS'_{ij} = \max \{SS_{ij} - d_i, SF_{ij} - d_i - d_j, FS_{ij}, FF_{ij} - d\}. \quad (12.18)$$

This eliminates the well-known anomalies (De Wit and Herroelen 1990; Moder *et al.* 1983) that can be entailed by the use of precedence diagramming (e.g. a project duration increase as the result of decreasing activity durations).

The *ready time* $g_i$ of an activity $i$ can easily be transformed into a finish–start relation between the dummy start activity 1, which starts and finishes at time 0, and activity $i$ itself:

$$FS'_{1i} = \max \{g_i, FS_{1i}\}. \quad (12.19)$$

Coping with *due dates* $h_i$ is somewhat more involved. For every activity $j$, a latest allowable start time $ls_j$ has to be computed such that whenever this activity $j$ is delayed to start later than $ls_j$, the due date of this activity or of one of its direct or indirect successors is exceeded, even if all subsequent activities were scheduled as soon as possible without considering the resource constraints. Consequently, if during the branch-and-bound procedure an activity $j$ is assigned an early start time $s_j$ that exceeds its latest allowable start time $ls_j$, backtracking can occur, since no feasible solution can be found by continuing the search from this partial schedule.

As before, we define $S_t$ as the *set of activities in progress* during the time interval $]t - 1, t]$, $PS_t$ as the *partial schedule* that contains the set of activities that have been assigned a finish time at time $t$, and the *cutset* $C_t$ as the set of all unscheduled activities whose predecessors all belong to the partial schedule $PS_t$. The *eligible set* $E_t$ then denotes the set of all activities that belong to the cutset $C_t$ and that can start at time $t$. The precise time instant at which these sets are defined will be clear from the context; hence the subscripts will be omitted for simplicity of notation.

The search process (Demeulemeester 1992; Demeulemeester and Herroelen 1992b) starts by adding the dummy start activity 1 to $S$ and $PS$ with a finish time $f_1 = 0$. All activities $i$ that have activity 1 as a single predecessor are added to the cutset and are assigned an early start time, based on the precedence relations $FS'_{1i}$ (which include the ready times). The next decision point $m$ is then computed as the smallest early start time of any activity in the cutset. The activities in the cutset that can start at time $m$ are added to the eligible set $E$. All activities in $S$ that do not finish later than at time $m$ are deleted from $S$, and all activities in $E$ are scheduled: they are added to $S$ and $PS$ and are assigned a finish time that equals the sum of the decision point $m$ and the duration of the activity involved. The cutset is updated. If, owing to resource constraints, it is impossible to schedule all activities in $E$ concurrently, a resource conflict occurs. Such a conflict will produce a new branching in the branch-and-bound tree at level $p$: the branches describe ways to resolve the resource conflict, i.e. decisions about which combinations of activities are to be delayed.

A *delaying alternative* $D_q$ is defined as the set of activities that belong to $S$, the delay of which would resolve the resource conflict that occurred at level $p$ of the solutions tree and for which it holds that if an activity belongs to $D_q$ then all its direct and indirect successors that belong to $S$ are also included in $D_q$. In order to simplify the construction of the delaying alternatives, a precedence relation is added for every activity that can be partially overlapped with one of its indirect successors. As such, only the direct successors need to be examined in order to satisfy the second condition. The *delaying set* $D(p)$ then consists of all possible delaying alternatives $D_q$ that resolve the resource conflict at level $p$ of the branching tree. For each delaying alternative $D_q$, the *delaying point* $w_q$ is computed as the earliest time at which either the resource availability changes, or an activity that belongs to $S - D_q$ finishes, or one of the unscheduled activities that has no predecessor in $D_q$ could finish if all unscheduled activities were scheduled as soon as possible. A *precedence-based lower bound* $L_q$ is then calculated by adding the maximal remaining critical path length of any of the activities that belong to $D_q$ to the delaying point $w_q$. The delaying alternative with the smallest lower bound is chosen (ties are broken arbitrarily), and these activities are removed from $S$ and $PS$ (as well as all completed successors of one of these activities). All other delaying alternatives are stored for backtracking purposes. The cutset is updated, and the process of constructing the eligible set, adding it to $S$ and $PS$ and branching whenever resource conflicts occur, is repeated until a solution to the problem is found or until it can be shown that by branching from this node only infeasible solutions or dominated solutions could be generated. When this happens, the procedure backtracks.

Demeulemeester and Herroelen (1992b) have proved that the partial schedules may be constructed by semi-active timetabling. In addition, they have shown that it is sufficient to consider only minimal delaying alternatives in order to resolve a resource conflict. Importantly, they have extended the left-shift and cutset dominance rules, and have also shown that the critical sequence bound and the extended critical sequence bound *cannot* be extended, leaving only the remaining critical path length as a possible lower bound.

As already mentioned, the literature on the GRCPSP is very limited and a standard set of test problems has not yet been established. Bartusch *et al.* (1988) have reported on computational results obtained with a branch-and-bound procedure that basically extends the Radermacher procedure originally developed for the RCPSP (Radermacher 1985; Möhring and Radermacher 1989). Various constellations of a bridge construction project with 46 activities and 1–3 resource types were solved on a CYBER 175 in CPU-times ranging from 0.2 s to 47.4 s for 28 of these problems, with an average of 3.8 s and a large out-runner of 1206.3 s for one problem. Bartusch *et al.* (1988) have also reported on computational experience obtained on a variant of the bridge construction project and a real-life pipeline project, reaching the conclusion that their branch-and-bound procedure works well when the number of reduced forbidden sets is not too large and these sets are small in size. This is usually the case when there are many temporal constraints. When the problem has a large degree of parallelism, i.e. if there are many (large) forbidden sets or if there are few temporal constraints, then establishing optimality of a solution will require too much time.

In order to test our procedure, it was coded in Turbo C Version 2.0 for IBM PS/2 Model 70 A21 personal computers or compatibles. The procedure was then tested on the 110 Patterson test problems as modified by Simpson (1991) to incorporate variable resource availabilities. Simpson's serial procedure was able to optimally solve 97 out of the 110 modified problems, when the procedure was allowed 600 CPU s on an IBM 3090. The average computation time for these 97 problems was 100.85 s, with a standard deviation of 199.62 s. The parallel procedure could solve 98 out of those 110 problems in an average CPU-time of 96.63 s with a standard deviation of 195.50 s. Our procedure could find the optimal solution for 107 problems within a 5 min CPU-time limit in an average CPU-time of 2.4975 s and a standard deviation of 4.6233 s on an IBM PS/2 with coprocessor and running at 16 MHz. As such, it seems to be a very efficient and effective exact solution procedure for the GRCPSP. In addition, the computational experience obtained indicates that moderate changes in the ready times or in the resource availabilities do not have a significant impact on the computation times. The introduction of due dates significantly reduces the solution time required. Allowing activity overlaps (negative $FS_{ij}$ values) causes a strong increase in the required computation time.

## 12.4 EXTENSION OF THE DH PROCEDURE TO THE PRCPSP

The DH procedure has been extended to the PRCPSP along the lines described by Demeulemeester (1992) and Demeulemeester and Herroelen (1992c). In order to do so, it is assumed that only two dummy activities exist in the project: the dummy start and the dummy end. This is caused by the time incrementing scheme used, which augments the decision points by one time unit at a time. In addition, we distinguish between *activities* and *subactivities*. At the start of the procedure, we create a new project network in which all activities are replaced by one or more subactivities. The dummy start and end activities are replaced by dummy start and end subactivities with a duration of 0. All other activities are split into subactivities, their number being equal to the duration of the original activity, each having a duration of 1 and resource requirements that are equal to those of the original activity. Demeulemeester (1992) and Demeulemeester and Herroelen (1992c) have proved that in order to solve the PRCPSP, it is sufficient to construct partial schedules by semi-active timetabling at the level of the subactivities.

We define an *eligible activity* as an activity for which one of the subactivities is eligible. An *unfinished activity* is an activity for which not all subactivities have been scheduled. Denote the $z$ unfinished subactivities of unfinished activity $i$ at time $t$ as

$i_1, i_2, \ldots, i_z$. We say that activity $i$ is scheduled immediately at time $t$ if all its remaining subactivities $i_x (x = 1, \ldots, z)$ are scheduled such that $f_{ix} = t + x$.

Theorems 12.1 and 12.2 for the RCPSP can now be extended in the following way.

**Theorem 12.5** *If for a partial schedule $PS_m$ at instant $m$ there exists an eligible activity $i$ that cannot be scheduled together with any other unfinished activity $j$ at any time instant $m' \geq m$ without violating the precedence or resource constraints then an optimal continuation of $PS_m$ exists with all remaining subactivities $i_1, i_2, \ldots, i_z$ of activity $i$ scheduled immediately at time $m$.*

It should be noted that we do not have to check whether or not an activity is in progress at time $m$. The scheduling of an activity at the previous decision point does not imply that if that activity was not completed, the same activity should also be scheduled at the current decision point. The preemption condition allows us to forget the scheduling decisions in previous periods and to consider only those possibilities implied by the set of eligible subactivities.

**Theorem 12.6** *If for a partial schedule $PS_m$ at time instant $m$ there exists an eligible activity $i$ that can be scheduled together with only one other unfinished activity $j$ at any time instant $m' \geq m$ without violating the precedence or resource constraints, and if activity $j$ is eligible, then an optimal continuation of $PS_m$ exists with all remaining subactivities of activity $i$ scheduled immediately and with as many subactivities of activity $j$ as possible scheduled concurrently with the subactivities of activity $i$.*

It will be noted that no test needs to be performed to check whether the remaining duration of activity $j$ is larger than that of activity $i$. Indeed, if the remaining duration of activity $j$ is larger, we shall schedule as many subactivities of activity $j$ as there are unscheduled subactivities in activity $i$. If, however, the remaining duration of activity $j$ is smaller or equal, we shall schedule all remaining subactivities of activity $j$ concurrently with those of activity $i$.

The left-shift dominance rule cannot be applied to the PRCPSP, since there is no need to delay a subactivity once it has been scheduled. Demeulemeester and Herroelen (1992c), however, were able to prove the following dominance rule, which very much resembles the cutset dominance rule stated earlier for the RCPSP.

**Theorem 12.7** *Consider a partial schedule $PS_m$ at time $m$. If there exists a partial schedule $PS_k$ that was previously saved at a similar time $m$, and if $PS_m$ is a subset of $PS_k$, then the current partial schedule $PS_m$ is dominated.*

Demeulemeester and Herroelen (1992c) have also shown that it is sufficient to consider only minimal delaying alternatives in order to resolve resource conflicts.

They have shown that all three lower bounds discussed earlier (remaining critical path length, critical sequence and extended critical sequence lower bound) remain applicable, at the trade-off of increased computational requirements. Therefore they only included the remaining critical path length bound in the code.

As already mentioned, the literature on the PRCPSP is almost void, and very little computational experience is available. Our procedure has been programmed in Turbo C Version 2.0 for a personal computer IBM PS/2 Model 70. On the same 41 Patterson test problems used by Kaplan (1988, 1991) and using a similar PC running at 16 MHz, our procedure finds the optimal solution in an average CPU-time of 4.9863 s with a standard deviation of 9.2932 s, while the Kaplan code requires an average of 425 s and a standard deviation of 713 s. Using a personal computer IBM PS/2 running at 25 MHz, we have tested our algorithm on all 110 Patterson test problems. All problems could be solved within 5 min of CPU-time, requiring an average of 6.8985 s and a standard deviation of 25.8149 s.

Demeulemeester (1992) has extended the code for the PRCPSP with variable resource availabilities. In that case, Theorems 12.5 and 12.6 no longer apply. A total of 107 out of the modified 110 Patterson test problems could be solved in an average computation time of 12.6321 s and a standard deviation of 36.9061 s.

## 12.5 CONCLUSIONS

In this chapter a new depth-first branch-and-bound methodology for solving the RCPSP, the GRCPSP and the PRCPSP has been reviewed. The overall search procedure generates a solutions tree in which nodes represent precedence and resource feasible partial schedules. Branches emanating from a parent node correspond to exhaustive and minimal combinations of activities, the delay of which resolves resource conflicts at each parent node. It has been indicated how the procedure can be equipped with dominance rules and lower bound calculations in order to restrict the growth of the solutions tree for each problem type.

The resulting computer codes have been written in Turbo C Version 2.0 for use on personal computers running under DOS. Computational experience with the RCPSP, GRCPSP and PRCPSP is very promising, and indicates that the procedure is one of the most efficient currently available. Research is in progress in order to upgrade the computer codes for personal computer environments running under OS/2 and workstations running under UNIX or AIX. In addition, a comparison paper is in preparation that should report on the computational experience obtained for the current best exact solution procedures on a standardized problem set.

## REFERENCES

Balas E. (1971) Project scheduling with resource constraints. In Beale E. M. L., editor, *Applications of Mathematical Programming* pp 187–200 The English University Press, London.

Bartusch M., Möhring R. H. and Radermacher F. J. (1988) Scheduling project networks with resource constraints and time windows *Ann. Oper. Res.*, **16**, 201–240.
Bell C. A. and Park K. (1990) Solving resource-constrained project scheduling problems by A* search *Nav. Res. Log.*, **37**, 61–84.
Blazewicz J., Lenstra J. K. and Rinnooy Kan A. H. G. (1983) Scheduling subject to resource constraints: classification and complexity *Discr. Appl. Math.*, **5**, 11–24.
Bowman E. H. (1959) The schedule-sequencing problem *Oper. Res.*, **7**, 621–624.
Brand J. D., Meyer W. L. and Schaffer L. R. (1964) The resource scheduling problem in construction. Civil Engineering Studies Report No. 5, Department of Civil Engineering, University of Illinois, Urbana.
Carlier J. (1987) Scheduling jobs with release dates and tails on identical machines to minimize makespan *Eur. J. Oper. Res.*, **29**, 298–306.
Carlier J. and Latapie B. (1991) Une méthode arborescente pour résoudre les problèmes cumulatifs *RAIRO*, **25**, 311–340.
Carruthers J. A. and Battersby A. (1966) Advances in critical path methods *Oper. Res. Quart.*, **17**, 359–380.
Christofides N., Alvares-Valdes R. and Tamarit J. M. (1987) Project scheduling with resource constraints: a branch and bound approach *Eur. J. Oper. Res.*, **29**, 262–273.
Davis E. W. (1966) Resource allocation in project network models—a survey *J. Ind. Engng*, **17**, 177–188.
Davis E. W. (1968) An exact algorithm for the multiple-constrained resource project scheduling problem. PhD Dissertation, Yale University.
Davis E. W. (1973) Project scheduling under resource constraints: historical review and categorization of procedures *AIIE Trans.*, **5**, 297–313.
Davis E. W. and Heidorn G. E. (1971) An algorithm for optimal project scheduling under multiple resource constraints *Management Sci.*, **27**, B803–B816.
Demeulemeester E. (1992) Optimal algorithms for various classes of multiple resource-constrained project scheduling problems. PhD Thesis, Department of Applied Economic Sciences, Katholieke Universiteit Leuven.
Demeulemeester E. and Herroelen W. (1992a) A branch-and-bound procedure for the multiple resource-constrained project scheduling problem *Management Sci.*, **38**, 1803–1818.
Demeulemeester E. and Herroelen W. (1992b) A branch-and-bound procedure for the generalized resource-constrained project scheduling problem. Research Report No. 9206, Department of Applied Economic Sciences, Katholieke Universiteit Leuven.
Demeulemeester E. and Herroelen W. (1992c) An efficient optimal solution procedure for the preemptive resource-constrained project scheduling problem. Research Report No. 9216, Department of Applied Economic Sciences, Katholieke Universiteit Leuven.
Demeulemeester E., Herroelen W., Simpson W. P., Baroum S., Patterson J. and Yang K.-K. (1994) On a paper by Christofides *et al.* for solving the multiple-resource constrained, single project scheduling problem *Eur. J. Oper. Res.*, **76**, 218–228.
De Wit J. and Herroelen W. (1990) An evaluation of microcomputer-based software packages for project management *Eur. J. Oper. Res.*, **49**, 102–139.
Elmaghraby S. E. (1967) The sequencing of $N$ Jobs on $M$ parallel processors. Unpublished Paper, North Carolina State University, Raleigh.
Herroelen W. S. (1972) Resource-constrained project scheduling—the state of the art *Oper. Res. Quart.*, **23**, 261–275.
Johnson T. J. R. (1967) An algorithm for the resource constrained project scheduling problem. PhD Dissertation, MIT.

Kaplan L. (1988) Resource-constrained project scheduling with preemption of jobs. PhD Dissertation, University of Michigan.
Kaplan L. (1991) Resource-constrained project scheduling with setup times. Unpublished paper, Department of Management, University of Tennessee, Knoxville.
Moder J. J., Phillips C. R. and Davis E. W. (1983) *Project Management with CPM, PERT and Precedence Diagramming* Van Nostrand Reinhold, New York.
Möhring R. H. and Radermacher F. J. (1989) The order-theoretic approach to scheduling: the deterministic case. In Slowinski R. and Weglarz J., editors, *Advances in Project Scheduling*. Part I, Chapter 2 Elsevier, Amsterdam.
Moodie C. L. and Mandeville D. E. (1966) Project resource balancing by assembly line balancing techniques *J. Ind. Engng*, **17**, 377–383.
Patterson J. (1984) A comparison of exact procedures for solving the multiple constrained resource project scheduling problem *Management Sci.*, **30**, 854–867.
Patterson J. H. and Huber W. D. (1974) A horizon-varying zero-one approach to project scheduling *Management Sci.*, **20**, 990–998.
Patterson J. H. and Roth G. (1976) Scheduling a project under multiple resource constraints: a zero-one programming approach *AIIE Trans.*, **8**, 449–456.
Petrovic R. (1968) Optimisation of resource allocation in project planning *Oper. Res.*, **16**, 559–586.
Pritsker A. B., Watters L. J. and Wolfe P. M. (1969) Multiproject scheduling with limited resources: a zero-one programming approach *Management Sci.*, **16**, 93–108.
Radermacher F. J. (1985) Scheduling of project networks *Ann. Oper. Res.*, **4**, 227–252.
Schrage L. (1970) Solving resource-constrained network problems by implicit enumeration—nonpreemptive case *Oper. Res.*, **10**, 263–278.
Simpson W. P. (1991) A parallel exact solution procedure for the resource-constrained project scheduling problem. PhD Dissertation, Indiana University.
Slowinski R. (1980) Two approaches to problems of resource allocation among project activities—a comparative study *J. Oper. Res. Soc.*, **31**, 711–723.
Stinson J. P., Davis E. W. and Khumawala B. M. (1978) Multiple resource-constrained scheduling using branch-and-bound *AIIE Trans.*, **10**, 252–259.
Talbot B. and Patterson J. H. (1978) An efficient integer programming algorithm with network cuts for solving resource-constrained scheduling problems *Management Sci.*, **24**, 1163–1174.
Weglarz J. (1981) Project scheduling with continuously-divisible, doubly constrained resources *Management Sci.*, **27**, 1040–1053.
Wiest J. D. (1964) Some properties of schedules for large projects with limited resources *Oper. Res.*, **12**, 395–418.

---

*Department of Applied Economic Sciences, Katholieke Universiteit Leuven, Naamsestraat 69, B-3000 Leuven, Belgium*

CHAPTER 13

# The Job Shop Scheduling Problem: A Concise Survey and Some Recent Developments

**E. Pinson**
*Institut de Mathématiques Appliquées, Université Catholique de l'Ouest*

**Abstract**

The job shop scheduling problem is one of the most well-known machine scheduling problems. It is strongly $\mathcal{NP}$-hard, and most methods proposed for solving it are of an enumerative type, and based on a disjunctive graph formulation. We present here a concise survey and also describe some recent developments regarding this important combinatorial optimization problem.

## 13.1 INTRODUCTION

In the job shop scheduling problem $n$ jobs have to be processed on $m$ different machines. Each job consists of a sequence of operations that have to be processed in a fixed order. An operation has to be executed on a given machine during a fixed processing time. The aim is to minimize the maximum of the job completion times or makespan under some additional constraints (specified in Section 13.2). Only a few particular cases have been proved to be polynomially solvable. The general problem is $\mathcal{NP}$-hard in the strong sense (Lenstra *et al.* 1977; Garey and Johnson 1979) and is probably one of the most computationally intractable combinatorial problems considered so far. A practical proof of this intractability comes from the fact that a small example with 10 jobs and 10 machines posed by Fischer and Thompson (1963) remained open for over 20 years. It was finally solved by Carlier

---

*Scheduling Theory and its Applications* Edited by P. Chrétienne, E. G. Coffman, Jr., J. K. Lenstra and Z. Liu
© 1995 John Wiley & Sons Ltd

and Pinson in 1980 (1989), as the culmination of a considerable amount of research.

Most methods proposed for solving this problem are of an enumerative type, and based on a disjunctive graph formulation. Other approaches have also been tested, using active schedule generation, mixed integer programming formulations and heuristic methods.

The aim of this chapter is, on the one hand, to present a concise survey and, on the other, to describe some recent developments regarding the job shop scheduling problem. It has no claim to exhaustivity. First, in Section 13.2, the problem statement is recalled and some related complexity results are reviewed. In Section 13.3 some developments for the general case are presented. For this purpose, instead of reviewing in a sequential way complete methods that have been proposed for solving the general case over the years, we have chosen to present important results in connection with the main components of enumerative methods: lower bounds, branching schemes and elimination rules. Finally, in Section 13.4 we briefly discuss the main approaches that have been tested in approximative techniques.

## 13.2 PROBLEM STATEMENT AND COMPLEXITY RESULTS

### 13.2.1 The job shop scheduling problem

In the job shop scheduling problem a set $\mathcal{J}$ of $n$ jobs $\mathcal{J}_1, \mathcal{J}_2, \ldots, \mathcal{J}_n$ have to be processed on a set $\mathcal{M}$ of $m$ different machines $\mathcal{M}_1, \mathcal{M}_2, \ldots, \mathcal{M}_m$. Job $\mathcal{J}_j$ consists of a sequence of $m_j$ operations $\mathcal{O}_{j1}, \mathcal{O}_{j2}, \ldots, \mathcal{O}_{jm_j}$, which have to be scheduled in this order. Moreover, each operation can be processed only by one machine among the $m$ available ones. Preemption of any operation is not allowed. Operation $O_{jk}$ has a processing time $p_{jk}$. The objective is to find an operating sequence for each machine such as to minimize a particular nondecreasing function of the job completion times, and in such a way that two operations are never processed on the same machine at any time instant. Note that in this chapter we focus on the particular criterion $C_{\max} = \max_{j=1,n} C_j$, where $C_j$ denotes the completion time of the last operation of job $\mathcal{J}_j (j = 1, \ldots, n)$.

### 13.2.2 $\mathcal{NP}$-hardness of the job shop scheduling problem

Only a few special cases of this problem can be solved in polynomial time, and they are briefly reviewed in the next section. Problems with two machines and $m_j \leq 3 (j = 1, \ldots, n)$, and problems with three machines and $m_j \leq 2 (j = 1, \ldots, n)$ are $\mathcal{NP}$-hard (Lenstra et al. 1977; Gonzalez and Sahni 1978). Problems with three machines and unit processing times have been proved to be strongly $\mathcal{NP}$-hard (Lenstra and Rinnooy Kan 1979). More recently, Sotskov (1991) proved that the job shop problem with three jobs is $\mathcal{NP}$-hard. Of course, the general job shop scheduling problem is also strongly $\mathcal{NP}$-hard. Note that these results hold even if preemption is allowed (Gonzalez and Sahni 1978).

### 13.2.3 Polynomially solvable particular cases

#### 13.2.3.1 Job shop scheduling problems with two machines

The job shop scheduling problem with two machines and at most two operations per job can be solved in $O(n \log n)$ time by a simple extension, due to Jackson (1955), of Johnson's algorithm for the two-machine flow shop scheduling problem. Let $p_{i1}$ (respectively $p_{i2}$) denote the processing time of job $\mathcal{J}_i$ on machine $\mathcal{M}_1$ (respectively $\mathcal{M}_2$), and $m_{i1}$ (respectively $m_{i2}$) the index of the first (respectively second) machine assigned to job $\mathcal{J}_i$ in its job sequence.

**Johnson's rule** *If $\min(p_{i1}, p_{j2}) \leq \min(p_{i2}, p_{j1})$ then there exists an optimal schedule for the two-machine flow shop scheduling problem in which job $\mathcal{J}_i$ precedes job $\mathcal{J}_j$.*

**Jackson's rule** Partition the set of jobs into the following four groups according to the number and the machine assignments of operations:

$J_{12} = \{\mathcal{J}_j : m_j = 2, m_{1j} = 1, m_{2j} = 2\}$, $\quad J_{21} = \{\mathcal{J}_j : m_j = 2, m_{1j} = 2, m_{2j} = 1\}$,

$J_1 = \{\mathcal{J}_j : m_j = 1, m_{1j} = 1\}$, $\quad J_2 = \{\mathcal{J}_j : m_j = 1, m_{1j} = 2\}$.

An optimal schedule for the job shop scheduling problem with two machines and $m_j \leq 2 (j = 1, \ldots, n)$ can be built by the following processing order:

- on machine $\mathcal{M}_1, J_{12}$–$J_1$–$J_{21}$,
- on machine $\mathcal{M}_2, J_{21}$–$J_2$–$J_{12}$,

where the jobs in $J_{12}$ and $J_{21}$ are ordered according to Johnson's rule, and the jobs in $J_1$ and $J_2$ are in an arbitrary order.

Note that the two-machine job shop scheduling problem with unit processing times can be solved in $O(\sum_{j=1}^{n} m_j)$ time (Hefetz and Adiri 1982).

#### 13.2.3.2 Job shop scheduling problems with two jobs

The job shop scheduling problem with two jobs can also be solved in polynomial time (Akers 1956; Brucker 1988). Brucker proposed an $O(m_1 m_2 \log m_1 m_2)$ time algorithm achieving this goal. As Akers did, he showed that the problem can be formulated as a shortest-path problem in the plane with rectangular objects as obstacles. These obstacles correspond to a common machine requirement of both jobs according to their job sequence. More precisely, denote by $m_{1j}$ (respectively $m_{2j}$ the number of the $j$th machine assigned to job $\mathcal{J}_1$ (respectively $\mathcal{J}_2$) in its job sequence, and by $p_{1j}$ (respectively $p_{2j}$) the corresponding processing times. If $m_{1i} = m_{2j} (i \in ]m_1], j \in ]m_2])$, we create an obstacle with south-west coordinates $(\sum_{k=1}^{i-1} p_{1k}, \sum_{k=1}^{j-1} p_{2k})$ and length $p_{1i}$ (respectively $p_{2j}$) on the $x$ axis (respectively $y$ axis). Figure 13.1 shows an example with $m_1 = 4$ and $m_2 = 3$.

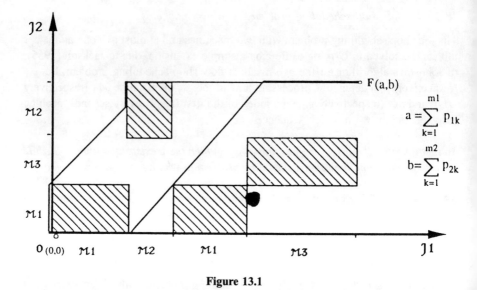

**Figure 13.1**

Any feasible path starts at the origin $O$ with coordinates $(0, 0)$ and ends at $F$ with coordinates $(\sum_{k=1}^{m_1} p_{1k}, \sum_{k=1}^{m_2} p_{2k})$, going either diagonally or parallel to the axes and avoiding the interior of the obstacles. This shortest-path problem can be reduced to the problem of finding a shortest path from $O$ to $F$ in a specific network that can be constructed in $O(p \log p)$, where $p$ is the number of obstacles. Moreover, Brucker (1988) showed that the shortest path on this network can be computed in $O(p)$ steps. (Note that $p = O(m_1 m_2)$.)

## 13.3 OPTIMIZATION ALGORITHMS

### 13.3.1 Introduction

As pointed out in Section 13.1, most methods proposed for solving the job shop scheduling problem are of an enumerative type, and use a disjunctive graph formulation proposed by Roy and Sussman (1964). Nevertheless, other approaches have been tested, most of them based on an active schedule generation or mixed integer programming (MIP) formulation. In this section, some of these methods are reviewed through illustrations of their main principles, including recent developments. Instead of reviewing in a sequential way complete methods that have been proposed for solving the general case over the years, we have preferred to present important results in connection with the main components of any enumerative method: lower bounds, branching schemes and elimination rules.

## 13.3.2 Problem formulation

### 13.3.2.1 Disjunctive graph formulation and connex notions

*Disjunctive graph*

Two operations $i$ and $j$, executed by the same machine, cannot be simultaneously processed. So we associate with them a pair of disjunctive arcs or disjunction $[i,j] = \{(i,j), (j,i)\}$. The problem is then modeled by a *disjunctive graph* $\mathcal{G} = (G, D)$, where $G = (X, U)$ is a conjunctive graph associated with the job sequences, in which $X$ is the set of operations associated with the $n$ jobs and $U$ is the set of conjunctive arcs (an arc connecting two consecutive operations in a job sequence), and $D$ is a set of disjunctions as defined before.

In the following it is assumed that the operations associated with the $n$ jobs are numbered in an arbitrary order. Figure 13.2 shows an example of disjunctive graph associated with a job shop with 3 jobs and 3 machines. In this instance, job $\mathcal{J}_1$ has to be processed on machine $\mathcal{M}_1$ with a processing time of 2 time units (operation 1), then on machine $\mathcal{M}_2$ with a processing time of 5 time units (operation 2), and finally on machine $\mathcal{M}_3$ with a processing time of 4 time units (operation 3). The job sequences for jobs $\mathcal{J}_2$ and $\mathcal{J}_3$ are respectively $\mathcal{M}_3 \mathcal{M}_1 \mathcal{M}_2$ with processing times 1,6,7 (operations 4,5,6), and $\mathcal{M}_3 \mathcal{M}_2 \mathcal{M}_1$ with processing times 6,8,3 (operations 7,8,9). o and * denote two dummy operations associated with the beginning and the end of the schedule.

*Basic notation*

In the following $p_i$ denotes the processing time of operation $i$, and $r_i$ (respectively $q_i$) its head (respectively its tail). Denoting the value of one of the longest paths from $i$ to $j$ in $G$ by $l(i,j)$, we define $r_i = l(\circ, i)$ and $q_i = l(i, *) - p_i$.

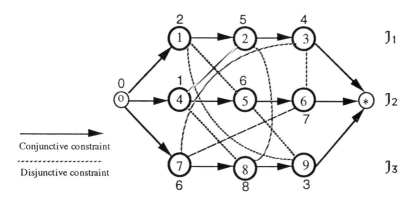

**Figure 13.2** An example of a 3 × 3 job-shop.

## Schedule

A *schedule* on a disjunctive graph $G = (G, D)$ is a set of starting times $T = \{t_i : i \in X\}$ such that

- the conjunctive constraints are satisfied:
$$t_j - t_i \geq p_i \quad \forall (i,j) \in U :$$
- the disjunctive constraints are satisfied:
$$t_j - t_i \geq p_i \quad \text{or} \quad t_i - t_j \geq p_j \quad \forall (i,j) \in D$$

## Selection

To build a schedule, we have to replace each disjunctive arc $[i,j]$ by either $(i,j)$ or $(j,i)$, and thus to choose an operating sequence for each machine.

A *selection* $A$ is a set of disjunctive arcs such that if $(i,j) \in A$ then $(j,i) \notin A$. The membership of $(i,j)$ in $A$ makes it necessary to process operation $i$ before operation $j$. We associate the conjunctive graph $G_A = (X, U \cup A)$ with selection $A$. Heads and tails can be dynamically adjusted on this conjunctive graph. By definition, a selection is *complete* if all the disjunctions of $D$ are selected. It is *consistent* if the associated conjunctive graph is acyclic. A schedule corresponds to a consistent complete selection. Its makespan is the value of a longest path in $G_A$.

## Clique of disjunctions

A *clique of disjunctions* is a set $K$ of operations such that every pair of operations of $K$ are in disjunction. Cliques are obtained for the job shop by considering a group of operations processed on a specific machine. $e \in K$ (respectively $s \in K$) is called the *input* (respectively *output*) of the clique $K$ if $e$ (respectively $s$) is sequenced in $T$ before (respectively after) all the other operations of $K$.

### 13.3.2.2 Mixed integer programming formulation

A large number of MIP formulations have been proposed over the years (for a review of some of them, see e.g. French 1982). The one we propose here has been recently used by Applegate and Cook (1991) in connection with polyhedral techniques and cutting plane generation. To the best of our knowledge, they were the first to obtain some interesting results on difficult instances using such an approach. Keeping the notation defined above, the problem can be formulated as follows:

minimize $C_{\max}$

subject to

$$\forall i \in X, \quad t_i \geq 0, \tag{13.1}$$

$$\forall i \in X, \quad C_{\max} \geq t_i + p_i, \tag{13.2}$$

$$\forall (i,j) \in U, \quad t_j \geq t_i + p_i \tag{13.3}$$

$$\forall [i,j] \in D, \quad t_i \geq t_j + p_j \text{ or } t_j \geq t_i + p_i. \tag{13.4}$$

This *disjunctive programming problem* leads to the following MIP formulation by introducing a binary variable $y_{ij}$ for each of the conditions (13.4) and setting the new constraints

$$\forall [i,j] \in D, \quad t_i \geq t_j + p_j - K y_{ij}, \tag{13.5}$$
$$t_j \geq t_i + p_i - K(1 - y_{ij}),$$
$$\forall [i,j] \in D, \quad y_{ij} \in \{0, 1\}, \tag{13.6}$$

where $K$ is some large constant, and $y_{ij} = 1$ if and only if $i$ is scheduled before $j$, and 0 otherwise.

### 13.3.3 Elimination rules: the concept of immediate selection

#### 13.3.3.1 Introduction

Elimination rules are of prime interest in branch-and-bound methods for efficiently pruning the associated search tree. To illustrate such a tool, we present in this section the key idea of immediate selection developed by Carlier and Pinson (1989, 1991, 1994).

Let $I$ be a set of operations that have to be processed on a machine. Our objective is to characterize particular partial operating sequences on subsets of operations of $I$ that cannot lead to a global solution for the job shop problem. So, in some cases, it will be possible either to fix some alternative sequences or to position some operations in relation to some others.

#### 13.3.3.2 Background and basic notation

*Solution*

Let UB be a given integer (in practice an upper bound on the optimal makespan). By definition, a *solution* of the job shop scheduling problem is a schedule with a makespan smaller than or equal to UB.

*Jackson's preemptive schedule: primal and dual version*

Let us consider a subset $I$ of operations that have to be processed on the same machine. Jackson's preemptive schedule is the list schedule associated with the most work remaining (MWR) priority dispatching rule. To build it, we schedule, at the first moment $t$ where the machine and at least one operation are available, the available operation with maximal tail: this operation is processed either up to its completion or until a more urgent operation becomes available. We update $t$, and iterate until all the operations are scheduled (Jackson 1955).

Because of the symmetric role of heads and tails, we can design in the same way a symmetric version of this algorithm we call the *dual Jackson's preemptive schedule*

*(DJPS)*, in contrast with the previous version, which we call for this reason the *primal Jackson's preemptive schedule (PJPS)*.

Using heap structures, both PJPS and DJPS can be computed in $O(n \log n)$ steps. Their makespans are lower bounds of the job shop problem (Carlier 1982), and are equal to $\max_{J \subseteq I} h(J)$, where

$$h(J) = \min_{j \in J} r_j + \sum_{j \in J} p_j + \min_{j \in J} q_j \qquad (13.7)$$

Figure 13.3 shows an example of PJPS and DJPS built on an instance with 6 operations.

### 13.3.3.3 Immediate selections on ascendant sets

By definition, a subset $J$ of operations is called an *ascendant set* of $c$ if $c \notin J$ and

$$\min_{j \in J} r_j + \sum_{j \in J} p_j + \min_{j \in J} q_j + p_c > UB, \qquad (13.8)$$

$$r_c + p_c + \sum_{j \in J} p_j + \min_{j \in J} q_j > UB. \qquad (13.9)$$

Clearly, from (13.8) and (13.9), if $J$ is an ascendant set of $c$ then $c$ is output of $K = J \cup \{c\}$, and we can put

$$r_c = \max\{r_c, C^*\},$$

| i | 1 | 2 | 3 | 4 | 5 | 6 |
|---|---|---|---|---|---|---|
| $r_i$ | 4 | 0 | 9 | 15 | 20 | 21 |
| $p_i$ | 6 | 8 | 4 | 5 | 8 | 8 |
| $q_i$ | 20 | 25 | 30 | 9 | 14 | 16 |

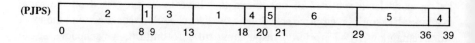

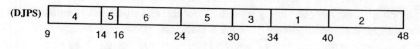

**Figure 13.3**

where

$$C^* = \max_{J' \subseteq J} \left\{ \min_{j \in J'} r_j + \sum_{j \in J'} p_j \right\}.$$

Let us introduce some additional notation. For $j \in I$, denote by $p_j^-$ the processed time of $j$ before $r_c$ in PJPS, $p_j^+$ the processed time of $j$ after $r_c$ in PJPS and $C_j$ its completion time in PJPS. Let $K_c^+ = \{j \in I : p_j^+ > 0\}$.

Carlier and Pinson (1991) proved that finding $C^*$ is equivalent to looking for a nonempty subset $K^+$ of $K_c^+$ satisfying

$$r_c + p_c + \sum_{j \in K^+} p_j^+ + \min_{j \in K^+} q_j > UB. \qquad (13.10)$$

Obviously, if such a set exists, there exists a maximal set $K_c^*$ satisfying (13.11), and we have

$$C^* = \max_{j \in K_c^*} C_j.$$

Figures 13.4 and 13.5 show a simple application of this result to the previous example with $c = 4$ and $UB = 52$.

For more details, we refer to Carlier and Pinson (1991), who proposed an $O(n^2)$ time algorithm for computing all the immediate selections on the set $I$. These results can of course be transposed to the symmetric case, that is to say, the search for descendant sets for operation $c$ (the symmetric relation of (13.9)).

### 13.3.3.4 Immediate selection on disjunctions

In the particular case where $J = \{j\}$, the condition (13.9) leads to the result that if $r_c + p_c + p_j + q_j > UB$ then $c$ is scheduled after $j$ in any solution.

Finding all immediate selections on disjunctions associated with operations of $I$ can be carried out using the simple $O(n^2)$ time procedure below:

*For all operations $i,j$ in $I, i \neq j$ do*
  *If $r_j + p_j + p_i + q_i > UB$ then*
    *fix $i \to j$*
  *Endif*
*Enddo*

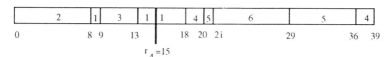

**Figure 13.4** Set $UB = 52$ and apply the previous results with $c = 4$. We obtain

$$r_4 + p_4 + p_1^+ + p_6^+ + q_5 = 15 + 5 + 3 + 8 + 8 + 14 = 53 > UB.$$

So $K_4^+ = \{1, 5, 6\}$, and we can adjust $r_4$ by setting $r_4 = \max\{C_1, C_5, C_6\} = 36$. We thus obtain the new preemptive schedule of Figure 13.5.

**Figure 13.5** The new preemptive schedule.

#### 13.3.3.5 Recent work

Carlier and Pinson (1994) presented some algorithmic improvements for computing heads and tails adjustments in relation with immediate selections, leading to two $O(n \log n)$ time algorithms for the cases detailed in Sections 13.3.3.3 and 13.3.3.4. Brucker et al. (1992) gave similar results, and, using a weaker characterization of immediate selections, presented an $O(n \log n)$ time algorithm for adjusting heads and tails.

### 13.3.4 Lower bounds

A natural way for computing lower bounds for scheduling problems is to relax some of their related constraints. Many relaxations have been tested for the job shop scheduling problem. Among the more usual ones are the following.

*Relaxation of capacity constraints*

The idea is to relax the capacity constraints of a subset $M \subset \mathcal{M}$ of machines (which is equivalent to ignoring the associated remaining disjunctive constraints). In that way, a smaller job shop scheduling problem with additional heads and tails constraints induced by the current conjunctive graph (Bratley et al. 1973) is obtained. In particular, if $|M| = m - 1$, we obtain a one-machine scheduling problem. This problem is strongly $\mathcal{NP}$-hard, but can be efficiently solved using Carlier's (1982) algorithm, even for quite large instances. Note also that the preemptive case can be solved in $O(n \log n)$ time.

*Relaxation of the precedence constraints and preemption allowed*

In this case the precedence constraints associated with each job sequence are relaxed, leading to a sequence of one-machine scheduling problems with additional head and tail constraints induced by the current conjunctive graph, and the preemption of any operation is allowed. Using such a relaxation, Carlier and Pinson (1994) proposed a tighter lower bound than all the classical ones. The main idea is to consider the procedure consisting, for a fixed quantity UB, in determining all immediate selections (primal and dual versions) on each machine, propagating head and tail adjustments through the current conjunctive graph. Obviously, such a procedure ends with one of the following two results:

(a) we can detect no more immediate selections;
(b) PJPS (or DJPS) has a makespan greater than UB on some machine.

In case (a) we can decrease UB by a given quantity $\delta$, and reapply the procedure. Necessarily, for a certain value UB' of UB, this procedure terminates with the second result, and UB' + 1 is a lower bound on the optimal makespan for the job shop scheduling problem. $\delta$ is computed in such a way that we are sure to detect at least one immediate selection at the next procedure run.

*Duality relaxations*

In this approach the capacity constraints of the machines or the precedence constraints on the operations of each job are weighted and aggregated in a single constraint (Fisher *et al.* 1983). Unfortunately, computational experiments on difficult instances seem to be quite disappointing in comparison with those obtained for other notorious combinatorial problems.

*MIP formulation and polyhedral techniques*

Computational results obtained with such techniques are also quite disappointing in view of those reported for other hard combinatorial problems (Balas 1985; Queyranne 1988; Dyer and Wolsey 1990). Nevertheless, Applegate and Cook (1991) have obtained some interesting results for difficult instances using such an approach. To illustrate their purpose, consider the disjunctive programming and MIP formulation described in Section 13.3.2. The basic idea is to start with relaxations of these problems (relaxation of the conditions (13.4) in the disjunctive programming formulation, and simple continuous relaxation, replacing the constraints (13.6) by

$$\forall [i,j] \in D, \quad 0 \leqslant y_{ij} \leqslant 1 \qquad (13.12)$$

in the MIP formulation), and to add successively inequalities (cutting planes) to the formulation that are valid for any schedule, but not satisfied by the current optimal solution to the linear programming relaxation. Of course, this optimal solution provides at each step a lower bound of the optimal makespan. Let us now give some examples of such cutting planes. In what follows, $I$ denotes a subset of operations that have to be processed on the same machine, and $J$ a subset of $I$.

*Basic cuts*  By considering the order on which the operations in $J$ are scheduled, it is easy to check, using for instance recursive arguments, that the inequality

$$\sum_{j \in J} p_j t_j \geqslant \left( \min_{j \in J} r_j \right) \sum_{j \in J} p_j + \sum_{i \neq j \in J} p_i p_j$$

is satisfied by all schedules.

*Two-jobs cuts*

This cut is a sharpened version of the previous basic cut. Let $[i,j]$ be a nonselected disjunctive constraint satisfying $r_j < r_i + p_i$ and $r_i < r_j + p_j$. It can be shown that the inequality

$$(p_i + r_i - r_j)t_i + (p_j + r_j - r_i)t_j \geq p_i p_j + r_i p_j + r_j p_i$$

is satisfied by any schedule. This cut can be generalized to a subset $J$ of operations with $|J|\geq 2$.

*Triangle cuts*
For any subclique of disjunctions $i,j,k \subseteq I$, we have obviously

$$y_{ij} + y_{jk} + y_{ki} \leq 2.$$

*Half cuts*
By considering each operation $k \in J$ as output of $J$, we have

$$\forall k \in J, \qquad t_k \geq \min_{j \in J} r_j + \sum_{j \in J \setminus \{k\}} y_{jk} p_j.$$

*Relaxation to job shop scheduling problems with two jobs*
Brucker et al. (1991) and Brucker and Jurisch (1993) proposed such a relaxation. In order to take into account the current selection, which induces heads, tails and precedence constraints on operations, they designed a polynomial algorithm for solving the two-job job shop scheduling problem with additional heads and tails on operations in the preemptive case, which is a nice generalization of the one proposed by Brucker for the two-job job shop scheduling problem.

### 13.3.5 Branching schemes

Branching schemes for the job shop scheduling problem have been widely studied. The usual ones are based on the following principles.

*Active schedule generation*
Recall that in an active schedule no global left-shift on any operation can be performed without delaying any other operation. The idea is to generate all active schedules in a forward way. At each step, the schedulable operations are those of which all predecessors have been processed. The machine on which the minimum of their completion time is achieved is used to branch by successively scheduling next on that resource all available operations according to their release dates (Giffler and Thompson 1960).

*Critical path rearrangement*
Such branching schemes consist in rearranging operations along the critical path associated with a complete and consistent selection $A$. Let $A$ be the current selection and $c$ a critical path in $G_A$ of length $l(c)$. A first result, due to Balas (1969), can be stated as follows.

*If a selection $A'$ exists with critical path $c'$ satisfying $l(c') < l(c)$ then at least one disjunctive arc of $A \cap c$ is reversed in $c'$. Moreover, any selection obtained by reversing a disjunctive arc of $A \cap c$ remains consistent.*

A related enumeration scheme can be the following: a node of the search tree corresponds to a pair $(A, c)$, where $A$ and $c$ are defined as above, and we branch by reversing each disjunctive arc of $A \cap c$.

A second result, due to Grabowski et al. (1986), can be viewed as an enhancement of the previous one. Let us define a block as a sequence of operations on the critical path that are scheduled consecutively on the same machine. We have the following.

*If a selection $A'$ exists with critical path $c'$ satisfying $l(c') < l(c)$ then at least one operation of at least one block is scheduled either after or before every other operation in its block in $A'$.*

The related enumeration schemes consist of generating all possible rearrangements of operations according to the previous result (Grabowski et al. 1986; Brucker et al. 1991).

*Selection on a subclique of disjunctions*

In such an enumeration scheme the main idea is to focus on a subclique of disjunctions not entirely selected at the current step. The simplest way to proceed is to choose a nonselected disjunctive constraint and to orient it in either of two ways. Carlier and Pinson (1991,1994) proposed a more sophisticated branching scheme based on this principle. As an illustration, we detail below the one proposed by Carlier and Pinson (1994), which relies on the idea developed in Section 13.3.3. They focused on a pair $(J, c)$ for which one of the three conditions ((13.8), (13.9) and the symmetric version of (13.9)) leading to an immediate selection holds. Then they branch by creating two subproblems corresponding to each of the two complementary conditions. The choice of the pair $(J, c)$ is made in order to improve the global lower bound as quickly as possible. If no pair $(J, c)$ satisfies the conditions above, they simply branch on a nonselected disjunctive constraint using a penalty function as defined by Carlier and Pinson (1989).

## 13.4 APPROXIMATION ALGORITHMS

A substantial effort has been devoted to design approximation algorithms for solving the job shop scheduling problem. The aim of this section is to review briefly the different main approaches that have been tested.

*List schedules*

Most approximation algorithms proposed up to now are based on list schedules associated with dispatching rules. Considerable effort has been expended on empirical tests of such rules, and we refer for example to Lawrence (1984) for intensive experiments on a large collection of priority dispatching rules. These experiments show the erratic behaviour of such an approach. Also it can be noted that nearly no results on performance guarantees are reported in the literature.

*Local search technique*

Simulated annealing (van Laarhoven *et al.* 1992) and taboo search techniques (Dell'Amico and Trubian 1993; Widmer 1989) are the main local search techniques that have been tested on the job shop scheduling problem. In both cases the neighborhood structure is based on critical path rearrangement (see Section 13.3.5). In van Laarhoven *et al.* (1992) and Widmer (1989) a transition is generated by reversing the sequencing order of two critical operations, while Dell'Amico and Trubian (1993) used a combination of three neighborhood structures including the previous one and the block approach designed by Grabowski. Note the nice result due to van Laarhoven *et al.* (1992) proving the asymptotic convergence in probability to a global minimal solution of a simulated annealing procedure using the first neighborhood mentioned above.

In comparison with other heuristic methods, both techniques yield quite consistently good solutions. Nevertheless, simulated annealing is comparatively much more time-consuming than taboo search on difficult instances.

*Truncated branch and bound methods*

One of the most efficient approximate methods proposed so far is probably the *shifting bottleneck procedure* developed by Adams *et al.* (1988). Its main idea is the following. Starting with the initial job shop scheduling problem, they optimally sequence one by one the machines, using Carlier's (1982) algorithm for the one-machine problem. At each optimization step, heads and tails adjustments are computed. The order in which the machines are sequenced depends on a bottleneck measure associated with them. Each time a new machine is sequenced, they attempt to improve the operating sequence of all previously scheduled machines in a reoptimization step. This procedure is embedded in a second heuristic of an enumerative type, for which each node of the search tree corresponds to a subset of sequenced machines.

Finally attempts to solve the job shop scheduling problem using genetic algorithms have recently been reported (Dellacroce *et al.* 1992).

## 13.5 CONCLUSIONS

This chapter has attempted to present a concise survey of the huge research effort devoted to the job shop scheduling problem over the last 30 years. Considerable progress has been made towards an efficient resolution of this problem in the 1980s. The $10 \times 10$ instance of Fischer and Thompson (1963) can now be solved to optimality within less than 2 min of CPU-time on a small workstation with recent methods (Carlier and Pinson 1991). Nevertheless, this remains one of the most difficult combinatorial problems to date, and always arouses new research interest. Larger instances proposed among others by Adams et al. (1988) still remain open. They will probably constitute one of the prime challenges for the 1990s.

## ACKNOWLEDGMENTS

We are grateful to J. K. Lenstra and anonymous referees for their helpful suggestions regarding the presentation of this chapter.

## REFERENCES

Aaarts, E, H. L., Lenstra, J. K. and Vaessens, R. J. M. (1994) Job shop scheduling by local search, Memorandum COSOR 94-05, Eindhoven University of Technology.
Akers S. B. (1956) A graphical approach to the production scheduling problem *Oper. Res.*, **4**, 244–245.
Adams J, Balas E. and Zawack D. (1988) The shifting bottleneck procedure for job shop scheduling *Management Sci.*, **34**, 391–401.
Applegate D. and Cook W. (1991) A computational study of job-shop scheduling *ORSA J. Comput.*, **3**, 149–156.
Barker J. R. and McMahon G. B. (1985) Scheduling the general job shop *Management Sci.*, **31**, 594–598.
Balas E. (1969) Machine sequencing via disjunctive graphs: an implicit enumeration algorithm *Oper. Res.*, **17**, 941–957.
Balas E. (1979) Disjunctive programming *Ann. Discr. Math.*, **5**, 3–51.
Balas E. (1985) On the facial structure of scheduling polyhedra *Math. Programming Study*, **24**, 179–218.
Bouma R. W. (1982) Job shop scheduling: a comparison of three enumeration schemes in a branch and bound approach. Master's Thesis. Faculty of Econometrics and Operations Research. Erasmus University, Rotterdam.
Bratley P., Florian M. and Robillard P. (1973) On sequencing with earliest starts and due dates with application to computing bounds for the $(n/m/G/F\mathrm{max})$ problem *Nav. Res. Log. Quart.*, **20**, 57–67.
Brucker P. (1988) An efficient algorithm for the job-shop problem with two jobs *Computing*, **40**, 353–359.
Brucker P. and Jurisch B. (1993) A new lower bound for the job-shop scheduling problem *Eur. J. Oper. Res.*, **64**, 156–167.
Brucker P., Jurisch B. and Sievers B. (1991) A branch & bound algorithm for the job-shop scheduling problem *Discr. Appl. Math.*, to appear.
Brucker P., Jurisch B. and Krämer A. (1992) The job-shop and immediate selection *Discr. Appl. Math.*, to appear.
Carlier J. (1982) The one-machine sequencing problem *Eur. J. Oper. Res.*, **11**, 42–47.
Carlier J. (1984) Problèmes d'ordonnancements à contraintes de ressources: algorithmes et complexité. Thèse d'état, Université Paris VI.
Carlier J. and Pinson E. (1989) An algorithm for solving the job shop problem *Management Sci.*, **35**, 164–176.

Carlier J. and Pinson E. (1991) A practical use of Jackson's preemptive schedule for solving the job-shop problem *Ann. Oper. Res.*, **26**, 269–287.

Carlier J. and Pinson E. (1994) Adjusting heads and tails for the job-shop problem *Eur. J. Oper. Res.* **78**, 146–161.

Dellacroce F., Tadei R. and Volta G. (1992) Genetic algorithm for the job-shop problem. In *Proceedings of 3rd International Workshop on PMS, Como, Italy.*

Dell' Amico M. and Trubian M. (1993) Applying taboo search to the job-shop scheduling problem *Ann. Oper. Res.*, **41**, 231–252.

Dyer M. and Wolsey L. A. (1990) Formulating the single machine sequencing problem with release dates as a mixed integer program *Discr. Appl. Math.*, **26**, 255–270.

Fischer J. F. and Thompson G. L. (1963) *Industrial Scheduling* Prentice-Hall, Englewood Cliffs, NJ.

Fisher M. L., Lageweg B. J., Lenstra J. K. and Rinnooy Kan A. H. G. (1983) Surrogate duality relaxation for job-shop scheduling *Discr. Appl. Math.*, **5**, 65–75.

French S. (1982) *Sequencing and Scheduling: An Introduction to the Mathematics of the Job Shop* Wiley, Chichester.

Garey M. R. and Johnson D. S. (1979) *Computers and Intractability: A Guide to the Theory of NP-completeness* Freeman, San Francisco.

Grabowski J., Nowicki E. and Zdrzalka S. (1986) A block approach for single machine scheduling with release dates and due dates *Eur. J. Oper. Res.*, **26**, 278–285.

Giffler B. and Thompson G. L. (1960) Algorithms for solving production-scheduling problems *Oper. Res.*, **8**, 487–503.

Gonzalez T. and Sahni S. (1978) Flow-shop and job-shop schedules: complexity and approximation *Oper. Res.*, **26**, 36–52.

Hefetz N. and Adiri I. (1982) An efficient optimal algorithm for the two-machines unit-time job-shop schedule-length problem *Math. Oper. Res.*, **7**, 354–360.

Jackson J. R. (1955) Scheduling a production line to minimize maximum tardiness. Research Report 43, Management Science Research Project, University of California. Los Angeles.

Lageweg B. J., Lenstra J. K. and Rinnooy Kan A. H. G. (1976) Minimizing maximum lateness on one machine: computational experience and some applications *Statistica Neerlandica*, **30**, 25–41.

Lawler E. L., Lenstra J. K., Rinnooy Kan A. H. G. and Shmoys D. B. (1993) *Sequencing and Scheduling: Algorithms and Complexity.* North-Holland, Amsterdam.

Lawrence S. (1984) Resource constrained project scheduling. An experimental investigation of heuristic scheduling techniques. GSIA, Carnegie-Mellon University.

Lenstra J. K. (1976) Sequencing by enumerative methods. Mathematical Centre Tract 69, Mathematisch Centrum, Amsterdam.

Lenstra J. K. and Rinnooy Kan A. H. G. (1979) Computational complexity of discrete optimization problems *Ann. Discr. Math.*, **4**, 121–140.

Lenstra J. K., Rinnooy Kan A. H. G. and Brucker P. (1977) Complexity of machine scheduling problems *Ann. Discr. Math.*, **1**, 343–362.

McMahon G. B. and Florian M. (1975) On scheduling with ready times and due dates to minimize maximum lateness *Oper. Res.*, **23**, 475–482.

Pinson E. (1988) Le problème de job-shop. Thèse de doctorat de l'Université Paris VI.

Potts C. N. (1980) An adaptative branching rule for the permutation flow-shop problem *Eur. J. Oper. Res.*, **5**, 19–25.

Queyranne M. (1988) Structure of a simple scheduling polyhedron. Working Paper 1277, Faculty of Commerce and Business Administration, University of British Columbia, Vancouver.

Rinnooy Kan A. H. G. (1976) *Machine Scheduling Problems: Classification, Complexity and Computations* Nijhoff, The Hague.

Roy B. and Sussman B. (1964) Les problèmes d'ordonnancement avec contraintes disjonctives. Note DS no 9 bis, SEMA, Paris.

Sotskov Y. N. (1991) The complexity of scheduling problems with 2 & 3 jobs *Eur. J. Oper. Res.*, **53**, 326–336.

van Laarhoven P. J. M., Aarts E. H. L. and Lenstra J. K. (1992) Job shop scheduling by simulated annealing *Oper. Res.*, **40**, 113–125.

Widmer M. (1989) Job-shop scheduling with tooling constraints: a tabu search approach. OR Working Paper 89/22, Département de Mathématiques, Ecole Polytechnique Fédérale de Lausanne.

*Institut de Mathématique Appliquées, Université Catholique de l'Ouest, BP808, 49008 Angers Cedex 01, France*

CHAPTER 14

# Application of Majorization to Control Problems in Queueing Systems

**D. Towsley**
*University of Massachusetts, Amherst Department of Computer Science*

**Abstract**

We give an overview of the use of *majorization* for proving the optimality of scheduling policies in queueing systems. Majorization is a method for comparing vectors with real- or integer-valued components. We define it and present a number of its properties. Applications to several problems in the control of queueing systems are then given. These include establishing the optimality of the smallest remaining processing time (SRPT) policy for the $G/G/1$ queue, establishing extremal properties of the first-come first-serve (FCFS) and last-come first-serve (LCFS) policies for the $G/G/s$ queue, and establishing the optimality of the shortest queue (SQ) policy for identical parallel queues. These applications illustrate how majorization can be coupled with sample path arguments. They also illustrate some of the subtleties that arise in coupling sample paths under different policies.

## 14.1 INTRODUCTION

Majorization ordering is a powerful method for comparing deterministic vectors. A number of stochastic orderings for random vectors have been developed based on this comparison method (Marshall and Olkin 1979; Chang 1992). Although the underlying theory has its roots in the 1930s (Hardy *et al.* 1952), it has only recently been applied to the control of queues (for some early examples, see Foss 1980; Daley 1987; Menich 1987). It is the purpose of this chapter to present an overview of this theory and to demonstrate its applicability to the control of queues. Applications will be made to a number of control problems, including routing to parallel queues, scheduling from parallel queues and scheduling multiple server and single-server queues. Several classical results, including the optimality of the *smallest remaining*

---

*Scheduling Theory and its Applications* Edited by P. Chrétienne, E. G. Coffman, Jr., J. K. Lenstra and Z. Liu
© 1995 John Wiley & Sons Ltd

processing time (SRPT) policy for the $G/G/1$ queue and the property that FIFO minimizes and LIFO maximizes the variance in the stationary response time of the $G/GI/1$ queue will be shown to be consequences of underlying majorizations. One of the objectives of this survey will be to show how sample path arguments are typically used, along with majorization, to establish such optimality results. Another objective is to illustrate the problems and approaches to coupling sample paths under different policies required to make comparisons.

There exist a number of different methodologies for studying control problems in queueing systems. In addition to sample path analysis, these include queueing theory, discrete event simulation and Markov decision processes. The first two of these techniques are primarily concerned with evaluating the performance of control policies. Their use in studying optimal control policies stems from their applicability to investigating the performance of a wide variety of policies. Markov decision processes (MDPs), on the other hand, are directly applicable to the control of queues. Majorization has been used with MDPs (see Menich 1987; Menich and Serfozo 1991) to solve a number of problems. However, in each of these problems majorization coupled with sample path arguments has led (or can lead) to more general results. This is because the Markovian assumptions typically required in MDPs are not always needed in sample path analysis. The reader is referred to excellent treatments of MDPs by Ross (1970) and Bertsekas (1987) for further details.

Sample path arguments have been widely used in research on control of queues, typically coupled with some comparison method other than majorization. These methods have included strong ordering and convex ordering (Ross 1983; Stoyan 1983; Shanthikumar and Sumita 1987; Vasicek 1977), along with many others. In addition, similar techniques have been used to derive first- and second-order monotonicity properties of discrete event systems (for examples, see Weber 1979; Shaked and Shanthikumar 1988; Bacelli et al. 1989; Bacelli and Liu 1991; Shanthikumar and Yao 1990). We shall only consider applications that use majorization as the underlying comparison.

The remainder of the chapter is organized as follows. Section 14.2 introduces several stochastic comparisons, including those based on majorization. The optimality of a generalized SRPT policy is presented in Section 14.3. Section 14.4 focusses on the $G/GI/s$ queue and establishes extremal properties of the first-come first-serve and last-come first-serve policies with respect to the response times of customers. Section 14.5 applies majorization to establish the optimality of the *join the shortest queue (SQ)* policy for identical parallel finite-capacity queues. The chapter concludes with a short summary in Section 14.6.

## 14.2 STOCHASTIC ORDERING AND MAJORIZATION

In this section we introduce the concept of majorization and several stochastic orderings. Let $X, Y \in \mathbb{R}^n$ be two random variables (r.v.s).

**Definition 14.1** *The random vector $X$ is stochastically less than the random vector $Y$ in the sense of strong stochastic ordering ($X \leq_{st} Y$), convex ordering, ($X \leq_{cx} Y$), and increasing convex ordering ($X \leq_{icx} Y$) if*

$$E[f(X)] \leq E[f(Y)] \quad \forall \text{ increasing } f : \mathbb{R}^n \to \mathbb{R},$$
$$E[f(X)] \leq E[f(Y)] \quad \forall \text{ convex } f : \mathbb{R}^n \to \mathbb{R},$$
$$E[f(X)] \leq E[f(Y)] \quad \forall \text{ increasing and convex } f : \mathbb{R}^n \to \mathbb{R},$$

*respectively, provided the expectations exist.*

**Definition 14.2** *The stochastic process $X(t)_{t \geq 0}$ is stochastically less than the stochastic process $Y(t)_{t \geq 0}$ in the sense of strong stochastic ordering $(X(t)_{t \geq 0} \leq_{st} Y(t)_{t \geq 0})$ if*

$$(X(t_1), \ldots X(t_k)) \leq_{st} (Y(t_1), \ldots, Y(t_k)) \quad \forall\, t_1, \ldots, t_k \geq 0, \quad k > 0.$$

See Stoyan (1983) for properties concerning these orderings.

We now focus on majorization and its associated orderings.

**Definition 14.3** *Let $x, y \in \mathbb{R}^n$ be two real vectors. Let $x_{[i]}$ denote the ith largest component of $x$. The vector $x$ is said to be majorized by the vector $y$ (written $x \prec y$) iff*

$$\sum_{i=1}^{k} x_{[i]} \leq \sum_{i} = 1^k y_{[i]}, \quad k = 1, \ldots, n-1$$

$$\sum_{i=1}^{n} x_{[i]} = \sum_{i} = 1^n y_{[i]}.$$

*The vector $x$ is said to be weakly submajorized by the vector $y$ (written $x \prec_w y$) iff*

$$\sum_{i=1}^{k} x_{[i]} \leq \sum_{i=1}^{k} y_{[i]}, \quad k = 1, \cdots, n,$$

*and weakly supermajorized by vector $y$ (written $x \prec^w y$) iff*

$$\sum_{i=k}^{n} x_{[i]} \geq \sum_{i=k}^{n} y_{[i]}, \quad k = 1, \cdots, n.$$

We shall refer to weak submajorization as *weak majorization* throughout the remainder of the chapter.

Another variation is *p-weak majorization*, first introduced in Sparaggis et al. (1993).

**Definition 14.4** *Let $x, y \in \mathbb{R}^n$ be two real vectors and $p \in \mathbb{R}$. The vector $y$ is said to p-weakly submajorize the vector $x$ (written $x \prec_{(p)w} y$) iff*

$$\sum_{i=1}^{k} x_{[i]} \leq p + \sum_{i=1}^{k} y_{[i]}, \quad k = 1, \cdots, n. \tag{14.3}$$

One can define a $p$-weak supermajorization in a similar manner. However, we shall only be interested in $p$-weak submajorization, which will henceforth be referred to as $p$-weak majorization.

In the case that $x$ and $y$ have integer components, and $p$ is an integer, we have the following useful properties (for the proof, see Sparaggis *et al.* 1994).

**Lemma 14.1** *If* $x \prec_{(p)w} y$ *then*

(1) $(x_{[1]},\cdots,(x_{[k]}-1)^+,\cdots,x_{[n]}) \prec_{(p)w} w(y_{[1]},\cdots,(y_{[j]}-1)^+,\cdots,y_{[n]})$, $1 \leqslant k \leqslant j \leqslant n$;
(2) $(x_{[1]},\cdots,x_{[k]}+1,\cdots,x_{[n]}) \prec_{(p)w} (y_{[1]},\cdots,y_{[j]}+1,\cdots,y_{[n]})$, $1 \leqslant j \leqslant k \leqslant n$;
(3) $(x_{[1]},\cdots,x_{[k]}+1,\cdots,x_{[n]}) \prec_{(p+1)w} y$, $1 \leqslant k \leqslant n$.

Consider two vectors $x, y \in \mathbb{R}$. We introduce the notation $y_\gamma$, $(y_{\gamma(1)},\ldots,y_{\gamma(n)})$ where $\gamma$ is a permutation on $(1, 2, \ldots, n)$. The following property, which is a special case of Corollary 3.2 in Day (1972), will prove useful later in this chapter.

**Lemma 14.2** *Let* $x, y \in \mathbb{R}^n$ *such that* $x_1 \leqslant x_2 \leqslant \cdots \leqslant x_n$ *and* $y_1 \leqslant y_2 \leqslant \cdots \leqslant y_n$. *Then, for all permutations* $\gamma$,

$$(y - x) \prec (y_\gamma - x) \prec (y_\chi - x),$$

*where $\chi$ is defined by* $\chi(i) = n - i + 1$, $i = 1, \ldots, n$.

See Marshall and Olkin (1979) for an excellent treatment of majorization and weak majorization and to Sparaggis *et al.* (1990) for a treatment of $p$-weak majorization.

Define the following classes of functions:

- $\mathscr{C}_1, (\mathscr{C}_1^\uparrow)$, the set of (increasing) Schur-convex functions, $f : \mathbb{R}^n \to \mathbb{R}$;
- $\mathscr{C}_2, (\mathscr{C}_2^\uparrow)$, the set of (increasing) symmetric and convex functions $f : \mathbb{R}^n \to \mathbb{R}$;
- $\mathscr{C}_3, (\mathscr{C}_3^\uparrow)$, the set of functions of the form $f(x) = \sum_{i=1}^n g(x_i)$, where $g : \mathbb{R} \to \mathbb{R}$ is a (increasing) convex function);
- $\mathscr{C}_4, (\mathscr{C}_4^\uparrow)$, the set of (increasing) functions of the form $f(x) = \sum_{i=1}^k x_{[i]}$ or $f(x) = -\sum_{i=k}^k x_{[i]}$, $k = 1, \ldots, n$;
- $\mathscr{C}_5, (\mathscr{C}_5^\uparrow)$, the set of (increasing) symmetric, L-subadditive functions $f : \mathbb{R}^n \to \mathbb{R}$ that are convex in each argument.

Here a function $f : \mathbb{R}^n \to \mathbb{R}$ is

(i) Schur-convex iff $f(x) \leqslant f(y)$ whenever $x \prec y$, $x, y \in \mathbb{R}^n$;
(ii) L-subadditive if for any $1 \leqslant i \leqslant j \leqslant n$ and any nonnegative $\epsilon_1$ and $\epsilon_2$ the following inequality holds:

$$f(\ldots, x_i+\epsilon_1, \ldots, x_j + \epsilon_2, \ldots) + f(\ldots, x_i, \ldots, x_j, \ldots)$$
$$\leqslant f(\ldots, x_i + \epsilon_1, \ldots, x_j, \ldots) + f(\ldots, x_i, \ldots, x_j + \epsilon_2, \ldots).$$

The following relations hold between these different classes of functions: $\mathscr{C}_4 \subset \mathscr{C}_2 \subset \mathscr{C}_1$, $\mathscr{C}_3 \subset \mathscr{C}_2$ and $\mathscr{C}_5 \subset \mathscr{C}_1$.

Examples of functions defined over $\mathbb{R}^n$ contained in both $\mathscr{C}_2$ and $\mathscr{C}_5$ are max $x$ and $\sum_{i=1}^n g(x_i)$ where $g : \mathbb{R} \to \mathbb{R}$ is convex. Another such function defined over $(\mathbb{R}^+)^n$ is $z(x) = \sum_{i=1}^n \mathbf{1}(x_i = 0)$.

We define the following orderings between r.v.s $X, Y \in \mathbb{R}^n$:

$$X \leqslant_{E_i} Y (X \leqslant_{E_i^\uparrow} Y) \quad \text{iff} \quad E[f(X)] \leqslant E[f(Y)], \quad \forall f \in \mathscr{C}_i(\mathscr{C}_i^\uparrow), \quad 1 \leqslant i \leqslant 5.$$

See Marshall and Olkin (1979) for properties of the $\leqslant_{E_i}$ and $\leqslant_{E_i^\uparrow}$ orderings for $1 \leqslant i \leqslant 4$ and Chang (1992) for properties of the $\leqslant_{E_5}$ and $\leqslant_{E_5^\uparrow}$ orderings. All but the $\leqslant_{E_4}$ and $\leqslant_{E_4^\uparrow}$ orderings have been regularly applied to problems concerning the control of queueing systems. The applications in this chapter will focus on the first three orderings. References to applications based on the $\leqslant_{E_5}$ and $\leqslant_{E_5^\uparrow}$ orderings will be given when appropriate.

The following classes of r.v.s will be of interest to us. Let $X \in \mathbb{R}$ be a nonnegative valued r.v. with probability density function $f_X(x)$ and cumulative distribution function $F_X(x) = \int_0^\infty f_X(y) \, dy$, $x \geqslant 0$. The r.v. $X$ is said to have *increasing failure rate (IFR)* if $f_X(x)/[1 - F_X(x)]$ is an increasing function of $x$, and to be *increasing in likelihood ratio (ILR)* if $f_X(x + c_2)/f_X(x + c_1)$ is increasing in $x$ for all $0 \leqslant c_1 \leqslant c_2$.

## 14.3 OPTIMALITY OF THE SMALLEST REMAINING PROCESSING TIME (SRPT) POLICY

In this section we state and prove a scheduling result for the $G/G/1$ queue with Bernoulli feedback. A simple consequence of this result is the optimality of the SRPT policy for the $G/G/1$ queue. Customers arrive at times $0 < a_1 < \ldots < a_n < \ldots$. At arrival, the $n$th customer (which arrives at $a_n$) generates a phase (the first phase of the $n$th customer) that requires $v_n > 0$ units of service. When the customer has completed executing its phase, it begins a new phase with probability $0 \leqslant p < 1$.

Let $\{u_n\}_{n=1}^\infty$ be a sequence of independent and identically distributed (i.i.d.) nonnegative real r.v.s such that for all $n \geqslant 1$, $\Pr[u_n > 0] = p$. If $u_n > 0$ then the $n$th completed phase initiates a new phase, requiring $u_n$ units of service, belonging to the same customer as the $n$th completed phase. This customer is placed into the waiting queue for execution. Otherwise, it has no successor phase and the customer to which it belongs is finished. Note that the service time requirements of the first phase need not have the same distribution as the service times of subsequent phases. Denote the $n$th generated phase by $\phi_n$.

Let $\Psi$ denote the class of policies that know the remaining service requirements of the phases of customers queued up for service and that allow preemption. We are interested in determining the policy from $\Psi$ that minimizes the number of customers in the system at any point in time. It should be clear that we need only consider non-idling policies. Let SP denote the policy that always schedules the customer in the queue whose phase requires the least amount of service to complete.

Let $\pi$ be an arbitrary policy in $\Psi$. We introduce the following notation.

- $N^\pi(t)$, the number of customers in the system at time $t \geq 0$;
- $S_n^\pi(t)$, the remaining service time for phase $\phi_n$, $n = 1, 2, \cdots$, in the system at time $t \geq 0$;
- $h^\pi(t)$, the number of phases generated (owing to either the arrival of a job or a phase completion) by time $t \geq 0$;
- $c_n^\pi$, the completion time of customer $n$;
- $R_n^\pi = c_n^\pi - a_n$, the response time of the $n$th customer;
- $R^\pi$, the stationary customer response time, provided that it exists;
- $n(t)$, the number of customers that arrive by time $t \geq 0$.

We now state and prove our result.

**Theorem 14.1** *Under the above assumptions, the smallest phase first policy stochastically minimizes the process of the number of jobs in the system:*

$$\{N^{SP}(t)\}_{t \geq 0} \leq_{st} \{N^\pi(t)\}_{t \geq 0} \quad \forall \pi \in \Psi.$$

**Proof** Fix the arrival times and the phase service times. Let $\pi \in \Psi$ be an arbitrary policy. Let $0 = t_0 < t_1 < \ldots < t_n < \ldots$ be the event times under $\pi$ and SP (i.e. arrival times, phase completion times and times at which preemptions occur). Define $Q^k(t) = (Q_1^k(t), \ldots, Q_{h^{SP}(t)}^k(t))$ to be the vector of remaining service times for phases $1, \ldots, h^{SP}(t)$ under policy $k = \pi$, SP at time $t > 0$. We shall establish the following relations through a forward-induction argument,

$$Q^\pi(t) \prec Q^{SP}(t), \qquad (14.4)$$
$$h^\pi(t) \leq h^{SP}(t)$$

for all $t \geq 0$. The latter inequality implies that $N^\pi(t) \geq N^{SP}(t)$. Although capital letters are usually reserved to denote random variables, within the proof of this and subsequent theorems, they also indicate the values of the variables at specific time instants on a sample path.

*Basis step* By the statement of the theorem, the above relations hold for $t = t_0$.

*Inductive step* Assume that the relations hold up through $t = t_i$. First, we establish the relation for $t_i < t < t_{i+1}$. By the inductive hypothesis,

$$\sum_{i=1}^k Q_{[i]}^\pi(t_i) \leq \sum_{i=1}^k Q_{[i]}^{SP}(t_i), \quad k = 1, \ldots, h^{SP}(t_i).$$

Now, SP serves the phase with the smallest residual service time. Let this be the $m$th largest component in $Q^{SP}(t)$ (i.e. $h^{SP}(t) - m$ phases have completed). Let $\pi$ execute the phase with the $l$th-largest service time. (Note that the phases in positions

$h^\pi(t_i)+1,\ldots,h^{SP}(t_i)$ cannot be chosen, since they have not arrived at the system under policy $\pi$.) Then at time $t$ we have two cases, depending on whether $l \leq m$ or not. If $l \leq m$ then

$$\sum_{j=1}^{k} Q_{[j]}^{\pi}(t) = \sum_{j=1}^{k} Q_{[j]}^{\pi}(t_i) \leq \sum_{j=1}^{k} Q_{[j]}^{SP}(t_i)$$

$$= \sum_{j=1}^{k} Q_{[j]}^{SP}(t), \quad k=1,\ldots,l-1,$$

$$\sum_{j=1}^{k} Q_{[j]}^{\pi}(t) = \sum_{j=1}^{k} Q_{[j]}^{\pi}(t_i) - (t-t_i) \leq \sum_{j=1}^{k} Q_{[j]}^{SP}(t_i), \quad k=l,\ldots,m-1,$$

$$\sum_{j=1}^{k} Q_{[j]}^{\pi}(t) = \sum_{j=1}^{k} Q_{[j]}^{\pi}(t_i) - (t-t_i) \leq \sum_{j=1}^{k} Q_{[j]}^{SP}(t_i) - (t-t_i)$$

$$= \sum_{j=1}^{k} Q_{[j]}^{SP}(t), \quad k=m,\ldots h^{SP}(t).$$

In the second case $l > m$ and

$$\sum_{j=1}^{k} Q_{[j]}^{\pi}(t) = \sum_{j=1}^{k} Q_{[j]}^{\pi}(t_i) \leq \sum_{j=1}^{k} Q_{[j]}^{SP}(t_i)$$

$$= \sum_{j=1}^{k} Q_{[j]}^{SP}(t), \quad k=1,\ldots,m-1,$$

$$\sum_{j=1}^{k} Q_{[j]}^{\pi}(t) = \sum_{j=1}^{k} Q_{[j]}^{\pi}(t_i) \leq \sum_{j=1}^{h^{SP}(t)} Q_{[j]}^{\pi}(t_i) - (t-t_i)$$

$$= \sum_{j=1}^{h^{SP}(t)} Q_{[j]}^{SP}(t_i) - (t-t_i)$$

$$= \sum_{j=1}^{k} Q_{[j]}^{SP}(t), \quad k=m,\ldots,l-1,$$

$$\sum_{j=1}^{k} Q_{[j]}^{\pi}(t) = \sum_{j=1}^{k} Q_{[j]}^{\pi}(t_i) - (t-t_i) \leq \sum_{j=1}^{h^{SP}(t)} Q_{[j]}^{SP}(t_i) - (t-t_i)$$

$$= \sum_{j=1}^{k} Q_{[j]}^{SP}(t), \quad k=l,\ldots h^{SP}(t).$$

Hence we conclude that

$$Q^{\pi}(t) \prec Q^{SP}(t), \quad t \geq 0.$$

If the event at time $t_{i+1}$ is either a departure under either policy or a preemption under $\pi$ then we are done, since $h^k(t_{i+1}) = h^k(t_i)$, $k = \pi, \text{SP}$ and $Q^\pi(t_{i+1}) \prec Q^{\text{SP}}(t_{i+1})$. In the case of a phase completion under $\pi$ followed by the initiation of a new phase it can be argued that $h^\pi(t_i) < h^{\text{SP}}(t_i)$; otherwise the relation (14.4) could not hold at $t_i$. Hence $h^\pi(t_{i+1}) = h^\pi(t_i) + 1 \leq h^{\text{SP}}(t_i) = h^{\text{SP}}(t_{i+1})$, and the relation (14.4) holds at time $t_{i+1}$ by the argument given above.

In the case of either an arrival at both systems or a phase completion under SP (and possibly $\pi$) followed by the initiation of a new phase at $t_{i+1}$ we have

$$Q^k(t_{i+1}) = (Q_1^k(t_{i+1}), \ldots, Q_{h^{\text{SP}}(t)}^k(t_{i+1}), \sigma), \quad k = \pi, \text{SP},$$

where $\sigma$ is the service time of the new phase. Clearly, $h^\pi(t_{i+1}) \leq h^{\text{SP}}(t_{i+1})$. An argument similar to that above can be used to establish that

$$Q^\pi(t_{i+1}) \prec Q^{\text{SP}}(t_{i+1}).$$

Removal of the conditioning on the arrival times and phase service times completes the proof. $\square$

**Remark** The proof of the last theorem illustrates the use of *forward induction*. We shall observe in the next section an example of an *interchange argument*. These are the proof techniques most commonly used in sample path analysis.

**Remark** In some sense the $\prec$ relation compares the balance of two vectors. In the previous theorem we compared workload vectors, where the components corresponded to the remaining service requirements of individual phases. The SP policy is the one that produces the most "unbalanced" such workload vector. As a consequence, it has completed more phases (produced more phases), which implies that it has completed more customers by time $t \geq 0$. In later applications we shall observe that the optimal policy is the one that produces the most "balanced" state vector.

If the system exhibits stationary behavior under SP and $\pi$ such that $N^\pi = \lim_{t \to \infty} N^\pi(t)$ and $N^{\text{SP}} = \lim_{t \to \infty} N^{\text{SP}}(t)$ exist then, by Little's formula,

$$E[R^{\text{SP}}] = E[N^{\text{SP}}]/\lambda,$$

where $\lambda$ is the arrival rate of jobs. Applying Theorem 14.1 yields the following result.

**Corollary 14.1** *Under the assumptions stated above, the SP policy minimizes the average job response time:*

$$E[R^{\text{SP}}] \leq E[R^\pi], \forall \pi \in \Psi.$$

**Remark** The optimality of the shortest remaining processing time (SRPT) policy in the case that service times are completely known follows as a consequence by setting $p = 0$. This result was first established by Schrage (1969). More recently, Walrand (1988) used a majorization argument to obtain the SRPT optimality result.

A similar result has been obtained for parallel processing systems (Liu and Towsley 1992c). Briefly, this work considers a branching fork/join processing model. In the case that all task service times are i.i.d. exponential r.v.s, the policy that schedules the customer with the smallest number of tasks is shown to minimize the process of the number of customers in the system.

## 14.4 EXTREMAL PROPERTIES OF QUEUEING DISCIPLINES

In this section we consider the $G/GI/s$ queue and establish stochastic majorization relationships between vectors of wait times and response times for different queueing disciplines. As before, let $0 < a_1 < \ldots < a_n < \ldots$ be the sequence of arrival times, i.e. the $n$th customer arrives at time $a_n$, $n = 1, 2, \ldots$. Customers arrive at a single queue that is served by $s$ servers. Let $\{\sigma_i\}_{i=1}^\infty$ be a sequence of i.i.d. r.v.s where $\sigma_i$ corresponds to the service time of the $i$th customer. We further assume that $\{\sigma_i\}_{i=1}^\infty$ is independent of the arrival times. Let $\Sigma_{np}$ denote the set of policies that are non-idling, non-preemptive and do not use service time information. Let FCFS and LCFS denote the first-come first-serve and last-come first-serve policies within $\Sigma_{np}$.

We define the following quantities.

- $\pi_n$, the identity of the customer scheduled in the $n$th position under $\pi \in \Sigma_{np}$;
- $i_n^\pi$, the identity of the $n$th customer to depart the system under $\pi \in \Sigma_{np}$;
- $s_n^\pi$, the time of the $n$th scheduling instance under $\pi \in \Sigma_{np}$.
- $d_n^\pi$, the time of the $n$th departure under $\pi \in \Sigma_{np}$.
- $W^\pi(n) = (W_1^\pi, \ldots, W_n^\pi)$, the vector of wait times of the first $n$ customers, where the wait time of the $n$th customer is $W_n^\pi = s_{\pi_n}^\pi - a_n$;
- $R^\pi(n) = (R_1^\pi, \ldots, R_n^\pi)$, the vector of response times of the first $n$ customers, where the response time of the $n$th customer is $R_n = d_{i_n^\pi}^\pi - a_n = W^\pi(n) + \sigma_n$.

The following result was established in Liu and Towsley (1992a).

**Theorem 14.2** *Consider a $G/G/s$ system with $n$ customer arrivals. If the service times form a sequence of i.i.d. r.v.s independent of the arrival process then*

$$W^{\text{FCFS}}(n) \leqslant_{E_1} W^\pi(n) \leqslant_{E_1} W^{\text{LCFS}}(n) \quad \forall \pi \in \Sigma_{np}. \tag{14.5}$$

**Proof** We condition on the arrival times and service times. Assume that the service times are assigned in the order of service. This is permitted because of the assumption that service times form an i.i.d. sequence of r.v.s independent of the arrival times. We focus on the relation for $\pi$ and FCFS and establish that

$$W^{\text{FCFS}}(n) \prec W^\pi(n) \quad \forall \pi \in \Sigma_{np}. \tag{14.6}$$

Choose an arbitrary policy $\pi \in \Sigma_{np}$ that differs from FCFS for the first time at the $m$th scheduling instance, i.e. $\pi_l = l$, $l = 1, 2, \ldots, m-1$, $\pi_m = m$, and $\pi_\nu = m$ for some $v > m$. We construct a new policy $\pi^1$ as

$$\pi_l^1 = \pi_l, \quad l \neq m, v,$$
$$\pi_m^1 = \pi_v,$$
$$\pi_v^1 = \pi_m.$$

We now apply Lemma 14.2 to obtain

$$W^{\pi^1}(n) \prec W^{\pi}(n).$$

The same procedure can now be repeated to construct a sequence of at most $n - m - 1$ policies, $\pi^2, \pi^3, \ldots \pi^{(n-m-1)} = $ FCFS such that

$$W^{\pi^i}(n) \prec W^{\pi^{(i-1)}}(n), \quad i = 2, \ldots n - m - 1.$$

This establishes (14.6). Since this holds *for every sample path*, removal of the conditioning yields the $E_1$ relationship between the wait time vectors under FCFS and $\pi$. A similar argument yields the result for LCFS. □

**Remark** The proof of this theorem illustrates the use of an interchange argument.
The following weaker result has also been established for response time vectors (see Liu and Towsley 1992a).

**Corollary 14.2** *Consider a $G/G/s$ system with $n$ arrivals. If the service times form a sequence of i.i.d. r.v.s. independent of the arrival process then*

$$R^{\text{FCFS}}(n) \leqslant {}_{E_2} R^{\pi}(n) \leqslant {}_{E_2} R^{\text{LCFS}}(n),; \quad \forall \pi \in \Sigma_{np} \quad (14.7)$$

**Proof** We have

$$R^{\text{FCFS}} = (W_1^{\text{FCFS}} + \sigma_1, \ldots, W_n^{\text{FCFS}} + \sigma_n),$$

where $\sigma_1, \ldots, \sigma_n$ are exchangeable r.v.s. If $f : \mathbb{R}^n \to \mathbb{R}$ is a symmetric convex function then $g : \mathbb{R}^n \to \mathbb{R}$ defined by

$$g(x_1, \ldots, x_n) = E[f(x_1 + \sigma_1, \ldots, x_n + \sigma_n)]$$

is also symmetric convex (see Marshall and Olkin 1979, Proposition B.5, p. 289). Since any symmetric convex function is a Schur-convex function and

$$E[f(R_1^{\pi}, \ldots, R_n^{\pi})] = E[g(W_1^{\pi} + \sigma_1, \ldots, W_n^{\pi} + \sigma_n)],$$

the proof follows from Theorem 14.2. □

**Remark** This last proof result was not based on sample path arguments. We (and others—Chang 1992) have observed that the $E_2$ ordering is not amenable to such arguments. Instead, such an ordering is typically established by invoking the property that $E_2$ is closed under convolution (under suitable assumptions). That is the idea behind the above proof.

**Remark** In the case of the $G/G/1$ and the $G/D/s$ queues this result can be extended to an $E_1$ ordering using the argument presented in Theorem 14.2 (for details, see Towsley and Baccelli 1991; Liu and Towsley 1992a). This is because $\pi_l = i_l^\pi, l = 1, 2, \ldots$. A more intricate argument based on an ingenious sample path construction has been used to extend the result to the $G/ILR/s$ queue as well (for details, see Liu and Towsley 1992b).

This result can be extended to the stationary regime, provided that it exists. Recall that the sequence of r.v.s $X_n \in \mathbb{R}$, $n \geq 1$ converge to the r.v. $X$ for the class of Borel mappings $\mathscr{F} : \mathbb{R} \to \mathbb{R}$ in the Cesaro sense (Feller 1971, p. 249) if

$$\lim_{N \to \infty} \frac{1}{N} \sum_{i=1}^{N} E[f(X_i)] = E[f(X)].$$

**Lemma 14.3** *Let $\{X_i\}_{i=1}^{\infty}$ and $\{Y_i\}_{i=1}^{\infty}$ be two sequences of r.v.s such that*

$$(X_1, \ldots, X_n) \leq_{E_3} (Y_1, \ldots, Y_n) \quad \forall n \geq 1.$$

*If the sequences $\{X_i\}_{i=1}^{\infty}$ and $\{Y_i\}_{i=1}^{\infty}$ converge weakly to $X$ and $Y$ with respect to the class of convex functions in the Cesaro sense as $n \to \infty$ then*

$$X \leq_{cx} Y.$$

**Proof** See Liu and Towsley (1992a). □

This last lemma implies that the stochastic orderings $\leq_{E_1}$ and $\leq_{E_2}$ established for the vectors of wait times and response times respectively reduce to the stochastic ordering $\leq_{cx}$ on the corresponding stationary performance metrics $W^\pi$ and $R^\pi$ respectively, provided that the weak convergence assumptions are satisfied. These results had been established previously in Vasicek (1977), Shanthikumar and Sumita (1987) and Liu and Towsley (1992a).

If we consider the class of nonpreemptive policies that allow idling then the results hold, with no change in the argument used above, provided that $E_1^\uparrow$, $E_2^\uparrow$ and icx replace $E_1$, $E_2$ and cx in the above results. If we consider the class of preemptive non-idling policies $\Sigma_p$ then we have the following result (Liu and Towsley 1992a).

**Theorem 14.3** *Consider a $G/M/s$ system with $n$ arrivals. If the service times are independent of the arrival process then*

$$R^{FCFS}(n) \leq_{E_1} R^\pi(n) \leq_{E_1} R^{LCFS}(n) \quad \forall \pi \in \Sigma_p. \tag{14.8}$$

**Proof** A sketch of the proof is provided here. The complete proof can be found in Liu and Towsley (1992a). First, we can restrict ourselves to policies that preempt only at arrival times and service completion times. This is a consequence of the memoryless property of the exponential distribution.

Second, we associate a Poisson process with each of the $s$ servers. The events correspond to times at which service completions may occur. In particular, a service completion occurs if there is a customer assigned to that server; otherwise there is no service completion. A customer assigned to service at an idle server always receives, as its service time, the remaining time until the next event. The memoryless property guarantees that the received service time is an exponentially distributed r.v. with the proper parameter.

Now, fix the arrival times, the service completion events, pick an arbitrary policy $\pi$ and focus on FCFS. We establish the following relation:

$$R^{\text{FCFS}}(n) \prec R^\pi(n).$$

If we consider the sequence of decision epochs (arrival and service completion times), an interchange argument can be used to show that the policy $\pi^1$ that differs from $\pi$ in that it makes one less non-FCFS decision yields a response time vector that is majorized by the response time vector associated with $\pi$. This step can then be repeated until the FCFS policy is produced. Removal of the conditioning on arrival times and service events yields the desired result for FCFS. A similar argument yields the result for LCFS. □

**Remark** In the case of the $G/IFR/1$ queue, combining arguments in Hirayama and Kijima (1989) with those presented above yield

$$R^{\text{FCFS}}(n) \leqslant_{E_1^\uparrow} R^\pi(n), \quad \forall \pi \in \Sigma_{\text{p}}.$$

This does not extend to the $G/IFR/s$ system for $s > 1$, since it is easy to construct a counterexample in the case of deterministic service times (for details, see Liu and Towsley 1992a).

Last, these results have been extended to in-forests (Liu and Towsley 1992b) and to a class of parallel processing systems modeled by fork/join queueing networks (Baccelli *et al.* 1990).

## 14.5 APPLICATIONS TO ROUTING

Consider a single arrival stream of packets feeding $s$ identical servers. Assume that the service times form an i.i.d. sequence of exponentially distributed r.v.s and that each server has a capacity to store $B$ customers (including the one in service). Let $0 < a_1 < \cdots < a_n < \cdots$ be the sequence of arrival times, i.e. the $n$th customer arrives at time $a_n$, $n = 1, 2, \cdots$. The customers arrive at a controller, which routes them to the different servers. We first consider the class of routing policies $\Sigma$ that have instantaneous queue length information available to them and that are always required to route customers to some queue if one or more of them have available space. Define SQ to be the policy that always routes a customer to the queue containing the least number of customers. In case of a tie, any rule can be used to choose the destination queue.

Let $N^\pi(t) = (N_1^\pi(t), \ldots, N_s^\pi(t))$ denote the joint queue lengths at time $t > 0$ under policy $\pi \in \Sigma$. Let $L^\pi(t)$ denote the number of customers lost owing to buffer overflow under policy $\pi$ by time $t$. The following theorem, taken from Towsley et al. (1992), states that under SQ, the number of customers that are rejected by any time $t$ is minimized (in a stochastic sense). Moreover, the vector $N^\pi(t)$ is shown to be larger than $N^{SQ}(t)$ in the sense of the $E_1^\uparrow$ order, for any $\pi \in \Sigma$ and all times $t$. Based on this last result, one immediately concludes that the total number of jobs present in the system at any time $t$ is minimized under the SQ policy.

**Theorem 14.4**
$$\{L^{SQ}(t)\}_{t \geq 0} \leq_{st} \{L^\pi(t)\}_{t \geq 0},$$
$$N^{SQ}(t) \leq_{E_1^\uparrow} N^\pi(t) \quad (14.9)$$

for all $\pi \in \Sigma$, $t > 0$, provided that $N^\pi(0) =_{st} N^{SQ}(0)$ and $L^{SQ}(0) =_{st} L^\pi(0)$.

**Proof** We condition on the arrival times, service times, and initial queue lengths. The proof is by induction on event times (i.e. arrival times and service completions), $t_0 = 0, t_1, t_2, \ldots$. Specifically, we shall show that

$$L^{SQ}(t) \leq L^\pi(t), \quad (14.10)$$
$$N^{SQ}(t) \prec_w N^\pi(t). \quad (14.11)$$

$\forall t \geq 0$ over all sample paths. We couple the systems at time $t = 0$ so that $L^\pi(0) = L^{SQ}(0)$ and $N^\pi(0) = N^{SQ}(0)$.

To carry out a forward induction, we need to couple the systems in the following manner. We couple the service completion times at the $k$th largest queue under both policies, $k = 1, \ldots, s$. Furthermore, if any queue is empty, we assume that the server serves a fictitious customer and that a customer that arrives at that queue receives the remainder of this service time. The exponential assumption is required here to guarantee that all true service times form a sequence of i.i.d. r.v.s.

*Basis step* By the statement of the theorem, the relations hold for $t = t_0$.

*Inductive step* Assume that the relations hold through $t = t_i$. Clearly they hold for $t_i < t < t_{i+1}$. For $t = t_{i+1}$, we treat service completion and arrival events separately.

*Service completion* Relation (14.10) clearly holds at $t = t_{i+1}$ since there are no losses at service completions. Hence we focus on the relation (14.11). This follows from property (1) of Lemma 14.1 with $p = 0$.

*Arrival* SQ routes the customer to the smallest queue and $\pi$ routes the customer to some arbitrary queue. Clearly the inductive hypothesis $\sum_{i=1}^{s} N^\pi_{[i]} \geq \sum_{i=1}^{s} N^{SQ}_{[i]}$ guarantees that $L^\pi(t_{i+1}) \geq L^{SQ}(t_{i+1})$. Thus if a job is admitted into the system under $\pi$, it will also be admitted into the system under SQ. In this case, the relation (14.11) is ensured by property (2) of Lemma 14.1. On the other hand, if a job is rejected under $\pi$ then the system is full at time $t_{i+1}$, which implies that (14.11) holds at $t_{i+1}$. This completes the inductive step.

Removal of the conditioning on arrival times and service times completes the proof. □

This result can be easily generalized using similar arguments to the case that the queue capacities vary from server to server (for details, see Towsley et al. 1992).

It is interesting to note that SQ maximizes throughput at the same time that it minimizes holding costs. This is not typical of control procedures for finite buffer systems. Typically, one has to tradeoff throughput and queue lengths, as we observe in the following variation of the routing problem. Let $\Sigma_a$ denote the class of policies in which policies have instantaneous queue length information and have the choice of rejecting a customer as well as routing it to a particular queue. Observe that $\Sigma \subset \Sigma_a$. We have the following result.

**Theorem 14.5**

$$\{L^{SQ}(t)\}_{t \geq 0} \leq_s t\{L^\pi(t)\}_{t \geq 0} \qquad (14.12)$$

for all $\pi \in \Sigma_a$, $t > 0$, provided that $N^\pi(0) =_{st} N^{SQ}(0)$ and $L^{SQ}(0) =_{st} L^\pi(0)$.

**Proof** This proceeds in a similar vein as that of Theorem 14.4, except that the following relation is established in place of (14.11):

$$N^{SQ}(t) \prec_{(L^\pi(t) - L^{SQ}(t))w} N^\pi(t). \qquad (14.13)$$

Observe that, unlike Theorem 14.4, it is no longer possible to establish any straightforward relationship between the queue length vectors under $\pi$ and SQ. The relation (14.13) emphasizes that the difference between the total number of customers in both systems can never exceed the difference in the number of lost customers for the two systems.

The proof is again by forward induction, and differs from that of Theorem 14.4 only in the possibility that, at the time of an arrival, $\pi$ may choose to reject a customer even though there is space in a queue. If this happens and SQ accepts the customer then the application of property (3) in Lemma 14.1 yields the desired result. □

The relation (14.9) was established in the case of infinite-capacity queues at all of the servers by Ephremides et al. (1980) using similar ideas. In that paper the authors also established the optimality of the *cyclic policy* (C) among the class of policies that use no information. One has to be very careful regarding how to couple the sample paths between C and an arbitrary policy so that a comparison can be made. For example, the simple coupling used in the preceding theorem does not allow one to make any kind of comparison between the joint queue lengths under C and those under some arbitrary policy. The authors make the key observation that it suffices to couple the service processes so that departures occur simultaneously *at all servers* (this is permitted by the exponential service time assumption). In doing so, the joint queue length statistics for the resulting modified system differ from the joint statistics in the real system. However, since the routing decisions do not depend on the joint queue length statistics, the statistics of the individual queue lengths are unaffected. Let $\tilde{N}^\pi(t)$ denote the queue lengths for the system where the service processes have been coupled. As mentioned above, the individual queue length statistics are unaffected by this coupling; hence

$$\tilde{N}_k^\pi(t) =_{st} N_k^\pi(t), \quad 1 \leq k \leq s; \quad t \geq 0.$$

The proof that the cyclic routing policy is the optimum policy consists of using the weak majorization arguments of Theorem 14.4 to establish

$$\tilde{N}^C(t) \leq_{E_1^\uparrow} \tilde{N}^\pi(t), \quad t \geq 0 \qquad (14.14)$$

which has the consequence that

$$N^C(t) \leq_{E_3^\uparrow} N^\pi(t), \quad t \geq 0. \qquad (14.15)$$

This result has been generalized in two ways. First, C has been shown to minimize the process of the number of losses at all times when all buffers have the same finite capacity and service times are exponentially distributed (Sparaggis *et al.* 1994). Second, C has been shown to be optimal in the case of infinite capacity buffers and i.i.d. service times with a distribution having increasing failure rate (Liu and Towsley 1992d).

Foss (1980) has applied similar ideas to prove that the optimum routing policy when the controller has *instantaneous information regarding the unfinished work* is one that routes to the server with the smallest amount of unfinished work in the case of an i.i.d. sequence of service times. Here the system state is taken to be $U^\pi(t) = (U_1^\pi(t), \ldots, U_s^\pi(t))$, where $U_k^\pi(t)$ is the unfinished work at server $k$. The proof is complicated in this application because the elements are real-valued and can be incremented by an arbitrary amount at an arrival. Consequently Lemma 1 no longer applies. Foss provides a novel way of coupling the sample paths that circumvents this problem. For details, see Foss (1980). A similar result can be found in Daley (1987).

Similar ideas have been applied to the problem of scheduling $n$ different classes of customers to a single server (Towsley *et al.* 1990; Sparaggis *et al.* 1993). The latter study establishes a duality result between routing problems and scheduling problems.

Last, Chang (1992) has used majorization arguments to establish the optimality of the balanced Bernoulli routing policy for identical servers and general service times out of the class of Markovian routing policies with positive correlation. This result is interesting, because it is based on an $\leq_{E_5^\uparrow}$ ordering.

## 14.6 SUMMARY

We have provided an overview of majorization and its application to the control of queueing systems. The focus of this chapter has been on its use in three different applications:

(i) establishing the optimality of the SRPT policy in the $G/G/1$ queue;
(ii) establishing extremal properties of the FCFS and LCFS policies for the $G/G/s$ queue;
(iii) establishing the optimality of the SQ policy in routing to identical servers. These applications have also attempted to illustrate how sample paths can be coupled together and the different proof techniques that are commonly used, namely *forward induction* (e.g. Theorem 14.4) and *interchange arguments* (e.g. Theorem 14.2).

## ACKNOWLEDGMENT

This work was partially supported by the National Science Foundation under Grants ASC 88-8802764 and NCR-9116183.

## REFERENCES

Baccelli F. and Liu Z. (1992) Comparison properties of stochastic decision free Petri nets. *IEEE Trans. Autom. Control*, **37**, 1905–1920.

Baccelli F., Massey W. and Towsley D. (1989) Acyclic fork-join queuing networks *J. Assoc. Comput. Mach.*, **36**, 615–642.

Baccelli F., Liu Z. and Towsley D. (1993) Extremal scheduling of parallel processing systems with and without real-time constraints *J. Assoc. Comput. Mach.*, **40**, 1209–1237.

Bertsekas D. P. (1987) *Dynamic Programming* Prentice-Hall, Englewood Cliffs, NJ.

Chang C. S. (1992) A new ordering for stochastic majorization: theory and applications *Adv. Appl. Prob.*, **24**, 604–634.

Daley D. J. (1987) Certain optimality properties of the first come first served discipline for $G/G/s$ queues *Stochastic Processes Applic.*, **25**, 301–308.

Day P. W. (1972) Rearrangement inequalities, *Can. J. Math.* **24**, 930–943.

Ephremides A., Varaiya P. and Walrand J. (1980) A simple dynamic routing problem *IEEE Trans. Autom. Control*, **25**.

Feller W. (1971) *An Introduction to Probability Theory and Its Applications*, Vol. 2, 2nd edition, Wiley, New York.

Foss S. (1980) Comparison of servicing strategies in multichannel queueing systems *Siberian Math. J.*, **21**, 851–857.

Hardy G. H., Littlewood J. E. and Polya G. (1952) *Inequalities*, Cambridge University Press.

Hordijk A. and Koole G. (1991) On the optimality of the generalized shortest queue policy *Prob. Eng. Info. Sci.*, **4**, 477–487.

Hirayama T. and Kijima M. (1989) An extremal property of FIFO discipline in $G/IFR/1$ queues *Adv. Appl. Prob.*, **21**, 481–484.

Liu Z. and Towsley D. (1994a) Effects of service disciplines in $G/GI/s$ queueing systems. COINS Technical Report TR 92-26; *Ann. Oper. Res.*, **48**, 401–409.

Liu Z. and Towsley D. (1994b) Stochastic scheduling in in-forest networks. *Adv. Appl. Prob.*, **26**, 222–241.

Liu Z. and Towsley D. (1994c) Scheduling a sequence of parallel programs containing loops within a centralized parallel processing system. COINS Technical Report, TR 92-34, May 1992.

Liu Z. and Towsley D. (1994d) Optimality of the round robin routing policy. *J. Appl. Prob.*, **31**, 466–475.
Marshall A. W. and Olkin I. (1979) *Inequalities: Theory of Majorization and Its Applications* Academic Press, New York.
Menich R. (1987) Optimality of shortest queue routing for dependent service stations. In *Proceedings of the 1987 IEEE CDC Conference*, Los Angeles, pp 1069–1072.
Menich R. and Serfozo R. F. (1991) Optimality of routing and servicing in dependent parallel processing systems *Queueing Syst.*, **9**, 403–418.
Ross S. M. (1970) *Applied Probability Models with Optimization Applications* Holden-Day, San Francisco.
Ross S. M. (1983) *Stochastic Processes*, Wiley, New York.
Schrage L. (1968) A proof of the optimality of the shortest remaining processing time discipline *Oper. Res.*, **16**, 687–690.
Shaked M. and Shanthikumar J. G. (1988) Stochastic convexity and its applications *Adv. Appl. Prob.*, **20**, 427–445.
Shanthikumar J. G. and Sumita U. (1987) Convex ordering of sojourn times in single-server queues: extremal properties of FIFO and LIFO service disciplines *J. Appl. Prob.*, **24**, 737–748.
Shanthikumar J. G. and Yao D. D. (1990) Second-order stochastic properties in queueing systems *Proc. IEEE*, **77**, 162–170.
Sparaggis P. D., Cassandras C. G. and Towsley D. (1993) On the duality between routing and scheduling systems with finite buffer space *IEEE Trans. Autom. Control*, **38**, 1440–1446.
Sparaggis P. D., Cassandras C. G. and Towsley D. (1994a) Routing with limited state information in queuing systems with blocking *IEEE Trans. on Autom. Control*, **39**, 1492–1497.
Sparaggis P. D., Towsley D. and Cassandras C. G. (1994b) Sample path criteria for weak majorization *Adv. Appl. Prob.*, **26**, 155–171.
Stoyan D. (1983) *Comparison Methods for Queues and Other Stochastic Models* Wiley, New York.
Towsley D. and Baccelli F. (1991) Comparison of service disciplines in tandem queueing networks with delay dependent customer behavior *Oper. Res. Lett.*, **10**, 49–55.
Towsley D., Fdida S. and Santoso H. (1990) Design and analysis of flow control protocols for metropolitan area networks. In *High-Capacity Local and Metropolitan Area Networks* (ed. G. Pujolle) Springer, Berlin, pp 471–492.
Towsley D., Sparaggis P. and Cassandras C. G. (1977) Optimal routing and buffer allocation for a class of finite capacity queueing systems *IEEE Trans. on Autom. Control*, **37**, 1446–1451.
Vasicek O. A. (1977) An inequality for the variance of waiting time under a general queueing discipline *Oper. Res.* **25**, 879–884.
Walrand J. (1988) *An Introduction to Queueing Networks*, Chapter 8, Prentice Hall, Englewood Cliffs, NJ.
Weber R. R. (1978) On the optimal assignment of customers to parallel queue, *J. of Appl. Prob.*, **15**, 406–413.
Weber R. R. (1979) On the marginal benefit of adding servers to $G/GI/m$ queues *Management Sci.*, **25**, 946–951.
Whitt W. (1986) Deciding which queue to join *Oper. Res.*, **34**, 55–62.
Winston W. (1977) Optimality of the shortest line discipline", *J. of Appl. Prob.*, **14**, 181–189.

*Department of Computer Science, University of Massachusetts, Amherst, MA 01003, USA*

CHAPTER 15

# Scheduling and Interchangeability in Tandem Queues

R. Weber
*Statistical Laboratory, University of Cambridge*

**Abstract**

We discuss optimal sequencing in stochastic flow shops. We find optimal sequences for some two machines flow shops, and note that there are $m$ machine shops in which the makespan is independent of the order in which the jobs are processed.

## 15.1 FUNDAMENTALS

The subject of performance analysis, scheduling and flow control in tandem queues is a rich one. Research has been motivated by a desire to model pathways between buffered switches in communication networks and by the fact that in the manufacturing context, a series of tandem queues may be used to model a flow shop. In this chapter we restrict attention to scheduling only. The generic problem throughout this chapter is one in which $n$ jobs are to be processed through a flow shop of $m$ machines. The processing time of job $i$ on machine $j$ is denoted by $X_{i,j}$, and this is typically a random variable, whose distribution is known. All jobs are present at the start, and follow the same order of processing at each machine. The aim is to sequence the jobs to minimize some performance measure, such as makespan or flow time.

### 15.1.1 Critical path diagrams

A critical path diagram, composed of nodes and directed arcs, is helpful in calculating and thinking about completion times in a flow shop. Consider the model

---

*Scheduling Theory and its Applications* Edited by P. Chrétienne, E. G. Coffman, Jr., J. K. Lenstra and Z. Liu
© 1995 John Wiley & Sons Ltd

in which $n$ jobs are processed in sequence $1, \ldots, n$ through a flow shop of $m$ stations with unlimited buffering between stations. An activity-on-node diagram can be drawn as a rectangular lattice of $n \times m$ nodes. Node $(i, j)$ is the node in the $i$th row and $j$th column, and it is associated with the activity of processing job $i$ on machine $j$ and with the processing time $X_{i,j}$. Nodes are connected by horizontal and vertical arcs that are directed to the right and downwards respectively. Let $D_{i,j}$ be the time at which job $i$ leaves station $j$. A fundamental recurrence is

$$D_{i,j} = \max \{D_{i,j-1}, D_{i-1,j}\} + X_{i,j}. \tag{15.1}$$

It follows that $D_{i,j}$ can be interpreted as a maximum-length path from $(1, 1)$ to $(i, j)$, taken over all paths between these node following the directed arcs of the lattice, and where the length of a path is computed as the sum of the values associated with the nodes through which it passes. So, for example, $D_{23}$ is the maximum of length of three paths, each of which passes through four nodes:

$$D_{23} = \max \{X_{11} + X_{12} + X_{13} + X_{23}, X_{11} + X_{12} + X_{22} + X_{23},$$
$$X_{11} + X_{21} + X_{22} + X_{23}\}.$$

The makespan is time at which all jobs are finished, $D_{n,m}$.

### 15.1.2 Finite buffers and types of blocking

We shall assume that when the buffer that follows station $j$ is of size $b_j$, no more than $b_j$ jobs may wait for processing at station $j + 1$, excluding any job presently occupying the machine. There are two principal types of blocking that can occur. Under *manufacturing blocking*, a job may finish processing at station $j$, but it may not leave the station nor another job begin processing at the station, until the buffer of the downstream station has room for the job to be received. In this case the $(i - b_j - 1)$th job must have left station $j + 1$ before the $i$th job can leave station $j$. So if jobs are sequenced in the order $1, \ldots, n$,

$$D_{i,j} = \max \{D_{i,j-1} + X_{i,j}, D_{i-1,j} + X_{i,j}, D_{i-b_j-1,j+1}\}. \tag{15.2}$$

Under *communications blocking*, a job may not start processing at station $j$ until there is room for it to be received in the buffer downstream. This models the notion that a packet should be transmitted only if it will not cause an overflow of the buffer at the destination node. In this case

$$D_{i,j} = \max \{D_{i,j-1}, D_{i-1,j}, D_{i-b_j-1,j+1}\} + X_{i,j}. \tag{15.3}$$

In both cases the makespan can be conveniently calculated from an activity-on-arc diagram. In this diagram there is a directed arc of length $X_{i,j}$ that is associated with the processing of job $i$ on machine $j$. The node at the start of the arc is associated with the event that job $i$ begins processing at station $j$, and the node at the end of the arc is associated with the event that the job leaves stations $j$. The constraints imposed by blocking are represented by directed arcs of length 0. For example, in manufacturing blocking there is a arc of length 0 from the end of arc

$(i - b_j - 1, j + 1)$ to the end of arc $(i,j)$. Again the makespan is the maximum-length path from the start of node $(1, 1)$ to the end of node $(n, m)$.

### 15.1.3 Reversibility, convexity and the bowl-shape phenomenon

By interpreting the makespan as a maximum-length path in a critical path diagram, it becomes apparent that the makespan is the same when jobs are processed in a reverse order through the reversed order of machines. The proof of this *reversibility* property is effected by noticing that the maximum-length path is unaltered when the diagram is turned through 180° and the directions of the arcs reversed.

The fundamental recurrences (15.1)–(15.3) are also useful in showing that for any fixed scheduling sequence, the departure times $D_{i,j}$ are convex functions of the processing times (since the maximum of convex functions is convex). For example, if servers of speeds $s_1, \ldots, s_m$ are available at the stations, so that the processing time of job $i$ at station $j$ is actually $X_{i,j}/s_j$, then the makespan is a convex function of $s$. This fact, combined with reversibility, means that to minimize expected makespan in a system in which all processing times are independent and identically distributed, there are equal sized buffers between machines and there is a constraint on total speed of $\sum_j s_j = 1$, the optimal allocation of speeds should be symmetrical about the middle of the line.

A long-standing conjecture for the above problem is that to minimize the makespan the speeds should be not only symmetrical about the middle, but bowl-shaped, in the sense that stations towards the middle of the line should be served at greater speed. Intuitively, stations towards the middle suffer from both starvation and blocking, whereas the first station is never starved, the last station is never blocked and we should expect that stations nearer the ends would be suffer less from either starvation or blocking than those towards the middle. Also, the paths in the activity-on-arc diagram, the maximum of whose lengths is the makespan, make more crossings of arcs in columns of the diagram corresponding to middle stations than they do arcs in columns at the sides. A similar conjecture is that any buffer space that is allocated ought also to be concentrated towards the middle.

### 15.1.4 Jobs whose service times are the same at each machine

Suppose the service time of job $i$ is the same at every machine, $X_{i,j} = X_i$. By thinking of the calculation of the makespan in terms of the activity-on-node diagram described above, it is clear that the maximum-length path runs from $(1, 1)$ to $(k, 1)$ to $(k, m)$ to $(n, m)$, where $k$ is the order of processing of the job with greatest service time. Thus

$$D_{n,m} = \sum_{i=1}^{n} X_i + (m - 1) \max_i X_i.$$

It follows that the makespan is independent of the order in which the jobs are processed. Consequently, the time at which the first $i$ jobs are complete is minimized by sequencing the $i$ jobs of smallest expected processing times first. Thus the flow time is minimized by the well-known *shortest expected processing time first* rule, *SEPT*. This minimization may be in expectation, in increasing convex ordering, or in distribution, depending upon whether processing times can be ordered in expectation, hazard rate, or likelihood ratio, respectively. These three forms of minimization will be met again in Theorem 15.4.

### 15.1.5 The dual system

For a flow shop model, in which there is infinite buffering between machines, there is a dual system in which the roles of the jobs and machines are interchanged and in which we visualize $m$ machines (which are now viewed as jobs) passing through $n$ jobs (which are now viewed as machines). The duality is also exposed by reflecting the activity-on-node diagram in the line joining $(1, 1)$ to $(n, m)$. In flow shops it is usually considered that processing times depend on the job; in tandem queues it is more usual that processing times should depend on the station. Thus there is a natural duality between flow shop and tandem queue models.

For the above model of a flow shop, in which any given job has equal processing times at all machines, the dual system is one in which identical jobs pass through a set of tandem stations and all jobs have equal processing times at any given station. The makespan is independent of the order of the stations, effectively being determined by the bottleneck station, where all but the first job to reach it must queue.

## 15.2 THE TWO MACHINE FLOW SHOP

We begin by considering scheduling models in which there are just two stations and an unlimited number of jobs may queue between the stations.

### 15.2.1 Minimizing makespan on two machines

Suppose $n$ jobs are to be processed in the same order through two machines. Suppose that on the first and second machines the processing times of job $i$ are $X_i$ and $Y_i$ respectively. If the processing times are deterministic then the makespan is minimized by Johnson's algorithm, which arranges the jobs in a sequence for which $i$ precedes $j$ only if min $\{X_i, Y_j\} \leq$ min $X_j, Y_i$.

If $X_i$ and $Y_i$ are distributed exponentially with parameters $\lambda_i$ and $\mu_i$ respectively then $E[\min \{X_i, Y_j\}] \leq E[\min \{X_j, Y_i\}]$ if and only if $\lambda_i + \mu_j \leq \lambda_j + \mu_i$, which suggests that expected makespan might be minimized by sequencing jobs in decreasing order of $\lambda_i - \mu_i$. This is the case, as shown by the following simple proof (adapted from Weiss 1982).

**Theorem 15.1** *For jobs with exponentially distributed processing times as above, the expected makespan in a two machine flow shop is minimized by sequencing in order of decreasing values of $\lambda_j - \mu_j$.*

**Proof** Suppose that machine 1 is idle, jobs $i$ and then $j$ are to be sequenced next, and the work remaining on the second machine is presently $z$. Consider the distribution of, $T$, the work remaining on the second machine when both these jobs have completed on the first machine. If $X_i + X_j \leq z$ then $T$ is independent of the order in which jobs $i$ and $j$ are ordered. So, for $X_i + X_j > z$, we have

$$P_{ij}(T > t | X_i + X_j > z) = \frac{\mu_i}{\mu_i + a_j} e^{-\mu_j t} + \frac{\lambda_j}{\lambda_i + \lambda_j} \frac{\mu_j e^{-\mu_i t} - \mu_i e^{-\mu_j t}}{\mu_j - \mu_i}.$$

Hence, considering the opposite order,

$$P_{ij}(T > t \mid X_i + X_j > z) - P_{ji}(T > t \mid X_i + X_j > z)$$

$$= \frac{\mu_i \mu_j}{(\lambda_i + \mu_j)(\lambda_j + \mu_i)} \frac{e^{-\mu_i t} - e^{-\mu_j t}}{\mu_j - \mu_i} [(\lambda_i - \mu_i) - (\lambda_j - \mu_j)].$$

This leads to the conclusion that sequencing the jobs in the order implied by the theorem stochastically minimizes the remainder on the second machine after any number of jobs have completed on the first machine. The theorem follows from the fact that $C_{\max} = T_n + \sum_i X_i$, where $T_n$ is the remainder on the second machine after all $n$ jobs have completed on the first machine. □

Ku and Nui (1986) proved by a simple interchange argument that a sequence minimizes the makespan in distribution if the order of processing times on the first and last machine can be simultaneously ordered in nondecreasing and nonincreasing order of likelihood ratio. The key condition, which as a special case is implied by $X_i \leq_{lr} X_j$ and $Y_i \geq_{lr} Y_j$, is stated in the theorem. Note, however, that Theorem 15.1 is not a special case of Theorem 15.2.

**Theorem 15.2** *Suppose that for all $i$,*

$$(\min\{X_i, Y_j\} \mid X_i + X_j = a \text{ and } Y_i + Y_j = b)$$
$$\leq (\min\{X_j, Y_i\} \mid X_i + X_j = a \text{ and } Y_i + Y_j = b)$$

*for all $a$ and $b$ for which the conditional distributions are defined. Then processing in the order $1, \ldots, n$ stochastically minimizes the makespan.*

**Proof** Suppose that $i$ and $j$ are the $(k-1)$th and $k$th jobs in order of processing. We condition on values $D_{k-2,1} = d_1$, $D_{k-2,2} = d_2$, $X_i + X_j = a$ and $Y_i + Y_j = b$. Then

$$D_{k,1} = d_1 + X_i + X_j$$
$$= d_1 + a$$
$$D_{k,2} = \max\{d_1 + X_i + X_j + Y_j, d_1 + X_i + Y_i + Y_j, d_2 + Y_i + Y_j\}$$
$$= d_1 + a + b - \min\{Y_i, X_j, d_2 - d_1 - a\}$$

The theorem follows by coupling realizations of the conditioned random variables so that if the order of $i$ and $j$ in the sequence is interchanged then $D_{k,1}$ is unchanged and $D_{k,2}$ is no less. □

### 15.2.2 Applications of stochastic ordering

Notions of stochastic ordering and their relation to bivariate functions can be used to study stochastic scheduling problems. We shall use them to obtain optimal sequences for certain two-machine flow shop problems. Some key results are contained in the following definitions and theorem.

For a bivariate function $g(x,y)$, define $\Delta g(x,y) = g(x,y) - g(y,x)$ and the following classes of function:

$$\mathcal{G}_{\text{lr}} = g(x,y) \; : \; \Delta g(x,y) \geqslant 0 \; \forall x \geqslant y$$
$$\mathcal{G}_{\text{hr}} = g(x,y) \; : \; g(x,y) \text{ and } \Delta g(x,y) \text{ are increasing in } x \; \forall x \geqslant y$$
$$\mathcal{G}_{\text{st}} = g(x,y) \; : \; \Delta g(x,y) \text{ is increasing in } x \; \forall x$$

**Theorem 15.3** *In each row of the following table the ordering between two random variables $X \geqslant_a Y$ that is stated in the first column holds $\iff$ the order $g(X,Y) \geqslant_b g(Y,X)$ given in the second column holds for every $g$ that satisfies the conditions stated in the third column.*

| $X \geqslant_a Y$ | $g(X,Y) \geqslant_b g(Y,X)$ | $g \in$ |
|---|---|---|
| $\geqslant$ lr | $\geqslant$ st | $\mathcal{G}_{\text{lr}}$ |
| $\geqslant$ hr | $\geqslant$ icx | $\mathcal{G}_{\text{hr}}^1$ |
| $\geqslant$ st | $\geqslant$ mean | $\mathcal{G}_{\text{st}}$ |

The following theorem is an application to the two-machine flow shop. We have already met a form of this result in Theorem 15.2.

**Theorem 15.4** *Suppose processing times on the first and second machines of jobs $1, \ldots, n$ are $X_1 \leqslant_a \ldots \leqslant_a X_n$ and $Y_1 \geqslant_a \ldots \geqslant_a Y_n$ respectively. Then the order $1, \ldots, n$ minimizes the makespan in the sense of $b$, where $(a,b) = (\text{lr, st}), (\text{hr, icx})$ or $(\text{st, mean})$.*

The fact that the processing times on the second machine are ordered in the reverse direction to those on the first machine is a form of *agreeability condition*.

**Proof** Suppose jobs are processed in the order $1, \ldots, n$. The makespan is

$$\max_{1 \leqslant k \leqslant n} \left\{ \sum_{i=1}^{k} X_i + \sum_{i=k}^{n} Y_i \right\}.$$

So define $f(x,y)$, the makespan for this order, by

$$f(x_1,\ldots,x_n,y_1,\ldots,y_n) = \max_{1 \leq k \leq n}\left\{\sum_{i=1}^{k} x_i + \sum_{i=k}^{n} y_i\right\}.$$

For fixed $x$ and $y$, define

$$g(a,b) = f(x_1,\ldots,x_{j-1},a,b,x_{j+2},\ldots,x_n,y_1,\ldots,y_n).$$

Then

$$g(a,b) - g(b,a) = \max\{M_1, s+a+y_j+t, M_2+a+b\}$$
$$- \max\{M_1, s+b+y_j+t, M_2+a+b\}$$

where

$$s = \sum_{i=1}^{j-1} x_i, \quad t = \sum_{i=j+1}^{n} y_i$$

$$M_1 = \max_{k<j}\left\{\sum_{i=1,\ldots,k} x_i + \sum_{i=k,\ldots,n} y_i\right\},$$

$$M_2 = \max_{k>j}\left\{\sum_{i=1,\ldots,k, i \neq j, j+1} x_i + \sum_{i=k,\ldots,n} y_i\right\}.$$

It therefore follows that both $g(a,b)$ and $(a,b)$ are increasing in $a$ for all $a$. Hence $g \in \mathcal{G}_{lr}, \mathcal{G}^1_{lr}, \mathcal{G}_{st}$. Pick random variables $A \leq_a B$. From Theorem 15.3, it follows that

$$g(A,B) \leq_b g(B,A)$$

in an appropriate sense. This proves that the processing times of jobs that are processed $j$th and $(j+1)$th at the first machine should be ordered by SEPT. This holds for all $j$, and so the jobs are best taken in SEPT order at the first machine. Similarly, by consideration of the reversed system, the processing times on the second machine are best taken in LEPT order. The agreeablility condition ensures that both of these achieved by the order $1,\ldots,n$, and the theorem follows. □

The theorem may be generalized by replacing the makespan by a function of the completion times, $f(C)$, where $f$ is increasing, increasing and supermodular, or increasing and separable, for each of the three cases of the theorem respectively.

## 15.3  INTERCHANGEABILITY

In certain cases the makespan turns out to be independent of the order of the jobs. In Section 15.1 we noted that this is true when each job has equal service times at all machines. In that case the makespan is determined by the bottleneck job, i.e. the one with greatest service time. More surprisingly, it is also true when $X_{i1},\ldots,X_{im}$ are independent exponential random variables with parameter $\mu_i$. In this case the set of $m$

instants at which the final job clears each of the $m$ machines form a set of random variables whose joint distribution is independent of the sequence in which the jobs are processed. A consequence of this fact is the expected flow time is minimized by sequencing the jobs in SEPT order.

The dual system to the above is one in which $m$ machines, now viewed as jobs, pass through $n$ jobs, now viewed as machines, where service times at machine $i$ are independent exponential random variables with parameter $\mu_i$. It is the set of $m$ departure times from the final station whose joint distribution is independent of the order of the stations. This result is known as the *interchangeability of $\cdot/M/1$ queues*, and a further generalization is the following theorem. In this the service times at machine $j$ are exponential random variables with parameters $\mu_1,\ldots,\mu_n$, but are conditioned on the fact that for customer $i$ their sum is fixed at $c_i$. The traditional interchangeablity of $\cdot/M/1$ queues is the special case in which $c_i$ has the distribution of the sum of exponential random variables with parameters $\mu_1,\ldots,\mu_n$. The dual system, appropriate to a flow shop, is one in which the sum of the processing times at machine $j$ is fixed. However, we shall state and prove the theorem in the context of tandem queues.

**Theorem 15.5** *Consider two stations in tandem with a finite buffer between stations, and at which the ith customer has service times of $X_i$ and $Y_i$, where these are exponential random variables with parameters $\lambda$ and $\mu$, that are conditioned on having sum $c_i$, i.e.*

$$f_i(x_i, y_i) = \frac{(\mu - \lambda)\lambda\mu e^{-\lambda x_i - \mu y_i}}{\lambda e^{-\lambda c_i} - \mu e^{-\mu c_i}}, \quad 0 \leq x_i, y_i, \quad x_i + y_i = c_i.$$

*Then the departure process from the second station is statistically the same when the order of the stations is reversed. This holds for any arrival process—even one that depends on the departure process from the last station.*

The truth of Theorem 15.5 under the assumption that there is only a finite buffer between stations is actually a consequence of the fact that the theorem holds even when the arrival process depends on the departure process. To model the effect of a finite buffer and communication blocking, we just consider a modified arrival process, for which the arrival of customer $n$ is taken as the maximum of its actual arrival time and $D_{n-b-1,2}$. This remark means that it suffices to prove Theorem 15.5 for the case that the intermediate waiting room is infinite.

In this chapter we shall prove a simple version of the theorem, in which it is assumed that all jobs are present at the start. The proof for an arbitrary arrival process is basically the same, but more technical, and implies that the result extends to the interchange of more than two stations with infinite intermediate buffers, and to the dual system with more than two jobs (Weber 1992). The simplified version is a consequence of the following lemma.

**Lemma 15.1** *Consider two stations in tandem with unlimited queueing between stations. Consider a given set of feasible departure times at the second station, say $D_{1,2} = d_1, \ldots, D_{n,2} = d_n$. Let $C_n = \sum_{j=1}^n c_j$. Then for each $s \in [C_n - d_n, d_n]$, there exists an unique corresponding set of service times at station 1, $x(s) = (x_1(s), \ldots, x_n(s))$, for which these departure times will be achieved and for which $\sum_{j=1}^n x_j(s) = s$. Moreover, these are the only ways the given set of departure times can be achieved.*

We shall prove Theorem 15.5 and then Lemma 15.1.

**Proof of Theorem 15.5** To prove the theorem, it suffices to show that the joint probability distribution of the departure times of the first $n$ customers is not altered by exchanging $\lambda$ and $\mu$. Condition on a set of possible values of the departure times, $d_1, \ldots, d_n$. Consider a feasible set of service times at the first station, for which these departure times are achieved. By the lemma, there is one such set, $x_1(s), \ldots, x_n(s)$, for each $s \in [C_n - d_n, d_n]$, with $\sum_{j=1}^n x_j(s) = s$. The joint probability of these service times is proportional (with respect to a multiplicative constant that is symmetric in $\lambda$ and $\mu$) to

$$\prod_{i=1}^n e^{-\lambda x_j(s)} e^{-\mu(c_i - x_j(s))} = e^{-\lambda s} e^{-\mu(C_n - s)}.$$

Integrating the above with respect to $s$ over the feasible range $s \in [C_n - d_n, d_n]$ leads to an expression that is symmetric in $\lambda$ and $\mu$. □

**Proof of Lemma 15.1** The proof is by induction on $n$. Clearly it is true for $n = 1$. So take as an inductive hypothesis that the lemma is true for $n-1$, and thus that $d_1, \ldots, d_{n-1}$ may be realized uniquely by some $(x_1(s), \ldots, x_{n-1}(s))$ for every $s \in [C_{n-1} - d_{n-1}, d_{n-1}]$. Suppose the sum of the service times of the first $n-1$ customers at the first station is $s$, and the departure times of the first $n-1$ customers from the second station are $d_1, \ldots, d_{n-1}$. Then

$$d_n = \max\{s + c_n, d_{n-1} + c_n - x_n\}. \tag{15.4}$$

It is possible for the maximum to be achieved by the second term in curly brackets on the right hand side of (15.4), because if $d_n$ is a feasible time at which the $n$th customer can depart the second station then $d_n \leq d_{n-1} + c_n$. Thus the value of $x_n$ for which the second term in curly brackets equals $d_n$, i.e. $\bar{x} = c_n + d_{n-1} - d_n$, is such that $\bar{x} \in [0, c_n]$. Since we have assumed that (15.4) has a solution, we must have $\bar{s} + c_n \leq d_n$ when $\bar{s}$ is the minimal value $s = C_{n-1} - d_{n-1}$ for which $d_1, \ldots, d_{n-1}$ can be achieved. Therefore we can satisfy (15.4) either by taking $x_n = \bar{x}$ and any $s \in [C_{n-1} - d_{n-1}, d_n - c_n]$, or by taking $s = d_n - c_n$ and any $x_n \in [\bar{x}, c_n]$. So there is exactly one way for $s + x_n$ to take each of the values between $\bar{s} + \bar{x} = C_n - d_n$ and $(d_n - c_n) + c_n = d_n$. As $s + x_n$ denotes $\sum_{j=1}^n x_j$, this completes the proof of the lemma. □

## 15.4 LITERATURE

We conclude with a brief review of the literature in this field.

*Reversibility, convexity and the bowl-shape phenomenon*

Muth (1979) considered tandem queues with blocking, and showed that if all customers are present at the start then the departure time is the same if the order of the stations is reversed. The observation that optimal server allocations are often bowl-shaped was made by Hillier and Boling (1979), and has been investigated by other authors, but it is still without proof in general. The longest line for which a bowl-phenomenon has been established is one studied by Weber and Weiss (1994), who considered a model with no buffer space and in which each job needs unit service at exactly one randomly chosen station, this being the $j$th station with probability $p_j$, independently of other jobs. They found that for up to nine stations the $p_j$ should be chosen so the middle stations are the less likely ones at which service will be required.

*The dual system*

Pinedo (1982) discussed the idea of a dual flow shop when considering flow shops both with infinite intermediate storage and with no intermediate storage and manufacturing blocking. Suppose that each job has identically distributed processing times on the $m$ machines. Pinedo showed that if only two of the jobs have stochastic processing times, and the processing times of the others are deterministic, then the stochastic jobs should be sequenced first and last. This is an application of the work of Tembe and Wolff (1974). Suresh *et al.* (1985) considered the same problem, with no intermediate waiting rooms, and showed that either this schedule or the schedule that puts the two stochastic jobs at the same end is optimal.

For processing times that are a.s. ordered, Pinedo showed that any SEPT–LEPT policy will minimize the expected makespan. This holds even if there is no intermediate storage. Also, for no intermediate storage, Pinedo considered jobs whose processing times are ordered in distribution, $X_1 \geqslant_{st} \cdots \geqslant_{st} X_n$. The expected makespan is minimized by $n, n-2, n-4, \ldots, 4, 2, 1, 3, 5, \ldots, n-3, n-1$ if $n$ is even and $n, n-2, n-4, \ldots, 3, 1, 2, 4, \ldots, n-3, n-1$ if $n$ is odd. The proof is by a reversal of a set of jobs in the middle of the sequence.

Foley and Suresh (1984, 1986) considered flow shops with non-overlapping jobs, i.e. ones whose processing times are ordered the same on all machines. They considered

(a) no buffers and non-overlapping jobs;
(b) non-overlapping jobs and one arbitrary job;
(c) stochastically identical jobs and two faster jobs;
(d) non-overlapping jobs and two arbitrary jobs.

In cases (a) and (b) SPT–LPT and SPT–Y are optimal. In (c) the faster jobs should go at the ends. In (d), SPT–Y–SPT–LPT–Z is optimal, where the processing times in SPT–LPT section are longer than those in the SPT section.

Kijima et al. (1989) extended these results to finite buffers. In case (d), with ample buffer space, SPT–Y–SPT–LPT–Z is optimal, where the processing times in the SPT–LPT section are longer than those in the SPT section.

*Minimizing makespan on two machines*

The generalization of Johnson's (1954) rule is the conjecture of Talwar's (1967) rule, proved by Cunningham and Dutta (1973), after a preliminary attempt by Bagga (1970). The proofs in this chapter are adapted from Weiss (1982) and Ku and Nui (1986). Frostig and Adiri (1985) generalized the result to a flow shop of three machines, and proved that the makespan is minimized stochastically by a job sequence in which processing times on the first and last machines are respectively in monotone nondecreasing and nonincreasing order of likelihood ratio, and those on the second machine are i.i.d.

*Applications of stochastic ordering*

The results in Section 15.2.2 are due to Shanthikumar and Yao (1991) and Chang and Yao (1990). Their applications to stochastic scheduling problems have been very well explained by Righter (1994).

*Interchangeability*

Friedman (1965) showed that, for arbitrary arrivals and deterministic service times, the departure time of every customer is the same if the order of the stations is reversed. This holds when there are $s_i$ identical servers working in parallel at each station.

Interchangeability of $\cdot/M/1$ queues has been studied by Weber (1979), Lehtonen (1986), Tsoucas and Walrand (1987) and Anantharam (1987). The proof given here is a simplified version of that in Weber (1992). A similar argument is used by Kijima and Makimoto (1990). The discrete-time version of the model includes interchangeability of stations for a model in which there are unlimited buffers and each job needs unit service at exactly one randomly chosen station, this being the station $j$ with probability $p_j$, independently of other jobs (Weber and Weiss 1994).

The method of incorporating blocking via an alteration of arrival times also leads to the generalizations proposed by Chao et al. (1989). For example, we may suppose that any customer who arrives to find more than a certain number of other customers in the system is lost.

*Further topics*

There are other performance measures that can be studied, such as job lateness, or end-to-end delay. It is not surprising that the *earliest due date* and *FCFS* rules are often optimal for such measures. See for example Towlsey and Baccelli (1991) and Liu and Towsley (1994).

We have only discussed problems in which jobs are sequenced in the same order at each machine. Some results are possible. For example, adopting a SEPT rule at the last station will minimize expected flow time when jobs have processing times that are exponentially distributed. However, as noted by Hordijk and Koole (1992), expected weighted flow time is not necessarily minimized by using the $c\mu$ rule at the last station.

## REFERENCES

Anantharam V. (1987) Probabilistic proof of interchangeability of $/M/1$ queues in series *Queueing Syst. Theory and Applic.* **2**, 387–392.

Bagga P. C. (1970) $N$-job, 2 machine sequencing problem with stochastic service times *Opsearch*, **7**, 184–199.

Chang C.-S. and Yao D. (1990) Rearrangement, majorization and stochastic scheduling. Technical Report RC 16259, IBM.

Chao X., Pinedo M. and Sigman K. (1989) On the interchangeability and stochastic ordering of exponential queues in tandem with blocking *Prob. Engng Inf. Sci.*, **3**, 223–236.

Cunningham A. A. and Dutta S. K. (1973) Scheduling jobs with exponentially distributed processing times on two machines in a flowshop *Nav. Res. Log. Quart.*, **16**, 69–81.

Foley R. D. and Suresh S. (1984) Stochastically minimizing the makespan in flowshops *Nav. Res. Log. Quart.*, **31**, 551–557

Foley R. D. and Suresh S. (1986) Scheduling $n$ non-overlapping jobs and two stochastic jobs in a flow shop *Nav. Res. Log. Quart.*, **33**, 123–128.

Friedman H. D. (1965) Reduction methods for tandem queueing systems *Oper. Res.*, **13**, 121–131.

Frostig E. and Adiri I. (1985) Stochastic flowshop no-wait scheduling *J. Appl. Prob.*, **22**, 240–246.

Hillier F. S. and Boling R. M. (1979) On the optimal allocation of work in symmetric balanced production line systems with variable operation times *Management Sci.*, **25**, 721–728.

Hordijk A. and Koole G. (1992) The $\mu c$-rule is not optimal in the second node of the tandem queue: a counterexample *Adv. Appl. Prob.*, **24**, 234–237.

Johnson S. M. (1954) Optimal two and three stage production schedules with setup times included *Nav. Res. Log. Quart.*, **1**, 61–74.

Kijima M. and Makimoto N. (1990) On interchangeability for exponential single-server queues in tandem *J. Appl. Prob.*, **27**, 459–464.

Kijima M., Makimoto N. and Shirakawa H. (1989) Stochastic minimization of the makespan in flow shops with identical machines and buffers of arbitrary size *Oper. Res.*, **38**, 924–928.

Ku P.-S. and Nui S.-C. (1986) On Johnson's two-machine flow shop with random processing times *Oper. Res.*, **34**, 130–136.

Lehtonen T. (1986) On the ordering of tandem queues with exponential servers *J. Appl. Prob.*, **23**, 115–129.

Liu Z. and Towsley D. (1994) Stochastic scheduling in in-forest networks *Adv. Appl. Prob.*, **26**, 222–241.

Muth E. J. (1979) The reversibility property of production lines *Management Sci.*, **25**, 152–158.

Pinedo M. (1982) Minimizing the expected makespan in stochastic flow shops *Oper. Res.*, **30**, 148–162.

Righter R. (1994) Scheduling. In: *Stochastic Orders and their Applications* (ed. M. Shaked and J. G. Shanthikumar), Academic, New York, pp. 381–432.

Shanthikumar J. G. and Yao D. J. (1991) Strong stochastic convexity: Closure properties and applications *J. Appl. Prob.*, **28**, 131–145.

Suresh S., Foley R. D. and Dickey S. E. (1985) On Pinedo's conjecture for scheduling in a stochastic flow shop *Oper. Res.*, **33**, 1146–1153.

Talwar P. P. (1967) A note on sequencing problems with uncertain job times *J. Oper. Res. Soc. Japan*, **9**, 93–97.

Tembe S. V. and Wolff R. W. (1974) The optimal order of service in tandem queues *Oper. Res.*, **30**, 148–162.

Towlsey D. and Baccelli F. (1991) Comparison of service disciplines in a tandem queueing network with delay dependent customer behavior *Oper. Res. Lett.* **10**, 49–55.

Tsoucas P. and Walrand J. (1987) On the interchangeability and stochastic ordering of $/M/1$ queues in tandem *Adv. Appl. Prob.*, **16**, 515–520.

Weber R. R. (1979) The interchangeability of $\cdot/M/1$ queues in series' *J. Appl. Prob.*, **16**, 690–695.

Weber R. R. (1992) The interchangeability of tandem queues with heterogeneous customers and dependent service times *Adv. Appl. Prob.*, **24**, 727–737.

Weber R. R. and Weiss G. (1994) The cafeteria process—tandem queues with 0–1 dependent service times and the bowl shape phenomenon *Oper. Res.*, **42**, 947–957.

Weiss G. (1982) Multiserver stochastic scheduling. In Dempster M. A. H., Lenstra J. K. and Rinnooy Kan A. H. G., editors, *Deterministic and Stochastic Scheduling* pp 157–179 Reidel, Dordrecht.

---

*Statistical Laboratory, University of Cambridge, 16 Mill Lane, Cambridge CB2 1SB, UK*

CHAPTER 16

# Dynamic Routing and Sequencing in Open Queueing Networks

**C. N. Laws**
*Department of Statistics, University of Oxford*

**Abstract**

We outline some recent work concerning open queueing networks in which the routing of customers through the network and the sequencing of service at the service-stations can be controlled. In many heavy traffic situations and under good control policies, networks with many single-server stations act like single station systems with multiple servers. We call this effect *resource pooling*, and describe how it can substantially reduce customer delay.

## 16.1 INTRODUCTION

Consider a queueing network such as that illustrated in Figure 16.1. This might represent a computer or communication network, or a manufacturing system. The nodes represent the service stations of the network, which are labeled by $i = 1, 2, \ldots, I$. Service times at station $i$ form an independent identically distributed sequence of random variables with mean $\mu_i^{-1}$ and variance $\sigma_i^2$. Different types of customers arrive at the network, customer types being labeled by $j = 1, 2, \ldots, J$, and pass through the system via one of several possible routes. Customers of type $j$ arrive according to a renewal process, interarrival times having mean $\lambda_j^{-1}$ and variance $\alpha_j^2$. All arrival and service processes are independent. A route $r$ is an ordered set of stations $(i_1, i_2, \ldots, i_n)$, and a customer using route $r$ queues for service at each of these stations in turn before leaving the network. Let $\mathcal{R}j$ be the set of routes available to type $j$ customers and let $a_{ir} \in \mathbb{Z}_+$ be the number of times that route $r$ visits station $i$; an important special case is where each $a_{ir}$ is either 0 or 1. In addition to there being customers on different routes that queue at station $i$, there may also be customers queueing at $i$ who are at different stages of the same route (since routes

---

*Scheduling Theory and its Applications* Edited by P. Chrétienne, E. G. Coffman, Jr., J. K. Lenstra and Z. Liu
© 1995 John Wiley & Sons Ltd

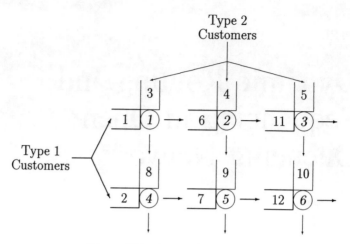

**Figure 16.1** A queueing network.

may visit $i$ more than once). At station $i$, distinguish between these different kinds of customers by having a different queue for each stage of each route that passes through $i$, queues being labeled by $k = 1, 2, \ldots, K$. In the network illustrated in Figure 16.1, type 1 customers pass through the system from left to right by using one of the two rows of the network, and type 2 customers move from top to bottom, using one of the three columns. Thus

$$\mathcal{R}_1 = \{(1,2,3), (4,5,6)\}, \quad \mathcal{R}_2 = \{(1,4), (2,5), (3,6)\},$$

and there are two queues at each station, these being labeled $1, 2, \ldots, 12$.

The control of such systems is important, since the variation in network performance with different control strategies can be substantial. Here we consider *dynamic* or *state dependent* control strategies, because, in comparison with less responsive strategies, this kind of control can offer significant improvements. The two methods of control we consider are the *routing* of customers through the network and the *sequencing* of service at the stations, and our aim is to minimize the mean delay per customer or equivalently, by Little's formula (Stidham 1974; Whitt 1991), to minimize the mean number of customers in the system. More precisely, a routing policy specifies the route $r \in \mathcal{R}_j$ along which a type $j$ customer is routed; such routing decisions are made at the instant when the customer arrives at the network and are irrevocable. A sequencing policy specifies which queue to serve at each station, at each point in time. Let $Q_k(t)$ be the number of customers in queue $k$ at time $t$, including the one in service if any, and define the queue length process $\mathbf{Q} = (Q_k, k = 1, 2, \ldots, K)$. Assume that the system starts from empty at time $t = 0$, so $\mathbf{Q}(0) = \mathbf{0}$. By dynamic control policies, we mean that routing and sequencing decisions can be based on any information available from observing the history of the process **Q**.

Since customers can be routed through the network using information about the current states of various queues, routing decisions may produce significant dependencies between different parts of the network. Such dependencies may be a major determinant of system performance and, further, make analysis of the system behavior difficult. For example, suppose customers arriving from a Poisson stream of rate $\lambda$ are to be routed to one of two exponential servers, each of rate $\mu$ ($\rho = \lambda/2\mu < 1$), and that customers leave after being served. While the optimal policy is for customers to join the shorter queue (Winston 1977; Weber 1978), in general the equilibrium behavior of the system is difficult to describe, although efficient computational methods are available (Adan et al. 1990). However, in heavy traffic, when $\rho$ is near 1, substantial information can be obtained about equilibrium and transient behavior (see e.g. Kingman 1961; Flatto and McKean 1977; Foschini 1977; Foschini and Salz 1978; Reiman 1983; Halfin 1985). To summarize, the lengths of the two queues are held approximately equal and there is a *resource pooling* effect: customer delay is distributed as in an $M/M/2$ queue with arrival rate $\lambda$ and servers each of rate $\mu$. This resource pooling reduces mean delay by a factor of 2 over a strategy that allocates customers to servers randomly on arrival (Foschini and Salz 1978).

A theme of much recent work has been the study of queueing systems using network models based on Brownian motion (Harrison and Williams 1987; Harrison 1988; Harrison and Nguyen 1990, 1993; Laws and Louth 1990; Wein 1990a,b; Dai and Harrison 1992). Such models are motivated by heavy traffic analysis, and in many cases have been shown to be exact in the heavy traffic limit (see e.g. Reiman 1984; Chen and Mandelbaum 1991; Peterson 1991). In Section 16.2 we discuss heavy traffic, and use a Brownian network model to describe the behavior of the system in heavy traffic under dynamic routing and sequencing. In Section 16.3 we consider some implications of the results for individual routing and sequencing decisions. We find that, in many respects, good dynamic routing and sequencing policies simplify the behavior of networks.

## 16.2 HEAVY TRAFFIC ANALYSIS

In this section and the next we outline and discuss some of the results of Laws (1992); see also the introductory review by Kelly and Laws (1993). To identify when the system is in heavy traffic, with each station $i$, associate a variable $c_i \geq 0$ and let

$$d_j = \min_{r \in \Re_j} \left( \sum_i a_{ir} c_i \right).$$

Call a routing strategy *stable* if, for all $i$, the average arrival rate of customers at station $i$ (from both inside and outside the network) is less than $\mu_i$. It is possible to find a stable routing strategy if and only if

$$\sum_j d_j \lambda_j < \sum_i c_i \mu_i \qquad (16.1)$$

as $\mathbf{c} = (c_i) \geq \mathbf{0}$) ranges over some finite set of vectors. This finite set of vectors for $\mathbf{c}$, and the associated vectors for $\mathbf{d}$, arise as dual variables in the natural linear programming problem that determines heavy traffic (Minoux 1981; Gondran and Minoux 1984). For the network illustrated in Figure 16.1, the 29 constraints of the form (16.1) include the following six constraints:

$$\lambda_1 < \mu_1 + \mu_4, \qquad \lambda_1 < \mu_1 + \mu_5$$

$$\lambda_2 < \mu_1 + \mu_2 + \mu_3, \qquad \lambda_1 + \lambda_2 < \mu_1 + \mu_2 + \mu_6$$

$$2\lambda_1 + \lambda_2 < \mu_1 + \mu_2 + 2\mu_6, \qquad 3\lambda_1 + 2\lambda_2 < 2\mu_1 + 2\mu_5 + \mu_3 + \mu_6.$$

The remaining 23 constraints can be obtained from those above by symmetry. The first four of these constraints are *cut constraints*. Consider the fourth constraint, for example. Stations 1, 2 and 6 form a cut for type 1 and type 2 traffic: their removal from the system prevents the flow of customers of types 1 and 2. Hence the minimum arrival rate at this cut, $\lambda_1 + \lambda_2$, must be less than the service rate of the cut, $\mu_1 + \mu_2 + \mu_6$, if the system is to be stable. The final constraint can be interpreted as follows. Customers are charged 2, 2, 1 and 1 units at stations 1, 5, 3 and 6 respectively; stations 2 and 4 have zero cost. To pay to get through the network, type 1 and 2 customers enter the system with 3 and 2 units respectively. If there is a feasible flow then the total arrival rate of revenue, $3\lambda_1 + 2\lambda_2$, must be less than the maximal charging rate, $2\mu_1 + 2\mu_5 + \mu_3 + \mu_6$. The constraint on $2\lambda_1 + \lambda_2$ has a similar interpretation with costs of 1, 1 and 2 units at stations 1, 2 and 6, and where type 1 and 2 customers arrive with 2 and 1 units respectively (route $(3, 6)$ is available to type 2 customers, but is too expensive). We call the constraints on $2\lambda_1 + \lambda_2$ and $3\lambda_1 + 2\lambda_2$, and more generally the constraints of the form (16.1), *generalized cut constraints*.

The general framework of Harrison (1988) can be used to obtain the Brownian model of the general network described above. Consider a regime in which $\boldsymbol{\lambda}$ is scaled by a factor $\rho$, where $\rho$ is increased from zero until one of the generalized cut constraints first hits equality, at $\rho = \rho^*$ say. In many contexts it will be natural for exactly one of this finite set of constraints to become tight in heavy traffic, that is, at $\rho = \rho^*$. (If $\boldsymbol{\mu}$ and $\boldsymbol{\lambda}$ are chosen from some continuous probability distribution over $\mathbb{R}_+^{I+J}$, for example, then there will be a single tight constraint with probability 1.) So assume that the constraint given by vectors $\mathbf{c}^* = (c_i^*)$ and $\mathbf{d}^* = (d_j^*)$,

$$\sum_j d_j^* \lambda_j < \sum_i c_i^* \mu_i,$$

is the only constraint close to equality and that the slack in this constraint is of order

$1/\sqrt{n}$ for some large integer $n$. For example, $\boldsymbol{\lambda} = (2\rho, 4\rho)$, $\boldsymbol{\mu} = (2, 6, 3, 6, 2, 3)$, where $1 - \rho$ is of order $1/\sqrt{n}$. More formally, we could consider a sequence of networks indexed by $n$, with arrival rates $\lambda_j(n)$ and service rates $\mu_i(n)$, in which

$$\sqrt{n}\left(\sum_j d_j^* \lambda_j(n) - \sum_i c_i^* \mu_i(n)\right) \to \theta \quad \text{as } n \to \infty.$$

Since it is not our aim to prove a heavy traffic limit theorem, we shall assume that

$$\sum_j d_j^* \lambda_j - \sum_i c_i^* \mu_i = O(1/\sqrt{n})$$

for the given arrival and service rate vectors $\boldsymbol{\lambda}$ and $\boldsymbol{\mu}$. Suppose that $c_i^* > 0$ for all $i$, that $d_j^* > 0$ for all $j$ and that, as we approach heavy traffic, it is possible for all routes $r \in \mathcal{R}_j$ to be in use. (If these conditions on $\mathbf{c}^*$, $\mathbf{d}^*$ and $\mathcal{R}_j$ do not hold then we should reduce the original network to its bottleneck subnetwork (Harrison 1988; Laws 1992) before proceeding with the heavy traffic analysis.)

The workload formulation of the Brownian model involves the scaled queue length process $\mathbf{Q}^*$ defined by

$$\mathbf{Q}^*(t) = \frac{\mathbf{Q}(nt)}{\sqrt{n}},$$

where the parameter $n$ is fixed. Since the system is initially empty $\mathbf{Q}^*(0) = \mathbf{0}$. Also define the scaled idleness process $\mathbf{I}^* = (I_i^*, i = 1, 2, \ldots, I)$ by

$$\mathbf{I}^*(t) = \frac{\mathbf{I}(nt)}{\sqrt{n}},$$

where $I_i(t)$ is the cumulative amount of time that server $i$ is idle over the time interval $[0, t]$. Observe that the components of $\mathbf{Q}^*$ are nonnegative, since they represent queue lengths, and the components of $\mathbf{I}^*$ are nondecreasing, since they measure cumulative idleness. Let $m_{ik}$ be the number of services (including the present one) that a customer currently in queue $k$ requires from station $i$ before leaving the network, and let

$$w_k = \sum_i m_{ik} c_i^*.$$

Now fixing the choice of routing and sequencing policies for the network determines the queue length and idleness processes $\mathbf{Q}$ and $\mathbf{I}$, and hence determines their scaled versions $\mathbf{Q}^*$ and $\mathbf{I}^*$. In Laws (1992) it is shown that, in the Brownian model, this choice of routing and sequencing policies is equivalent to choosing $\mathbf{Q}^*$ nonnegative and $\mathbf{I}^*$ nondecreasing subject to

$$\sum_k w_k Q_k^*(t) = B(t) + \sum_i c_i^* \mu_i I_i^*(t) \tag{16.2}$$

where $B$ is a Brownian motion, starting at 0, with drift $\theta = \sqrt{n}(\sum_j d_j^* \lambda_j - \sum_i c_i^* \mu_i)$

and variance $\sum_j (d_j^*)^2 \lambda_j^3 \alpha_j^2 + \sum_i (c_i^*)^2 \mu_i^3 \sigma_i^2$. Since $w_k$ is a weighted average of the remaining number of services of a customer in queue $k$, we can regard $w_k Q_k^*(t)$ as the total workload for the system in queue $k$ at time $t$: a service at station $i$ is regarded as being $c_i^*$ units of work. Then $\sum_k w_k Q_k^*(t)$ is the total work in the system at time $t$. Further, $\mu_i I_i^*(t)$ is the amount of lost service effort of server $i$ in $[0, t]$, and so $\sum_i c_i^* \mu_i I_i^*(t)$ measures the total lost service effort in the network, lost service effort at station $i$ being weighted by a factor $c_i^*$. Thus the Brownian model depends critically on the vectors $\mathbf{c}^*$ and $\mathbf{d}^*$ that define heavy traffic, and our control policies can effect any choice of queue lengths and idleness satisfying equation (16.2), which links the system workload to the total lost service effort.

Here, optimal control in the Brownian model produces what is termed a *pathwise solution* (Harrison 1988): rather than simply minimizing the mean number of customers in the system, the total number in the system, $\sum_k Q_k^*(t)$, can actually be minimized for all $t \geq 0$. Let

$$\bar{w} = \max_k (w_k),$$

and define the set of customer classes $\mathscr{K} = \{k : w_k = \bar{w}\}$. To minimize lost service effort over $[0, t]$ and hence minimize the system workload, for all $t \geq 0$, choose $\mathbf{I}^*$ nondecreasing such that

$$\sum_i c_i^* \mu_i I_i^*(t) = - \inf_{0 \leq s \leq t} B(s). \qquad (16.3)$$

Then the system population, $\sum_k Q_k^*(t)$, is minimized for all $t$ by choosing $\mathbf{Q}^*$ nonnegative such that

$$\sum_{k \in \mathscr{K}} Q_k^*(t) = \frac{1}{\bar{w}} b(t),$$
$$Q_k^*(t) = 0, \quad k \notin \mathscr{K}, \qquad (16.4)$$

where

$$b(t) = B(t) - \inf_{0 \leq s \leq t} B(s).$$

The interpretation of the above pathwise solution, (16.3) and (16.4), is in terms of a single station with multiple servers. Since the system workload is constrained by equation (16.2), the number of customers in the system is minimized by maximizing the workload per customer; that is, $Q_k^* > 0$ only if $k \in \mathscr{K}$. Also, the solution shows that $I_i^*$ can only increase when the system workload is zero: whenever one server is working, all servers are working, and service effort is only lost when the entire network is empty. If the network were operated without dynamic routing then the Brownian model would have a separate constraint on the workload of each server (Harrison 1988). So observe the important effect of dynamic routing: it has the effect of combining the workloads of individual servers and the network behaves as if servers $1, 2, \ldots, I$ are combined to form a *single pooled resource*.

## 16.3 ROUTING AND SEQUENCING POLICIES

The pathwise solution, (16.3) and (16.4), describes $\mathbf{Q}^*$ and $\mathbf{I}^*$ under the optimal dynamic routing and sequencing controls for the Brownian model. However, the solution does not directly indicate which individual routing and sequencing decisions will result in these optimal choices of $\mathbf{Q}^*$ and $\mathbf{I}^*$. For example, the pathwise solution does not immediately suggest how a single arriving customer of type $j$ should be routed. In this section we briefly consider the behavior of particular dynamic routing and sequencing policies based on delay estimates. (Routing algorithms based on delay estimates are of considerable practical importance in communication networks: see Bertsekas and Gallager (1987) and Schwartz (1987).)

Suppose initially that all sequencing is on a first-come first-served (FCFS) basis and let

$$N_i(t) = \sum_{k \in \mathscr{C}_i} Q_k(t),$$

where $\mathscr{C}_i$ is the set of queues at station $i$. Thus $N_i(t)$ is the total number of customers at station $i$ at time $t$. Suppose a type $j$ customer arriving at time $t$ uses the route $r \in \mathscr{R}_j$ that minimizes $D_r(t)$, where

$$D_r(t) = \sum_i a_{ir} \frac{N_i(t)}{\mu_i}. \tag{16.5}$$

We regard this policy as a shortest expected delay routing (SDR) policy. Although, for a customer arriving at time $t$, the number of customers at station $i$ may not be $N_i(t)$ when the customer reaches $i$, the *snapshot relation* (Foschini 1982; Reiman 1982; Reiman 1984) suggests that $D_r(t)$ will be an accurate estimate in heavy traffic. As a heavy traffic approximation, SDR equalizes $D_r(t)$ over the routes available to type $j$ customers so that

$$D_r(t) = D_{r'}(t) \quad \text{for all} \quad r, r' \in \mathscr{R}_j. \tag{16.6}$$

Equations (16.5) and (16.6) imply (Laws 1992) that

$$N_i(t) \propto c_i^* \mu_i, \tag{16.7}$$

and so, under SDR and FCFS, all servers are kept busy whenever there are customers anywhere in the system. Hence the first requirement of the pathwise solution, the condition (16.3), is satisfied and the network behaves as if all of the servers are combined to form a single pooled resource. Observe that SDR actually has the effect of calculating the (possibly unknown) dual cost vector $\mathbf{c}^*$.

As an alternative and an improvement to FCFS sequencing, consider the following dynamic priority sequencing policy used with SDR. Let

$$\hat{w}_k(t) = \sum_i m_{ik} \frac{N_i(t)}{\mu_i}.$$

Suppose that, on completion of a service, server $i$ calculates $\hat{w}_k(t)$ for each nonempty queue $k \in \mathscr{C}_i$ and then serves a customer in the queue with the smallest value of $\hat{w}_k(t)$. Given the relation (16.7),

$$\hat{w}_k(t) \propto w_k,$$

and observe that $\hat{w}_k(t)$ can also be regarded as an estimate of the remaining delay of a customer in queue $k$ at time $t$ if all future sequencing is FCFS. Heavy traffic limit theorems for single stations (Whitt 1971; Reiman 1983) and networks (Johnson 1983; Peterson 1991) in which routing and sequencing decisions are made without knowledge of $\mathbf{Q}$ show that only customers with the lowest priority level at a station have to queue there. So here we make the corresponding heavy traffic approximation that, under dynamic priority sequencing,

$$Q_k(t) = 0 \quad \text{if } w_k < \bar{w}_i, \quad k \in \mathscr{C}_i \tag{16.8}$$

where

$$\bar{w}_i = \max_{k \in \mathscr{C}_i}(w_k).$$

Although the condition (16.8) is similar to the second requirement of the pathwise solution (16.4), they are not equivalent (Laws 1992): the condition (16.4) implies (16.8) but not vice versa. However the conditions (16.4) and (16.8) are the same whenever $\bar{w}_i = \bar{w}$ for all $i$; that is, whenever there is a queue of maximum workload $\bar{w}$ at each station.

So, under our heavy traffic approximations, the above dynamic routing and sequencing policies satisfy the resource pooling property (16.3) of the pathwise solution (all servers being kept busy whenever there are customers in the system), but they do not satisfy the additional requirement of the solution (16.4) that the entire system population is concentrated in queues $k \in \mathscr{K}$ (maximizing the workload per customer). While SDR succeeds in satisfying the condition (16.3), it might appear that dynamic priority sequencing fails to satisfy the condition (16.4). However this is not the case: SDR leads to the relation (16.7), which in turn implies that the condition (16.4) cannot hold in general.

We give comparisons (from Kelly and Laws (1993) and based on the above approximations) of the mean delay per customer in heavy traffic under pathwise optimal routing and sequencing, SDR and dynamic priority sequencing, and an optimal nondynamic policy. Suppose arrival processes are Poisson and service times are exponential. Consider again the network in Figure 16.1 with service rate vector $\boldsymbol{\mu} = (2, 6, 3, 6, 2, 3)$ and with a sequence of arrival rate vectors, indexed by $n$, $\boldsymbol{\lambda}(n) = (2\rho(n), 4\rho(n))$ where $\rho(n) \uparrow 1$ (at a suitable rate) as $n \to \infty$. The nondynamic policy we consider uses FCFS sequencing and router arrivals according to the roll of an optimally biased die. In the heavy traffic limit ($n \to \infty$) the improvement of SDR and dynamic priority sequencing over this optimal random routing policy is by a factor of $f = 4.69$, while the improvement of the pathwise solution over the optimal random routing policy is by a factor of $f = 5.97$. Although simulation evidence (Laws 1992) suggests that the factor of improvement in the $n$th network, $f_n$, approaches its limit $f$ relatively slowly as $n \to \infty$, these simulations also indicate that dynamic controls can offer significant improvements in performance even at moderate system loads.

## REFERENCES

Adan I. J. B. F., Wessels J. and Zijm W. H. M. (1990) Analysis of the symmetric shortest queue problem *Stoch. Models*, **6**, 691–713.

Bertsekas D. P. and Gallager R. G. (1987) *Data Networks* Prentice-Hall, Englewood Cliffs, NJ.

Chen H. and Mandelbaum A. (1991) Stochastic discrete flow networks: diffusion approximations and bottlenecks *Ann. Prob.*, **19**, 1463–1519.

Dai J. G. and Harrison J. M. (1992) Reflected Brownian motion in an orthant: numerical methods for steady-state analysis *Ann. Appl. Prob.*, **2**, 65–86.

Flatto L. and McKean H. P. (1977) Two queues in parallel *Comm. Pure Appl. Math.*, **30**, 255–263.

Foschini G. J. (1977) On heavy traffic diffusion analysis and dynamic routing in packet switched networks. In Chandy K. M. and Reiser M., editors, pp 499–513 *Computer Performance* North-Holland, Amsterdam.

Foschini G. J. (1982) Equilibria for diffusion models of pairs of communicating computers—symmetric case *IEEE Trans. Inf. Theory*, **28**, 273–284.

Foschini G. J. and Salz J. (1978) A basic dynamic routing problem and diffusion *IEEE Trans. Comm.*, **26**, 320–327.

Gondran M. and Minoux M. (1984) *Graphs and Algorithms* Wiley-Interscience, New York.

Halfin S. (1985) The shortest queue problem *J. Appl. Prob.*, **22**, 865–878.

Harrison J. M. (1988) Brownian models of queueing networks with heterogeneous customer populations. In Fleming W. and Lions P. L., editors, *Stochastic Differential Systems, Stochastic Control Theory and Applications*, pp 147–186 Springer, New York.

Harrison J. M. and Nguyen V. (1990) The QNET method for two-moment analysis of open queueing networks *Queueing Syst.*, **6**, 1–32.

Harrison J. M. and Nguyen V. (1993) Brownian models of multiclass queueing networks: current status and open problems *Queueing Syst.*, **13**, 5–40.

Harrison J. M. and Williams R. J. (1987) Brownian models of open queueing networks with homogeneous customer populations *Stochastics*, **22**, 77–115.

Johnson D. P. (1983) Diffusion approximations for optimal filtering of jump processes and for queueing networks. PhD Thesis, Department of Mathematics, University of Wisconsin, Madison.

Kelly F. P. and Laws C. N. (1993) Dynamic routing in open queueing networks: Brownian models, cut constraints and resource pooling *Queueing Syst.*, **13**, 47–86.

Kingman J. F. C. (1961) Two similar queues in parallel *Ann. Math. Statist.*, **32**, 1314–1323.

Laws C. N. (1992) Resource pooling in queueing networks with dynamic routing *Adv. Appl. Prob.*, **24**, 699–726.

Laws C. N. and Louth G. M. (1990) Dynamic scheduling of a four-station queueing network *Prob. Engng. Inf. Sci.*, **4**, 131–156.

Minoux, M. (1981) Optimum synthesis of a network with non-simultaneous multicommodity flow requirements. In Hansen P., editor, *Studies on Graphs and Discrete Programming*, pp 269–277. North-Holland, Amsterdam.

Peterson W. P. (1991) A heavy traffic limit theorem for networks of queues with multiple customer types. *Math. Oper. Res.*, **16**, 90–118.

Reiman M. I. (1982) The heavy traffic diffusion approximation for sojourn times in Jackson networks. In Disney R. L. and Ott T. J., editors, *Applied Probability—Computer Science: The Interface*, Vol. 2, pp 409–422. Birkhäuser, Boston,

Reiman M. I. (1983) Some diffusion approximations with state space collapse. In *Lecture Notes in Control and Information Science*, No. 60, pp 209–240 Springer, Berlin.

Reiman M. I. (1984) Open queueing networks in heavy traffic *Math. Oper. Res.*, **9**, 441–458.

Schwartz M. (1987) *Telecommunication Networks* Addison-Wesley, Reading, MA.

Stidham S. (1974) A last word on $L = \lambda W$ *Oper. Res.*, **22**, 417–421.

Weber R. R. (1978) On the optimal assignment of customers to parallel servers *J. Appl. Prob.*, **15**, 406–413.

Wein L. M. (1990a) Optimal control of a two-station Brownian network *Math. Oper. Res.*, **15**, 215–242.

Wein L. M. (1990b) Scheduling networks of queues: heavy traffic analysis of a two-station network with controllable inputs *Oper. Res.*, **38**, 1065–1078.

Whitt W. (1971) Weak convergence theorems for priority queues: preemptive-resume discipline *J. Appl. Prob.*, **8**, 74–94.

Whitt W. (1991) A review of $L = \lambda W$ and extensions *Queueing Syst.*, **9**, 235–268.

Winston, W. (1977) Optimality of the shortest line discipline *J. Appl. Prob.*, **14**, 181–189.

---

*Department of Statistics, University of Oxford, 1 South Parks Road, Oxford OX1 3TG, UK*

CHAPTER 17

# Bandit Processes: Control, Analysis and Characterization

**Haya Kaspi and Avi Mandelbaum**
*Faculty of Industrial Engineering and Management, Technion, Israel*
**Robert J. Vanderbei**
*Department of Civil Engineering and Operations Research, Princeton University*

## Abstract

This chapter is about multi-armed bandit problems and bandit processes, with a focus on research in which we have been involved. We start with a description of previous work, proceed with a formulation of the multi-armed bandit problem and its solution, and then conclude with some open problems and research directions that we feel are worth pursuing. These include bandit problems with special structures, analysis of the stochastic processes that arise from the solutions, characterization of the structure of bandits that is amenable to a relatively simple solution, and, finally, relations of the above to ongoing research activities in other fields. Solving these problems is likely to encompass areas such as multiparameter processes, stochastic control (of semimartingales, specialized to diffusions), excursion theory (of Markov processes, generalized to semimartingales), stochastic waves, and more.

## 17.1 INTRODUCTION

The multi-armed bandit problem is a prototype of an important and challenging class of dynamic allocation problems, concerned with optimal division of resources among competing projects. What makes these problems important, as well as difficult, is that they embody the conflict between taking actions that yield immediate reward and pursuing those whose benefit will be realized in the future (the latter might require, for example, acquisition of information and learning).

The importance and scope of bandit problems is amply manifested by the appearance of three recent books on the subject: Presman and Sonin (1990), Berry and Fristedt (1985), and Gittins (1989). Interestingly, (quoting freely from Presman

---

*Scheduling Theory and its Applications* Edited by P. Chrétienne, E. G. Coffman, Jr., J. K. Lenstra and Z. Liu
© 1995 John Wiley & Sons Ltd

and Sonin 1990) "while these three books have equal rights to use the term bandit, the intersection of the content of any pair out of the three is either empty or is rather small." The models here are closest to those in Gittins (1989), with some overlap with Berry and Fristedt (1985) (especially Chapter 8).

## 17.2 PRIOR RESEARCH

The evolution of our research on bandits is described in the introduction to the survey by Mandelbaum (1988), from which the following paragraph has been extracted:

> In recent years, considerable effort has been devoted to the development of a theory for multiparameter processes. These are stochastic processes that evolve in time that is only partially ordered. The multiparameter theory provides a natural way to formulate problems in dynamic allocation of resources, including discrete- and continuous-time multi-armed bandits as special cases. Multiparameter processes that describe a game played by a gambler against a multi-armed bandit are called *bandit processes*. This article [i.e. Mandelbaum (1988)] surveys two control problems for bandit processes. The first problem, the *optimal stopping* problem, is that of a gambler who can stop playing at any time. The reward from the game depends only on the state of affairs at the time of stopping, and the gambler's problem is to choose an optimal stopping time. In the second problem, the *optimal navigation* problem, the gambler plays forever and seeks to maximize total discounted reward over an infinite horizon.

Roughly divided, optimal stopping problems are analyzed in Mandelbaum and Vanderbei (1981) and Mandelbaum *et al.* (1990), optimal navigation is the subject of Mandelbaum (1986,1987), and combining the two is the subject of Lawler and Vanderbei (1982). This whole line of research began in Mandelbaum and Vanderbei (1981), as a mathematical exercise in generalizing the classical optimal stopping problem. It has turned out to be useful, beyond a mere generalization, in the following directions.

*Control theory* The research by Mazziotto and Millet (1987) and Mandelbaum *et al.* (1990) added concreteness, which we believe had been lacking, to the developing abstract theory of multiparameter processes. New research has also been stimulated, as exemplified by Mazziotto and Millet (1987), Dalang (1990) and Chapter 13 of Friedman (1989) (devoted to the subject of Mandelbaum *et al.* 1990).

*Foundations* The technical machinery developed in Mandelbaum and Vanderbei (1981) enables one to formulate rigorously *continuous*-time models of dynamic allocation of resources and acquisition of information. (The introduction to Mandelbaum (1987) highlights the differences between discrete- and continuous-time formulations.)

*Switched processes* Continuous-time stochastic processes that arise from bandits seem novel and are interesting in their own right, and their relatives have been emerging elsewhere (see Section 17.4.3 below, as well as Benes and Rishel (1988), Karatzas and Ocone (1991), Barlow *et al.* (1989) and references therein).

## 17.3 THE MULTI-ARMED BANDIT MODEL AND ITS SOLUTION

The primitives for the $d$-armed bandit model are $d$ real-valued adapted stochastic processes $Z_i, \mathscr{F}_i$, $i = 1, \ldots, d$, on a common probability space. Here $Z_i = Z_i(t), t \geq 0$ is the *reward process* and $\mathscr{F}_i = \{F_i(t), t \geq 0\}$ the *information process* associated with arm $i$. It is assumed that the $\mathscr{F}_i(\infty)$ are independent and the $Z_i$ are bounded. Further technical restrictions must be imposed, but they are omitted, here and in the sequel, for brevity and clarity of presentation.

An *allocation strategy* $T$ is a $d$-dimensional stochastic process

$$\{T(t) = (T_1(t), \ldots T_d(t)), t \geq 0\}$$

that satisfies

(1) $T_i = \{T_i(t), t \geq 0\}$ is nondecreasing, with $T_i(0) = 0$ for each $i$;
(2) $T_1(t) + \ldots + T_d(t) = t$ for all $t \geq 0$;
(3) $\{T_1(t) \leq s_1, \ldots, T_d(t) \leq s_d\} \in \mathscr{F}_1(s_1) \vee \ldots \vee \mathscr{F}_d(s_d)$, for all $(s_1, \ldots, s_d) \geq 0$.

The random variable $T_i$ models the total amount of time allocated to arm $i$ over the interval $[0, t]$. Properties (1) and (2) are thus self-explanatory. Property (3) is a mathematical articulation of the non-anticipative nature of $T$: for all $i$, the event "no more than $s_i$ units of time have been allocated to arm $i$" does not depend on information beyond $\mathscr{F}_i(s_i)$. (In multiparameter terminology, an allocation strategy is called an *optional increasing path* (Walsh 1981).)

The present value $R(T)$ of future rewards associated with a strategy $T$ is the random variable

$$R(T) = \int_0^\infty e^{-\beta t} Z[T(t)] \, \mathrm{d}T(t),$$

where the discount factor $\beta$ is a given positive real number, and $Z[T(t)] \, \mathrm{d}T(t)$ is an abbreviation for $\sum_{i=1}^d Z_i[T_i(t)] \, \mathrm{d}T_i(t)$. The *bandit problem* is to find optimal strategies that maximize over all allocation strategies $T$ the *value function* $V(T) = \mathrm{E}R(T)$.

Optimal strategies are described in terms of dynamic allocation *index processes*. The index process $\Gamma_i = \{\Gamma_i(t), t \geq 0\}$ associated with arm $i$ is given by

$$\Gamma_i(t) = \operatorname*{ess\,sup}_{\tau > t} \frac{\mathrm{E}^{\mathscr{F}_i(t)} \int_t^\tau e^{-\beta t} Z_i(u) \, \mathrm{d}u}{\mathrm{E}^{\mathscr{F}_i(t)} \int_t^\tau e^{-\beta t} \, \mathrm{d}u}, \quad t \geq 0,$$

where the essential supremum is over all $\mathscr{F}_i$ stopping times $\tau$, not necessarily finite, with $\tau > t$ almost surely. An *index strategy* is one that follows the leader among the index processes. Formally, a strategy $I = \{I(t), t \geqslant 0\}$ is an index strategy if $I_i(\cdot)$ increases at $t$ only when $\Gamma_i[I_i(t)] = \vee_{i=1}^d \Gamma_j[I_j(t)]$, for all $t \geqslant 0$. (Such strategies are simple to implement in discrete time, but rather subtle if time is continuous; see Section 3.1 in Mandelbaum et al. (1990) for an explanation of how to follow the leader between two independent Brownian motions.)

The solution to the bandit problem is always an index strategy. To see this, we introduce the *index field* $\Gamma(s), s \in S = \mathbb{R}_+^d$ given by

$$\Gamma(s) = \max_{1 \leqslant i \leqslant d} \Gamma_i(s_i), \quad s = (s_1, \ldots, s_d) \in S,$$

and its lower envelope

$$\underline{\Gamma}(s) = \inf_{0 \leqslant r \leqslant s} \Gamma(r) = \max_{1 \leqslant j \leqslant d} \underline{\Gamma}_j(s_j),$$

where $\underline{\Gamma}_j$ is the lower envelope of $\Gamma_j$:

$$\underline{\Gamma}_j(t) = \inf_{0 \leqslant u \leqslant t} \Gamma_j(u), \quad t \geqslant 0.$$

Then for any two strategies $T$ and $U$, one can show that

$$V(T) \leqslant \mathbf{E} \int_0^\infty e^{-\beta t} \underline{\Gamma}[U(t)] \, dt, \tag{17.1}$$

while for any index strategy $I$,

$$V(I) = \mathbf{E} \int_0^\infty e^{-\beta t} \underline{\Gamma}[I(t)] \, dt. \tag{17.2}$$

The relations (17.1) and (17.2) clearly imply the optimality of index strategies.

## 17.4 OPEN PROBLEMS AND FUTURE DIRECTIONS OF RESEARCH

We describe three possible avenues of research, in concert with the three directions mentioned at the end of Section 17.2.

### 17.4.1 Control theory

#### 17.4.1.1 General bandits

The bandit problem in Section 17.3 is very general, and the proof seems very simple. This is indeed so for reward processes that correspond to discrete-time models (roughly $Z_i$ with piecewise-constant paths). Otherwise, a proof is available only under significant restrictions (Mandelbaum 1987), notably continuity of the reward processes and their indices, and it is based on cumbersome discrete approximations, especially for (17.1) and (17.2).

*Open problems*
Generalize Section (17.3) to reward processes that are allowed to have random jumps, for example bounded semimartingales. The approach in Mandelbaum (1987) is inadequate. A natural direction to start from is a direct proof of (17.1) and (17.2).

## 17.4.1.2 Special bandits

Solutions to the bandit problem have been obtained for many special bandits, notably discrete-time models (Gittins 1989; Berry and Fristedt 1985; Keener 1985, 1986; Varaiya et al. 1985; Lawler and Vanderbei 1982), diffusion bandits (Karatzas and Ocone 1991; Karatzas 1984; Mandelbaum 1987; Eplett 1986) and some discrete-space Markov processes (Eplett 1986). The analysis of such special bandits is important, since they are amenable to explicit calculations, and their solutions provide insight that stimulates further research. There is much to be done on the explicit solutions of special bandits, specifically Brownian, diffusion, Lévy and Markovian bandits. (Naturally, the more assumptions, the more explicit is the solution.) Here are some details.

Call arm $i$ *Markovian* if $Z_i(t) = r_i[X_i(t)]$, where $X_i$, the state of arm $i$, is a strong Markov process, and $r_i[x]$, the reward function, is a real-valued function on its state space. A Markovian bandit is one whose arms are all Markovian. In view of existing results (Karatzas 1984; Mandelbaum 1987), one expects (though a proof does not exist at that level of generality) that the index process $\Gamma_i$ of a Markovian arm has the representation $\Gamma_i(t) = \gamma_i[X_i(t)]$, $t \geq 0$, where $\gamma_i$, the *index function* associated with arm $i$, is given by

$$\gamma_i(x_i) = \sup_{\tau > 0} \frac{E_{x_i} \int_0^\tau e^{-\beta u} r_i[X_i(u)]\, du}{E_{x_i} \int_0^\tau e^{-\beta u} du}. \tag{17.3}$$

For $X_i$ that is a 1-dimensional diffusion process, and with a reward function that is monotone increasing and sufficiently smooth, Karatzas (1984) represented the index function in terms of resolvents and Green's functions. We have computed this representation to be

$$\gamma_i(x_i) = \frac{{}^*P^i \int_0^R e^{-\beta u} r_i(x_i + X_i(u))\, du}{{}^*P^i \int_0^R e^{-\beta u}\, du}, \tag{17.4}$$

where ${}^*P^i$ is the excursion law from the minimum set $\{t \geq 0 : X_i(t) = \underline{X}_i(t)\}$. When $X_i$ is a standard Brownian motion and the reward function is increasing, this formula reduces to

$$\gamma_i(x) = \sqrt{\frac{2}{\beta}} \int_0^\infty r_i(x+z) e^{-\sqrt{2\beta} z}\, dz. \tag{17.5}$$

For $r_i$ that is first decreasing and then increasing, we can show that

$$\gamma_i(x) = \gamma_i^+(x) \vee \gamma_i^-(x), \tag{17.6}$$

where $\gamma_i^+$ is identical to $\gamma_i$ in (17.5), and

$$\gamma_i^-(x) = \sqrt{\frac{2}{\beta}} \int_0^\infty r_i(x-z) e^{-\sqrt{2\beta} z}\, dz.$$

A similar result holds for sufficiently smooth $r_i$ that is increasing then decreasing.

A *Lévy* bandit is a Markovian bandit with $X_i$ that are independent Lévy processes. We have been working on this model with $r_i$ continuous, bounded and *increasing*, while the $X_i$ are arbitrary Lévy processes (which, of course, may have discontinuous sample paths). As might be anticipated from previous research (after all, Lévy processes are Markov), such Lévy bandits are optimally controlled by index strategies that are determined by the index functions $\gamma_i$ in (17.3). What is less anticipated, however, is the generalization of (17.4) to

$$\gamma_i(x_i) = \frac{\ell r_i(x_i) + {}^*P \int_0^R e^{-\beta u} r_i(u + X_i(u))\, du}{\ell + {}^*P \int_0^R e^{-\beta u}\, du}, \qquad (17.7)$$

where $l$ is the derivative of the Lebesgue measure of the minimum set with respect to its local time $L$, namely

$$\text{Lebesgue } \{0 \leqslant u \leqslant t : X_u = \underline{X}_u\} = \ell L_t,$$

$R$ is the first hitting time of this minimum ladder set of $X_i$ and $^*P$ is the excursion law. Furthermore, one can use the Wiener–Hopf factorization of the Lévy exponent of the arms to obtain the characteristic function of the above excursion laws, through which the indices are defined. With this representation one can explicitly calculate index functions of some interesting Levy arms, calculate optimal rewards, and also rediscover along the way that local time naturally quantifies continuous-time switching.

*Open Problems* Extend (17.5) and (17.7) to reward functions that are not necessarily monotone increasing. One should probably first establish the analogue of (17.6) to piecewise-monotone $r$. (Our proofs resemble those of Karatzas': they are based on the "principle of smooth fit". This requires smooth data, but the hope is to ultimately cover nonsmooth rewards, in the spirit of $r(x) = |x|$.) With the insight gained, it might then be possible to extend (17.7). This might entail further exploration of the relevant excursion theory, which does not seem easy. (Indeed, monotonic rewards give rise to excursions from the lower envelope of the index processes. We anticipate that general rewards will require delicate concatenation of excursions from lower and upper envelopes; bandits where the $X_i$ are more general than Markov, say semimartingales, might require a non-Markovian excursion theory.

### 17.4.2 Foundations

#### 17.4.2.1 Characterization of index strategies

*Open problem* Characterize the circumstances in which index strategies are optimal, and quantify performance deterioration due to the use of index strategies when such circumstances are violated.

It is natural to pursue this important problem within the multiparameter framework of Mandelbaum (1986,1987), in the spirit of Lawler and Vanderbei

(1982). It might be necessary first to extend this framework to cover superprocesses, as in Varaiya *et al.* (1985).

*A more specialized open problem*   Along the lines of Lawler and Vanderbei (1982) and given $d$ Markovian arms, is the optimal strategy an index strategy for all bounded reward functions if and only if the arms are independent (in some sense yet to be understood)?

*Related work*   The first breakthrough is due to Whittle (1980), later complemented by Glazebrook (1982). Whittle introduced a condition that is sufficient (and in some sense necessary) for the following to prevail: given a Markov decision process (MDP) consisting of several MDPs in parallel (an example of which is a Markovian multi-armed bandit), optimal strategies take actions in a way that is consistent with the optimal strategies for the individual MDPs. For lack of a better term, call such strategies *separable*. Separable strategies are attractive because of their simplicity. They need not be optimal in general, but it is possible to quantify the loss in reward due to their use, at least when Whittle's condition is violated (Glazebrook 1983,1988; Glazebrook and Fay 1990).

### 17.4.2.2   Bandits with switching costs

The introduction of switching costs into optimal control problems is notorious for making problems intractable. It is not hopeless, however, as work by Davis and Norman (1990) has beautifully demonstrated.

*Open problem*   Incorporate switching costs into existing bandit theory. A reasonable guiding principle is to start with simple models, specifically Brownian bandits similar to those in Mandelbaum *et al.* (1990). Switching costs can then be incorporated in alternative ways. For example, one can introduce a cost $c$ per switch, or perhaps incur costs that are proportional to the local time on some switching curve. (A "proper" formulation for the latter version is still being sought—"proper" in the sense that switching costs do indeed change the nature of the solution; this does not happen in symmetric examples, which are the ones we are able to analyze.) We have implemented on a computer a discretized version of the "cost $c$ per switch" formulation with Brownian motions, and this gave rise to the following.

*Conjecture*   The switching curve from the problem with $c = 0$ splits into two curves, each residing on a different side of the original curve. The buffer region between the two curves is a "continue with whatever you've been doing" region. Switching only occurs when the process actually leaves this buffer zone. One should first prove this conjecture, which might enable generalizing beyond Brownian bandits.

### 17.4.3 Switched processes

The stochastic processes that arise from bandits are interesting in their own right. We understand them best in discrete time or when sample paths are continuous, but the following examples amply demonstrate how interesting they might turn out in general.

Let $Z_i, i = 1, \ldots, d$, be $d$ stochastic processes. We say that a strategy $I = \{I(t), t \geq 0\}$ *follows the leader* among $Z_i, i = 1, \ldots, d$, if $I_i(\cdot)$ increases at $t$ only when $Z_i(I_i(t)) = \vee_{j=1}^{d} Z_j(I_j(t))$ for all $t \geq 0$ (to *follow the loser* is to follow the leader among the $-Z_i$'s).

#### 17.4.3.1 Lévy bandits

Let $Z_i, i = 1, \ldots, d$, be i.i.d. Lévy processes. Using basic tools from the theory of multiparameter processes, we have shown that, for any strategy $T$, the process $X^T$ given by

$$X^T(t) = Z_1(T_1(t)) + \ldots + Z_d(T_d(t)), \quad t \geq 0,$$

has the same distribution as the $Z_i$'s.

Suppose now that the $Z_i$'s do not have downward jumps. Let $I(t)$ be a strategy that follows the leader among the $Z_i$'s and define

$$\underline{Z}_i(t) = \inf_{0 \leq s \leq t} Z_i(t), \quad \xi_i(t) = Z_i(I_i(t)) - \underline{Z}_i(I_i(t)).$$

Then

$$\underline{Z}_1(I_1(t)) = \cdots = \underline{Z}_d(I_d(t)) = \frac{1}{d} \underline{X}^I(t).$$

The process $-\underline{X}^I(t)/d = L(t)$ is a continuous process that increases only when $X^I(t)$ assumes its minimum; in fact, it is a version of the local time at the minimum. At each time $t$, there is at most one $i$ for which $\xi_i(t) > 0$, and when this is the case, arm $i$ is pulled. We have derived the path decomposition

$$Z_i(I_i(t)) = \xi_i(t) - L(t),$$

where $L$ is the local time at the minimum of $X^I$ (local time of switching), and the $\xi_i$ correspond to the excursions from the minimum of $X^I$ (excursions associated with pulled arms).

#### 17.4.3.2 Nonlinearly-skewed Brownian motion

Consider the stochastic functional equation

$$X_t = B_t + \phi(L_t^X(0)), \quad t \geq 0,$$

where $B$ is a standard 1-dimensional Brownian motion, $X$ is the unknown process, and $L^X(0)$ is the (symmetric) local time of $X$ at 0. For $\phi$ linear, say $\phi(s) = \beta s$,

$s \geqslant 0$, Harrison and Shepp (1981) proved that a solution $X$ exists if and only if $|\beta| \leqslant 1$, the solution is then a strong solution adapted to $B$, and its distribution is that of a skew Brownian motion. Skew Brownian motion also arises from following the leader between $Z_1 = a_1 B_1$ and $Z_2 = a_2 B_2$, where the $B_i$ are independent standard Brownian motions. Indeed, let $S$ denote the "distance" of the switched process $(Z_1(T_1), Z_2(T_2))$ from the line $a_1 z_1 = a_2 z_2$, where distance is positive and measured vertically below the line, and is negative and measured horizontally above it. Then $S$ can be shown to solve the above equation with $\phi(s) = [(a_2 - a_1)/(a_2 + a_1)]s$; that is, $S$ is a skew Brownian motion with $\beta = (a_2 - a_1)/(a_2 + a_1)$. Similarly, following the leader between $Z_1 = a_1(B_1)$ and $Z_2 = a_2(B_2)$, for *non-linear* functions $a_1$ and $a_2$, gives rise to a solution to the above, with a $\phi$ that is related to the $a_i$ in a rather intricate fashion. This provides a weak solution, but gives rise to the natural question of existence and uniqueness of a strong solution. Recently (with help from M. Barlow), we have made significant progress on this problem, which confirms existence and uniqueness for a wide class of $\phi$. However, a definite resolution is still lacking. Indeed, it remains to characterize the $\phi$ for which a unique strong solution exists. One could also attempt to extend the analysis beyond $B$ that are Brownian motion, or perhaps to $d$-dimensional Brownian motion, with $d \geqslant 2$.

### 17.4.3.3  *Thin diffusions and the arcsine law*

The generalizations of the arcsine laws, by Pitman and Yor (1990), can be recast naturally in a bandit framework. Indeed, the excursion process $(\xi_1, \ldots, \xi_d)$ in the path decomposition above, with Brownian data, is precisely Walsh's (1978) diffusion, which they use. Since the Brownian nature of the path plays no role in the bandit framework (except for giving rise to a unique index strategy), we see what it all means in the context of arbitrary semimartingales, for example.

### 17.4.3.4  *Stochastic waves*

There exists a connection between switched processes and stochastic waves (Dynkin and Vanderbei 1983). Two examples which suggest this connection are as follows.

**Example 17.1**  Construct a martingale $X$ whose absolute value satisfies

$$|X_t| = t \quad \forall t \geqslant 0.$$

(This a special case of a result by D. Gilat, who proved that every nice submartingale is, in distribution, an absolute value of a martingale.) We now demonstrate one way of solving this problem. Let $B$ be a standard one-dimensional Brownian motion with $B(0) = 0$. Introduce

$$\tau_a = \inf \{s > 0 : |B_s| \geqslant a\},$$

for all $a \geqslant 0$. Then $\tau_0 = 0$, $\tau_a$ is increasing in $a$ and $\tau_{a+b} = \tau_a + \tau_b \circ \theta_{\tau_a}$, implying that $\tau$ is a random time change. It follows that $X_t \equiv B_{\tau_t}$ is a martingale that satisfies

$|X_t| = t$, $t \geqslant 0$. The martingale $X$ is a very simple stochastic wave (Vanderbei 1983; Dynkin and Vanderbei 1983), and it turns out that one can construct it by following the loser between a uniform motion to the right ($Z_1(t) = t$, $t \geqslant 0$) and the absolute value of a standard Brownian motion ($Z_2(t) = |B(t)|$, $t \geqslant 0$).

**Example 17.2** Follow the loser between the radial part of a 2-dimensional Markov process and the uniform motion to the right. It turns out that this yields exactly the stochastic wave in Vanderbei (1983).

## REFERENCES

Barlow M., Pitman J. and Yor M. (1989) On Walsh's Brownian motion *Séminaire de Probabilités XXIII. Springer Lecture Notes in Math.*, **1372**, 275–293.
Benes V. E. and Rishel R. W. (1988) Optimality of full-bang-to-reduce-predicted-miss for some partially observed stochastic control problems *IMA Lecture Notes*, **10**, 1–15.
Berry D. A. and Fristedt D. (1985) *Bandit Problems: Sequential Allocation of Experiments* Chapman Hall, London.
Dalang R. C. (1990) Randomization in the two-armed bandit problem *Ann. Prob.*, **18**, 218–225.
Davis M. H. A. and Norman A. R. (1990) Portfolio selection with transaction costs *Math. Oper. Res.*, **15**, 676–713.
Dynkin E. B. and Vanderbei R. J. (1983) Stochastic waves *Trans. AMS*, **275**, 771–779.
Eplett W. J. R. (1986) Continuous-time allocation indices and their discrete-time approximation *Adv. Appl. Prob.*, **18**, 724–746.
Friedman A. (1989) *Mathematics in Industrial Problems*, Part 2 Springer, Berlin.
Gittins J. C. (1989) *Multi-armed Bandit Allocation Indices* Wiley, New York.
Glazebrook K. D. (1982) On a sufficient condition for superprocesses due to Whittle *J. Appl. Prob.*, **19**, 99–110.
Glazebrook K. D. (1983) The role of dynamic allocation indices in the evaluation of suboptimal strategies for families of bandit processes. In Herkenrath U., Kalin D. and Vogel W., editors, *Mathematical Learning Models—Theory and Algorithms: Lecture Notes in Statistics*, No. 20, pp 68–77 Springer, Berlin.
Glazebrook K. D. (1988) On a reduction principle in dynamic programming *Adv. Appl. Prob.*, **20**, 836–851.
Glazebrook K. D. and Fay N. A. (1990) Evaluating strategies for Markov decision processes in parallel *Math. Oper. Res.*, **15**, 17–32.
Harrison J. M. and Shepp L. A. (1981) On skew Brownian motion *Ann. Prob.*, **9**, 309–313.
Karatzas I. (1984) Gittins indices in the dynamic allocation problem for diffusion processes *Ann. Prob.*, **12**, 173–192.
Karatzas I. and Ocone D. L. (1991) The resolvent of a degenerate diffusion on the plane, with applications to partially-observed stochastic control. Technical Report.
Keener R. (1985) Further contributions to the "two-armed bandit" problem *Ann. Statist.*, **13**, 418–422.
Keener R. (1986) Multi-armed bandits with simple arms, *Adv. Appl. Math.*, **7**, 199–204.
Lawler G. F. and Vanderbei R. J. (1982) Markov strategies for optimal control problems indexed by a partially ordered set *Ann. Prob.*, **11**, 642–647.
Mandelbaum A. (1986) Discrete multiarmed bandits and multiparameter processes *Prob. Theory Rel. Fields*, **71**, 129–147.

Mandelbaum A. (1987) Continuous multi-armed bandits and multi-parameter processes *Ann. Prob.,* **15**, 1527–1556.
Mandelbaum A. (1988) Navigating and stopping multi-parameter bandit processes. In Fleming W. and Lions P. L., editors, *Stochastic Differential Systems, Stochastic Control Theory and Applications* pp 339–372 Springer, Berlin.
Mandelbaum A. and Vanderbei R. J. (1981) Optimal stopping and supermartingales over partially ordered sets *Z. Warsch. verw. Gebiete,* **57**, 253–264.
Mandelbaum A., Shepp L. A. and Vanderbei R. J. (1990) Optimal switching between a pair of Brownian motions *Ann. Prob.,* **18**, 1010–1033.
Mazziotto G. and Millet A. (1987) Stochastic control of two-parameter processes: the two-armed bandit problem *Stochastics,* **22**, 251–288.
Pitman J. and Yor M. (1990) Arcsine laws and interval partitions derived from a stable subordinator. Technical Report 189, Department of Statistics, University of California, Berkeley.
Presman E. L. and Sonin I. N. (1990) *Sequential Control with Incomplete Information: The Bayesian Approach to Multi-armed Bandit Problems* Academic, New York.
Vanderbei R. J. (1983) Toward a stochastic calculus for several Markov processes *Adv. Appl. Math.,* **4**, 125–144.
Varaiya P. P., Walrand J. C. and Buyukkoc C. (1985) Extensions to the multi-armed bandit problem: the discounted case *IEEE Trans. Autom. Control,* **30**, 426–439.
Walsh J. B. (1978) A diffusion with discontinuous local time *Astérique,* **52–54**, 37–45.
Walsh J. B. (1981) Optional increasing paths. In *Colloque ENST-CNET: Lecture Notes in Mathematics,* No. 863, pp 172–201 Springer, Berlin.
Whittle P. (1980) Multi-armed bandits and the Gittins index *J. R. Statist. Soc.,* B **42**, 143–149.

*Faculty of Industrial Engineering and Management, Technion—Israel Institute of Technology, Haifa 32000, Israel*

*Department of Civil Engineering and Operations Research, School of Engineering/Applied Science, Princeton University, Princeton, NJ 08544, USA*

# Author Index

Aarts, E. H. L. *See* van Laarhoven, P. J. M. *et al.*
Adam, T. *et al.* 125
Adams, J. *et al.* 33, 290
Adan, I. J. B. F. *et al.* 329
Adelsberger, H. H. *See* Kanet, J. J. and Adelsberger, H. H.
Adiri, I. *See* Frostig, E. and Adiri, I.; Hefetz, N. and Adiri, I.
Adler, L. *et al.* 34
Agrawala, A. K. *et al.* 43
Aiken, A. and Nicolau, A. 210
Akers, S. B. 279
Alvares-Valdes, R. *See* Christofides, N. *et al.*
Anantharam, V. 323
Anger, F.D. *See* Hwang, J. J. *et al.*
Applegate, D. and Cook, W. 282, 287
Aragon, C.R. *See* Johnson, D. S. *et al.*
Arndt, U. *See* Franken, P. *et al.*
Asmussen, S. 19

Babai, L. *et al.* 12
Baccelli, F. *See* Towsley, D. and Baccelli, F.
Baccelli, F. and Brémaud, P. 19
Baccelli, F. and Liu, Z. 296
Baccelli, F. *et al.* 296, 306
Bagga, P. C. 323
Baker, K. R. 33
Balas, E. 263, 264, 287, 288
Balas, E. *See* Adams, J. *et al.*
Bárány, I. 49
Bárány, I. and Fiala, T. 10
Barlow, M. *et al.* 339
Baroum, S. *See* Demeulemeester, E. *et al.*
Bartholdi, J. J. *et al.* 49
Bartholdi, J. J. III. *See* Ramudhin, A. *et al.*
Barton, D.E. *See* David, F. N. and Barton, D. E.
Bartusch, M. *et al.* 99, 261, 265, 271

Battersby, A. *See* Carruthers, J. A. and Battersby, A.
Bell, C. A. and Park, K. 263, 266
Bellman, R. 41
Benes, V. E. and Rishel, R. W. 339
Bentley, J. L. 24
Berge, C. 228, 229
Berry, D. A. and Fristedt, D. 337, 338, 341
Berryman, H. *See* Saltz, J. *et al.*
Bertsekas, D. P. 296
Bertsekas, D. P. and Gallager, R. G. 333
Billingsley, P. 25
Blazewicz, J. *See* de Werra, D. and Blazewicz, J.
Blazewicz, J. *et al.* 16, 228, 260
Bokhari, S. H. 131
Boling, R. M. *See* Hillier, F. S. and Boling, R. M.
Bondy, J. A. and Murty, U. S. R. 11
Bowman, E. H. 263
Boxma, O. J. 18, 22
Brand, J. D. *et al.* 263
Bratley, P. *et al.* 286
Brémaud, P. *See* Baccelli, F. and Brémaud, P.
Browne, J. C. *See* Kim, S. J. and Browne, J. C.
Browne, S. and Yechiali, U. 40
Brucker, P. 279
Brucker, P. and Jurisch, B. 288
Brucker, P. *et al.* 99, 100, 286, 288, 289
Brucker, P. *See* Lenstra, J. K. *et al.*
Bruno, J. 43
Bruno, J. and Hofri, M. 34
Bruno, J. *et al.* 43
Bruno, J. L. 95, 105
Bruno, L. L. and Downey, P. J. 18
Burton, F. W. 146
Burton, F. W. and Rayward-Smith, V. J. 151

Burton, F. W. *et al.* 146
Buyukkoc, C. *See* Varaiya, P. P. *et al.*; Varaiya, P. P. *et al.*

Callahan, D. and Kennedy, K. 123
Calvin, J. *See* Bartholdi, J. J. *et al.*
Calvin, J. M. *See* Ramudhin, A. *et al.*
Carlier, J. 267, 286, 290
Carlier, J. and Chrétienne, P. 194, 210
Carlier, J. and Latapie, B. 263, 267
Carlier, J. and Pinson, E. 208, 213, 218, 267, 277, 283, 285, 286, 289
Carruthers, J. A. and Battersby, A. 263
Cassandras, C. G. *See* Sparaggis, P. D. *et al.*
Cellary, W. *See* Blazewicz, J. *et al.*
Chandy, K. M. *See* Adam, T. *et al.*
Chandy, K. M. and Reynolds, P. F. 43, 105, 106
Chang, C. S. 188, 295, 304, 309
Chang, C. S. and Yao, D. 323
Chao, X. *et al.* 51, 323
Chaouiya, C. 167, 173, 186, 188
Chaouiya, C. *See* Lefebvre-Barbaroux, S. *et al.*
Chen, H. and Mandelbaum, A. 58, 329
Chen, H. and Yao, D. D. 58
Chow, Y. C. *See* Hwang, J. J. *et al.*
Chrétienne, P. 73, 77, 112, 118, 130, 199, 205
Chrétienne, P. and Picouleau, C. 65
Chrétienne, P. *See* Carlier, J. and Chrétienne, P.; Colin, J.-Y. and Chrétienne, P.
Christofides, N. 5
Christofides, N. *et al.* 260, 263, 265
Çinlar, E. 21
Claver, J. F. and Jackson, P. 194
Cobham, A. 34, 37
Coffman, E. G. 15, 241
Coffman, E. G. *See* Agrawala, A. K. *et al.*
Coffman, E. G. and Denning, P. J. 112, 125
Coffman, E. G. and Liu, Z. 44
Coffman, E. G. and Whitt, W. 36, 49
Coffman, E. G. *et al.* 43
Coffman, E. G. Jr. 153
Coffman, E. G. Jr. *See* Muntz, R. R. and Coffman, E. G. Jr.
Coffman, E. G. Jr. and Gilbert, E. N. 18, 20
Coffman, E. G. Jr. and Graham, R. L. 94, 97, 98

Coffman, E. G. Jr. and Liu, Z. 105, 107
Coffman, E. G. Jr. and Lueker, G. S. 17, 22, 24, 36
Coffman, E. G. Jr. and Wright, P. E. 21
Coffman, E. G. Jr. *et al.* 17, 18, 20, 22
Cohen, G. *et al.* 194, 199, 201, 204
Colin, J.-Y. and Chrétienne, P. 67
Conway, R. W. *et al.* 33, 37
Cook, S. A. 6
Cook, W. *See* Applegate, D. and Cook, W.
Cormen, T. H. *et al.* 247, 254
Cox, D. R. and Smith, W. L. 34
Cox, J. *See* Goldratt, E. M. and Cox, J.
Crowley, K. *See* Saltz, J. *et al.*
Cruz, R. L. 168, 188
Cunningham, A. A. and Dutta, S. K. 34, 323
Cytron, R. 208

Dai, J. G. and Harrison, J. M. 329
Dai, J. G. and Wang, Y. 34
Dalang, R. C. 338
Daley, D. J. 295, 309
David, F. N. and Barton, D. E. 161
Davidson, E. S. and Patel, J. H. 208, 211
Davies, E. W. *See* Stinson, J. P. *et al.*
Davies, M. H. A. and Norman, A. R. 343
Davis, E. W. 260, 263
Davis, E. W. *See* Moder, J. J. *et al.*
Davis, E. W. and Heidorn, G. E. 262, 263, 264
Dellacroce, F. *et al.* 290
Dell'Amico, M. and Trubian, M. 290
Dell'Olmo, P. *et al.* 28
Demeulemeester, E. 259, 260, 262, 266, 267, 269, 270, 272
Demeulemeester, E. and Herroelen, W. 260, 262, 263, 267, 268, 269, 270, 271, 272, 273
Demeulemeester, E. *et al.* 266
Denning, P. J. *See* Coffman, E. G. and Denning, P. J.
deWerra, D. 227, 232
deWerra, D. *See* Mahadev, N. V. R. *et al.*
deWerra, D. and Blazewicz, J. 238
deWerra, D. and Solot, P. 222, 223, 231, 237
deWerra, D. *et al.* 222, 223
deWit, J. and Herroelen, W. 270
Dickey, S. E. *See* Suresh, S. *et al.*
Dickson, J. R. *See* Adam, T. *et al.*

# AUTHOR INDEX

Dolev, D. and Warmuth, M. K.  92, 95, 96
Dongarra, J. J. and Sorensen, D. C.  137
Downey, P. See Bruno, J. et al.
Downey, P. J. See Bruno, L. L. and Downey, P. J.
Dunigan, T. H.  117
Dutta, S. K. See Cunningham, A. A. and Dutta, S. K.
Dyer, M. and Wolsey, L. A.  287
Dynkin, E. B. and Vanderbei, R. J.  345

Eager, D. L. et al.  146
Ecker, K. See Blazewicz, J. et al.
Efe, K.  87
Eisenbeis, C.  194
Eisenbeis, C. and Wang, J.  220
Eisenbeis, C. and Windheiser, D.  217
Elmaghraby, S. E.  263
El-Rewini, H. and Lewis, T. G.  113, 137
Ephremides, A. et al.  308
Eplett, W. J. R.  341
Ethier, S. N. and Kurtz, T. K.  25

Fay, N. A. See Glazebrook, K. D. and Fay, N. A.
Fdida, S. See Towsley, D. et al.
Feige, U. et al.  12
Feller, W.  18, 161, 305
Fiala, T. See Bárány, I. and Fiala, T.
Fischer, J. F. and Thompson, G. L.  277, 290
Fisher, M. L. et al.  287
Flatto, L.  17
Flatto, L. See Coffman, E. G. et al.; Coffman, E. G. Jr. et al.
Flatto, L. and McKean, H. P.  329
Florian, M. See Bratley, P. et al.
Foley, R. D. See Suresh, S. et al.
Foley, R. D. and Suresh, S.  322
Fortnow, L. See Babai, L. et al.
Foschini, G. J.  329, 333
Foschini, G. J. and Salz, J.  329
Foss, S.  295, 309
Fraiman, N. See Adler, L. et al.
Fraiman, N. et al.  34
Franken, P. et al.  19
Frederickson, G. N. See Bruno, J. et al.; Coffman, E. G. Jr. et al.
French, S.  282
Frenk, J. B. G. See Rinnooy Kan, A. H. G. and Frenk, J. B. G.

Frenk, J. B. G. and Rinnooy Kan, A. H. G.  22, 42
Friedman, A.  338
Friedman, H. D.  51, 323
Fristedt, D. See Berry, D. A. and Fristedt, D.
Frostig, E.  43, 105
Frostig, E. and Adiri, I.  323

Gajski, D. See Wu, M. Y. and Gajski, D.
Gallager, R. G. See Bertsekas, D. P. and Gallager, R. G.
Gao, G. and Ning, Q.  210
Garey, M. See Coffman, E. G. et al.
Garey, M. R. See Agrawala, A. K. et al.; Brucker, P. et al.
Garey, M. R. and Johnson, D. S.  5, 16, 69, 73, 74, 80, 99, 151, 277
Garey, M. R. et al.  26, 92, 95
Gasperoni, F.  241
Gasperoni, F. and Schwiegelshohn, U.  219, 220, 242
Gaubert, S.  194, 204, 207
Geist, G. A. and Heath, M. T.  123, 130
George, A. et al.  130
Gerasoulis, A.  111
Gerasoulis, A. See Yang, T. and Gerasoulis, A.
Gerasoulis, A. and Nelken, I.  130
Gerasoulis, A. and Yang, T.  82, 121, 122, 129
Getzler, A. See Roundy, R. O. et al.
Giffler, B. and Thompson, G.L.  288
Gilat, D.  345
Gilbert, E. N. See Coffman, E. G. Jr. and Gilbert, E. N.
Girkar, M. See Polychronopoulos, C. et al.
Gittins, J. See Nash, P. and Gittins, J.
Gittins, J. C.  40, 41, 337, 338, 341
Glazebrook, K. D.  343
Glazebrook, K. D. and Fay, N. A.  343
Glynn, P. W. and Whitt, W.  17, 26
Goldratt, E. M. and Cox, J.  34
Goldwasser, S. See Feige, U. et al.
Gondran, M. and Minoux, M.  197, 201, 202, 330
Gonzalez, T. See Sahni, S. and Gonzalez, T.
Gonzalez, T. and Sahni, S.  10, 151, 228, 278
Gonzalez, T. et al.  8
Grabowski, J. et al.  289
Graham, R. L.  5, 6, 16, 87

Graham, R. L. *See* Coffman, E. G. Jr. and Graham, R. L.
Graham, R. L. *et al.* 94, 241
Graves, S. C. *et al.* 194
Greenberg, A. G. *et al.* 17, 26
Greenberg, B. S. and Wolff, R. W. 49, 51
Haghighat, M. *See* Polychronopoulos, C. *et al.*
Hajek, B. 34
Halfin, S. 329
Hall, L. A. *See* Williamson, D. P. *et al.*
Han, S. *et al.* 18
Hanen, C. 193, 194, 208, 211, 212, 217
Hanen, C. and Munier, A. 211, 212, 214, 216, 218
Hardy, G. H. *et al.* 295
Harrison, J. M. 34, 40, 53, 329, 330, 331, 332
Harrison, J. M. *See* Dai, J. G. and Harrison, J. M.
Harrison, J. M. and Nguyen, V. 329
Harrison, J. M. and Shepp, L. A. 345
Harrison, J. M. and Wein, L. M. 49
Harrison, J. M. and Williams, R. J. 329
Heath, M. T. *See* Geist, G. A. and Heath, M. T.; George, A. *et al.*
Heath, M. T. and Romine, C. H. 120
Hefetz, N. and Adiri, I. 279
Heidorn, G. E. *See* Davis, E. W. and Heidorn, G. E.
Herer, Y. T. *See* Roundy, R. O. *et al.*
Herroelen, W. *See* De Wit, J. and Herroelen, W.; Demeulemeester, E. and Herroelen, W.; Demeulemeester, E. *et al.*
Herroelen, W. S. 259, 260
Hillier, F. S. and Boling, R. M. 322
Hillion, H. 205
Hillion, H. and Proth, J.-M. 207, 212
Hiranandani, S. *et al.* 137
Hirayama, T. and Kijima, M. 306
Hochbaum, D. S. and Shmoys, D. B. 5, 8
Hofri, M. *See* Bruno, J. and Hofri, M.
Hong, D. *See* Han, S. *et al.*
Hoogeveen, J. A. *See* Williamson, D. P. *et al.*
Hoogeveen, J. A. *et al.* 2, 7, 8, 75, 132, 133
Hu, T. C. 44, 95, 155, 209
Huang, C. C. and Weiss, G. 52
Huber, W. D. *See* Patterson, J. H. and Huber, W. D.
Hurkens, C. A. J. *See* Williamson, D. P. *et al.*
Hwang, J. J. *et al.* 82

Ibarra, O. H. *See* Gonzalez, T. *et al.*
Jackson, J.R. 279, 283
Jackson, P. *See* Claver, J. F. and Jackson, P.
Jaffe, J. M. 149, 153
Jakoby, A. and Reischuk, R. 80
Janacek, G. J. 145
Jean-Marie, A. 167
Jean-Marie, A. *See* Lefebvre-Barbaroux, S. *et al.*
Johnson, D. P. 334
Johnson, D. S. 3
Johnson, D. S. *See* Brucker, P. *et al.*; Garey, M.R. and Johnson, D.S.; Garey, M.R. *et al.*
Johnson, D. S. and Papadimitriou, C. H. 5
Johnson, D. S. *et al.* 24
Johnson, S. M. 26, 323
Johnson, T. J. R. 263
Jung, H. *et al.* 71
Jurisch, B. *See* Brucker, P. and Jurisch, B.; Brucker, P. *et al.*

Kailath, T. *See* Varvarigou, T. A. *et al.*
Kämpke, T. 43
Kanet, J. J. and Adelsberger, H. H. 34
Kaplan, L. 262
Karatzas, I. 341
Karatzas, I. and Ocone, D. L. 339, 341
Karmarkar, N. and Karp, R. 42
Karmarkar, N. and Karp, R. M. 4, 17, 22
Karmarkar, N. *et al.* 24, 42
Karp, R. *See* Karmarkar, N. and Karp, R.; Karmarkar, N. *et al.*
Karp, R. M. 3, 5
Karp, R. M. *See* Karmarkar, N. and Karp, R. M.; Karmarkar, N. *et al.*
Kaspi, H. *et al.* 36, 40
Kawaguchi, T. and Kyan, S. 42
Keener, R. 341
Kelly, F. P. 34
Kelly, F. P. and Laws, C. N. 34, 329, 334
Kennedy, K. *See* Callahan, D. and Kennedy, K.; Hiranandani, S. *et al.*
Khumawala, B. M. *See* Stinson, J. P. *et al.*
Kijima, M. *See* Hirayama, T. and Kijima, M.
Kijima, M. and Makimoto, N. 323
Kijima, M. *et al.* 323
Kim, K. H. and Naghibzadeh, M. 168
Kim, S. J. and Browne, J. C. 123, 127
Kingman, J. F. C. 329
Kirousis, L. M. *See* Jung, H. *et al.*

Kleinrock, L. 34, 43, 58
Kleitman, D. J. and Rothschild, B. L. 28
Klimov, G. P. 53
Kobacker, E. See Adler, L. et al.
Koelbel, C. and Mehrotra, P. 137
Kogge, P. M. 208
König, D. See Franken, P. et al.
Korst, J. 194, 223
Krämer, A. See Brucker, P. et al.
Kruatrachue, B. and Lewis, T. 87
Ku, P.-S. and Nui, S.-C. 317
Kumar, P. R. 34
Kumar, P. R. See Lin, W. and Kumar, P. R.
Kung, S. Y. 123
Kurtz, T. K. See Ethier, S. N. and Kurtz, T. K.
Kyan, S. See Kawaguchi, T. and Kyan, S.

Labetoulle, J. 168, 176
Laftit, S. et al. 208
Lageweg, B. J. See Fisher, M. L. et al.; Veltman, B. et al.
Lai, T. H. and Sprague, A. 146
Lam, S. and Sethi, R. 153
Latapie, B. See Carlier, J. and Latapie, B.
Lawler, E. L. 100, 103, 104, 253
Lawler, E. L. See Graham, R. L. et al.; Valdes, J. et al.
Lawler, E. L. et al. 28, 33, 92, 167
Lawler, G. F. and Vanderbei, R. J. 338, 341, 342, 343
Lawrence, S. 290
Laws, C. N. 34, 327, 329, 331, 333, 334
Laws, C. N. See Kelly, F. P. and Laws, C. N.
Laws, C. N. and Louth, G. M. 34, 329
Layland, J. W. See Liu, C. L. and Layland, J. W.
Lazowska, E. D. See Eager, D. L. et al.
Leachman, R. C. 34
Leadbetter, M. R. et al. 21, 25
Lee, C. See Polychronopoulos, C. et al.
Lee, C. Y. See Hwang, J. J. et al.
Lefebvre-Barbaroux, S. 167, 168, 182
Lefebvre-Barbaroux, S. et al. 173, 182
Lehtonen, T. 51, 323
Leiserson, C. E. See Cormen, T. H. et al.
Leitstand 34
Lenstra, J. K. 1
Lenstra, J. K. See Blazewicz, J. et al.; Fisher, M. L. et al.; Graham, R. L. et al.; Hoogeveen, J. A. et al.; Lawler, E. L. et al.; O'kEigearthaich, M. et al.; van Laarhoven, P. J. M.

et al.; Veltman, B. et al.; Williamson, D. P.
Lenstra, J. K. and Rinnooy Kan, A. H. G. 2, 6, 112, 278
Lenstra, J. K. et al. 8, 277, 278
Leung, B. See Polychronopoulos, C. et al.
Leung, J. Y. -T. See Han, S. et al.
Leung, J. Y. -T. and Merril, M. L. 167
Lewis, T. See Kruatrachue, B. and Lewis, T.
Lewis, T. G. See El-Rewini, H. and Lewis, T. G.
Liggett, T. M. 26
Lin, W. and Kumar, P. R. 43
Lindgren, G. See Leadbetter, M. R. et al.
Littlewood, J. E. See Hardy, G. H. et al.
Liu, C. L. and Layland, J. W. 168, 176, 178, 182
Liu, J. See George, A. et al.
Liu, Z. 91
Liu, Z. See Baccelli, F. and Liu, Z.; Baccelli, F. et al.; Coffman, E. G. and Liu, Z.; Coffman, E. G. Jr. and Liu, Z.
Liu, Z. and Sanlaville, E. 97, 99, 102, 103, 105
Liu, Z. and Towsley, D. 303, 304, 305, 306, 309, 323
Loulou, R. 18
Louth, G. M. See Laws, C. N. and Louth, G. M.
Lovász, L. See Feige, U. et al.
Luecker, G. S. See Karmarkar, N. et al.
Lueker, G. S. 22, 42
Lueker, G. S. See Coffman, E. G. Jr. and Lueker, G. S.; Coffman, E. G. Jr. et al.; Karmarkar, N. et al.
Lund, C. See Babai, L. et al.

McCormick, S. T. et al. 194
McGeoch, L. A. See Johnson, D. S. et al.
McKean, H. P. See Flatto, L. and McKean, H. P.
McKeown, G. P. See Burton, F. W. et al.
McNaughton, R. 101
Mahadev, N. See de Werra, D. et al.
Mahadev, N. V. R. et al. 230
Makimoto, N. See Kijima, M. and Makimoto, N.; Kijima, M. et al.
Mandelbaum, A. 338, 340, 341, 342
Mandelbaum, A. See Chen, H. and Mandelbaum, A.; Kaspi, H. et al.
Mandelbaum, A. and Vanderbei, R. J. 338

Mandelbaum, A. *et al.* 338, 340, 343
Mandeville, D. E. *See* Moodie, C. L. and Mandeville, D. E.
Marshall, A. W. and Olkin, I. 295, 298, 299, 304
Massey, W. *See* Baccelli, F. *et al.*
Matloff, N. 37
Matsuo, H. 194, 211
Maxwell, W. L. *See* Conway, R. W. *et al.*; Roundy, R. O. *et al.*
Mazziotto, G. and Millet, A. 338
Meal, H. C. *See* Graves, S. C. *et al.*
Mehrotra, P. *See* Koelbel, C. and Mehrotra, P.
Meilijson, I. and Weiss, G. 40, 41, 53
Menich, R. 295, 296
Menich, R. and Serfozo, R. F. 296
Merril, M. L. *See* Leung, J. Y. -T. and Merril, M. L.
Messerschmitt, D. G. *See* Parhi, K. K. and Messerschmitt, D. G.
Meyer, W. L. *See* Brand, J. D. *et al.*
Meyn, S. P. and Tweedie, R. L. 20
Miller, L. W. *See* Conway, R. W. *et al.*; Schrage, L. and Miller, L. W.
Millet, A. *See* Mazziotto, G. and Millet, A.
Minoux, M. 330
Minoux, M. *See* Gondran, M. and Minoux, M.
Mirchandaney, R. *See* Saltz, J. *et al.*
Mirchandani, P. B. *See* Xu, S. H. *et al.*
Mirchandani, P. B. and Xu, S. H. 43
Mitra, D. *See* Weiss, A. and Mitra, D.
Moder, J. J. *et al.* 270
Möhring, R. H. *See* Bartusch, M. *et al.*
Möhring, R. H. and Radermacher, F. J. 264, 271
Moler, C. 138
Moller, P. *See* Cohen, G. *et al.*
Monma, C. L. and Rinnooy Kan, A. H. G. 49
Moodie, C. L. and Mandeville, D. E. 263
Munier, A. 193, 194, 205, 206, 210, 211
Munier, A. *See* Hanen, C. and Munier, A.
Munshi, A. A. and Simons, B. 209
Muntz, R. R. and Coffman, E. G. Jr. 100, 102, 104
Murty, U. S. R. *See* Bondy, J. A. and Murty, U. S. R.
Muth, E. J. 49, 322

Naghibzadeh, M. *See* Kim, K. H. and Naghibzadeh, M.

Nash, P. and Gittins, J. 53
Nelken, I. *See* Gerasoulis, A. and Nelken, I.
Newell, G. F. 58
Nguyen, V. *See* Harrison, J. M. and Nguyen, V.
Nicolau, A. *See* Aiken, A. and Nicolau, A.
Ning, Q. *See* Gao, G. and Ning, Q.
Norman, A. R. *See* Davies, M. H. A. and Norman, A. R.
Nowicki, E. *See* Grabowski, J. *et al.*
Nui, S. -C. *See* Ku, P. -S. and Nui, S. -C.

Ocone, D. L. *See* Karatzas, I. and Ocone, D. L.
Odlyzko, A. M. *See* Karmarkar, N. *et al.*
O'kEigearthaich, M. *et al.* 241
Olkin, I. *See* Marshall, A. W. and Olkin, I.
Orlin, J. B. *See* Young, N. E. *et al.*
Ortega, J. M. 123, 130

Papadimitriou, C. and Yannakakis, M. 97, 112, 118, 130
Papadimitriou, C. H. *See* Johnson, D. S. and Papadimitriou, C. H.
Papadimitriou, C. H. and Tsitsiklis, J. N. 105
Papadimitriou, C. H. and Yannakakis, M. 69, 71, 85
Parhi, K. K. and Messerschmitt, D. G. 210, 217
Park, K. *See* Bell, C. A. and Park, K.
Patel, J. H. *See* Davidson, E. S. and Patel, J. H.
Patterson, J. 260, 264
Patterson, J. *See* Demeulemeester, E. *et al.*
Patterson, J. H. *See* Talbot, B. and Patterson, J. H.
Patterson, J. H. and Huber, W. D. 263
Patterson, J. H. and Roth, G. 263
Peterson, W. P. 329, 334
Petrovic, R. 263
Phillips, C. R. *See* Moder, J. J. *et al.*
Phillips, S. and Westbrook, J. 16
Picouleau, C. 7, 75, 76, 80, 85, 122
Picouleau, C. *See* Chrétienne, P. and Picouleau, C.
Pinedo, M. 322
Pinedo, M. *See* Adler, L. *et al.*; Chao, X. *et al.*
Pinedo, M. and Weiss, G. 105
Pinedo, M. L. 34, 49

Pinedo, M. L. *See* McCormick, S. T. *et al.*;
 Weiss, G. and Pinedo, M. L.; Wie, S. H.
 and Pinedo, M. L.
Pinedo, M. L. and Ross, S. L.  34
Pinedo, M. L. and Weiss, G.  34, 43
Pinedo, M. L. and Wie, S. -H.  49
Pinson, E.  277
Pinson, E. *See* Carlier, J. and Pinson, E.;
 Carlier, J. and Pinson, E.
Pitman, J. *See* Barlow, M. *et al.*
Pitman, J. and Yor, M.  345
Pliska, S. R. *See* Tcha, D. W. and Pliska, S. R.
Plotnicoff, J. C. *See* Adler, L. *et al.*
Polya, G. *See* Hardy, G. H. *et al.*
Polychronopoulos, C. *et al.*  136
Presman, E. L. and Sonin, I. N.  337, 338
Pritsker, A. B. *et al.*  263
Pritsker Corporation  34
Proth, J. -M. *See* Hillion, H. and Proth, J.-M.; Laftit, S. *et al.*

Quadrat, J. -P. *See* Cohen, G. *et al.*
Queyranne, M.  287

Radermacher, F. J.  263, 264, 271
Radermacher, F. J. *See* Bartusch, M. *et al.*;
 Möhring, R. H. and Radermacher, F. J.
Ramamoorthy, C. V. *See* Reddi, S. S. and
 Ramamoorthy, C. V.
Ramchandani, C.  197
Ramudhin, A. *See* Bartholdi, J. J. *et al.*
Ramudhin, A. *et al.*  17, 27
Rayward-Smith, V. J.  7, 85, 86, 145
Rayward-Smith, V. J. *See* Burton, F. W. and
 Rayward-Smith, V. J.; Burton, F. W. *et al.*
Reddi, S. S. and Ramamoorthy, C. V.  230
Reiman, M. I.  329, 333, 334
Reischuk, R. *See* Jakoby, A. and Reischuk, R.
Reiss, R. -D.  21
Resnick, S. I.  25
Reynolds, P. F. *See* Chandy, K. M. and
 Reynolds, P. F.
Righter, R.  34, 43, 323
Rinnooy Kan, A. H. G. *See* Blazewicz, J. *et al.*; Fisher, M. L. *et al.*; Frenk, J. B. G.
 and Rinnooy Kan, A. H. G.; Graham, R.
 L. *et al.*; Lawler, E. L. *et al.*; Lenstra, J.
 K. and Rinnooy Kan, A. H. G.; Lenstra,
 J. K. *et al.*

Rinnooy Kan, A. H. G. *See* Monma, C. L.
 and Rinnooy Kan, A. H. G.;
 O'kEigearthaich, M. *et al.*
Rinnooy Kan, A. H. G. and Frenk, J. B.
 G.  22
Rishel, R. W. *See* Benes, V. E. and Rishel, R. W.
Rivest, R. L. *See* Cormen, T. H. *et al.*
Robillard, P. *See* Bratley, P. *et al.*
Röck, H. and Schmidt, G.  12
Romine, C. H. *See* Heath, M. T. and Romine, C. H.
Rootzén, H. *See* Leadbetter, M. R. *et al.*
Ross, S. L. *See* Pinedo, M. L. and Ross, S. L.
Ross, S. M.  296
Roth, G. *See* Patterson, J. H. and Roth, G.
Rothschild, B. L. *See* Kleitman, D. J. and
 Rothschild, B. L.
Roundy, R.  211, 212, 217
Roundy, R. O. *et al.*  34
Roy, B. and Sussman, B.  280
Roychowdhury, V. P. *See* Varvarigou, T. A. *et al.*

Saad, Y.  123, 130, 138
Safra, S. *See* Feige, U. *et al.*
Sahni, S. *See* Gonzalez, T. and Sahni, S.;
 Gonzalez, T. *et al.*
Sahni, S. and Gonzalez, T.  5
Sakasegawa, H. *See* Shanthikumar, J. G. *et al.*; Yamazaki, G. and Sakasegawa, H.;
 Yamazaki, G. *et al.*
Saltz, J. *et al.*  117, 137
Salz, J. *See* Foschini, G. J. and Salz, J.
Sanlaville, E.  91, 97, 98, 100, 104, 108
Sanlaville, E. *See* Liu, Z. and Sanlaville, E.
Santoso, H. *See* Towsley, D. *et al.*
Sarkar, V.  82, 113, 114, 117, 118, 119, 127, 130
Schaffer, L. R. *See* Brand, J. D. *et al.*
Schevon, C. *See* Johnson, D. S. *et al.*
Schlunk, O. *See* Greenberg, A. G. *et al.*
Schmidt, G. *See* Blazewicz, J. *et al.*; Röck,
 H. and Schmidt, G.
Schmidt, J. P. *et al.*  12
Schmidt, V. *See* Franken, P. *et al.*
Schouten, D. *See* Polychronopoulos, C. *et al.*
Schrage, L.  34, 263, 302
Schrage, L. and Miller, L. W.  34, 37, 39
Schwartz, M.  333
Schwiegelshohn, U.  241

Schwiegelshohn, U. *See* Gasperoni, F. and Schwiegelshohn, U.
Serafini, P. and Ukovich, W.   194, 221, 223, 241
Serfozo, R.   34
Serfozo, R. F. *See* Menich, R. and Serfozo, R. F.
Serlin, O.   168, 176, 182
Sethi, R. *See* Garey, M. R. *et al.*; Lam, S. and Sethi, R.
Sevcik, K. C.   40, 53
Shaked, M. and Shanthikumar, J. G.   34, 296
Shanthikumar, J. G. *See* Shaked, M. and Shanthikumar, J. G.; Yamazaki, G. *et al.*
Shanthikumar, J. G. and Sumita, U.   296, 305
Shanthikumar, J. G. and Yao, D.   34
Shanthikumar, J. G. and Yao, D. D.   296
Shanthikumar, J. G. and Yao, D. J.   323
Shanthikumar, J. G. *et al.*   52
Shenker, S. *See* McCormick, S. T. *et al.*
Shepp, L. A. *See* Harrison, J. M. and Shepp, L. A.; Mandelbaum, A. *et al.*
Shirakawa, H. *See* Kijima, M. *et al.*
Shmooys, B. B. *See* Lawler, E. L. *et al.*
Shmoys, D. B.   1
Shmoys, D. B. *See* Hochbaum, D. S. and Shmoys, D. B.; Lawler, E. L. *et al.*; Lenstra, J. K. *et al.*; Williamson, D. P. *et al.*
Siegel, A. *See* Schmidt, J. P. *et al.*
Sievers, B. *See* Brucker, P. *et al.*
Sigman, K. *See* Chao, X. *et al.*
Silly, M.   168
Simons, B. *See* Munshi, A. A. and Simons, B.
Simpson, W. P.   262, 272
Simpson, W. P. *See* Demeulemeester, E. *et al.*
Sleep, M. R. *See* Burton, F. W. *et al.*
Slowinski, R.   262
Slowinski, R. *See* Blazewicz, J. *et al.*
Smith, W. L. *See* Cox, D. R. and Smith, W. L.
Solot, P.   227
Solot, P. *See* de Werra, D. and Solot, P.; de Werra, D. *et al.*; Mahadev, N. V. R. *et al.*
Sonin, I. N. *See* Presman, E. L. and Sonin, I. N.
Sorensen, D. C. *See* Dongarra, J. J. and Sorensen, D. C.
Sotskov, Y. N.   278
Sparaggis, P. D. *et al.*   297, 298, 309

Speranza, M. G. *See* Dell'Olmo, P. *et al.*
Spirakis, P. *See* Jung, H. *et al.*
Sprague, A. *See* Lai, T. H. and Sprague, A.
Srikantakumar, P. S. *See* Xu, S. H. *et al.*
Srinivasan, A. *See* Schmidt, J. P. *et al.*
Srinivasan, R.   17, 26, 27
Stefek, D. *See* Graves, S. C. *et al.*
Stidham, S.   34, 328
Stidham, S. and Weber, R. R.   34
Stinson, J. P. *et al.*   263, 264, 266, 269
Stoyan, D.   34, 296
Sumita, U. *See* Shanthikumar, J. G. and Sumita, U.
Suresh, S. *See* Foley, R. D. and Suresh, S.
Suresh, S. *et al.*   322
Sussman, B. *See* Roy, B. and Sussman, B.
Szegedy, M. *See* Feige, U. *et al.*

Tadei, R. *See* Dellacroce, F. *et al.*
Takagi, H.   40
Talbot, B. and Patterson, J. H.   263, 264
Talwar, P. P.   323
Tamarit, J.M. *See* Christofides, N. *et al.*
Tardos, É. *See* Lenstra, J. K. *et al.*
Tarjan, R. E. *See* Garey, M. R. *et al.*; Valdes, J. *et al.*; Young, N. E. *et al.*
Tayur, S. R. *See* Roundy, R. O. *et al.*
Tcha, D. W. and Pliska, S. R.   53
Tembe, S. V. and Wolff, R. W.   49, 51, 322
Thompson, G. L. *See* Fischer, J. F. and Thompson, G. L.; Giffler, B. and Thompson, G. L.
Towsley, D.   295
Towsley, D. *See* Baccelli, F. *et al.*; Liu, Z. and Towsley, D.; Sparaggis, P. D. *et al.*
Towsley, D. and Baccelli, F.   305, 323
Towsley, D. *et al.*   307, 308, 309
Tripathi, S. K. *See* Agrawala, A. K. *et al.*
Trubian, M. *See* Dell'Amico, M. and Trubian, M.
Tseng, C. W. *See* Hiranandani, S. *et al.*
Tsitsiklis, J. N. *See* Papdimitriou, C. H. and Tsitsiklis, J. N.
Tsoucas, P. and Walrand, J.   323
Tuza, Zs. *See* Dell'Olmo, P. *et al.*
Tweedie, R. L. *See* Meyn, S. P. and Tweedie, R. L.

Ukovich, W. *See* Serafini, P. and Ukovich, W.
Ullman, J. D.   91, 92, 94

Valdes, J. *et al.* 79
Van der Heyden, L. 43
Vande Vate, J. H. *See* Bartholdi, J. J. *et al.*; Ramudhin, A. *et al.*
Vanderbei, R. J. *See* Kaspi, H. *et al.*; Lawler, G. F. and Vanderbei, R. J.; Mandelbaum, A. and Vanderbei, R. J.; Mandelbaum, A. *et al.*
Van de Velde, S. L. *See* Hoogeveen, J. A. *et al.*
van Laarhoven, P. J. M. *et al.* 290
Varaiya, P. *See* Ephremides, A. *et al.*; Weber, R. R. *et al.*
Varaiya, P. P. *et al.* 53, 341, 343
Varvarigou, T. A. *et al.* 85
Vasicek, O. A. 296, 305
Veltman, B. 85
Veltman, B. *See* Hoogeveen, J. A. *et al.*
Veltman, B. *et al.* 66
Venderbei, R. J. *See* Dynkin, E. B. and Vanderbei, R. J.
Viot, M. *See* Cohen, G. *et al.*
Volta, G. *See* Dellacroce, F. *et al.*

Walrand, J. *See* Ephremides, A. *et al.*; Tsoucas, P. and Walrand, J.; Varaiya, P. P. *et al.*; Weber, R. R. *et al.*
Walrand, J. C. *See* Varaiya, P. P. *et al.*
Walsh, J. B. 339, 345
Wang, J. *See* Eisenbeis, C. and Wang, J.
Wang, Y. *See* Dai, J. G. and Wang, Y.
Warmuth, M. K. *See* Dolev, D. and Warmuth, M. K.
Warren Burton, F. 145
Watters, L. J. *See* Pritsker, A. B. *et al.*
Weber, R. R. 40, 43, 49, 51, 296, 313, 320, 323, 329
Weber, R. R. *See* Coffman, E. G. *et al.*; Stidham, S. and Weber, R. R.; Xu, S. H. *et al.*
Weber, R. R. and Weiss, G. 51, 52, 58, 322, 323
Weber, R. R. *et al.* 43
Weglarz, J. 262
Weglarz, J. *See* Blazewicz, J. *et al.*
Wein, L. M. 34, 49, 52, 329
Wein, L. M. *See* Harrison, J. M. and Wein, L. M.
Weiss, A. 17
Weiss, A. and Mitra, D. 58
Weiss, G. 33, 34, 44, 46, 48, 53, 55, 316, 323

Weiss, G. *See* Bartholdi, J. J. *et al.*; Huang, C. C. and Weiss, G.; Meilijson, I. and Weiss, G.; Pinedo, M. and Weiss, G.; Pinedo, M. L. and Weiss, G.; Ramudhin, A. *et al.*; Weber, R. R. and Weiss, G.
Weiss, G. and Pinedo, M. L. 34, 43
Wessels, J. *See* Adan, I. J. B. F. *et al.*
Westbrook, J. *See* Phillips, S. and Westbrook, J.
Whitt, W. 15, 21, 52, 328, 334
Whitt, W. *See* Coffman, E. G. and Whitt, W.; Coffman, E. G. Jr. *et al.*; Glynn, P. W. and Whitt, W.; Greenberg, A. G. *et al.*
Whittle, P. 40, 41, 343
Widmer, M. 290
Wie, S. H. *See* Pinedo, M. L. and Wie, S. -H.
Wie, S. H. and Pinedo, M. L. 49
Wiest, J. D. 263
Williams, R. J. *See* Harrison, J. M. and Williams, R. J.
Williamson, D. P. *et al.* 2, 10, 11, 12
Windheiser, D. *See* Eisenbeis, C. and Windheiser, D.
Winkler, P. 28
Winston, W. 329
Wolf, B. *See* McCormick, S. T. *et al.*
Wolfe, P. M. *See* Pritsker, A. B. *et al.*
Wolff, R. W. *See* Greenberg, B. S. and Wolff, R. W.; Tembe, S. V. and Wolff, R. W.
Wolsey, L. A. *See* Dyer, M. and Wolsey, L. A.
Wright, P. E. 17
Wright, P. E. *See* Coffman, E. G. Jr. and Wright, P. E.
Wu, M. Y. and Gajski, D. 113, 124, 126, 137
Wu, T. -P. *See* Adler, L. *et al.*

Xie, X. *See* Laftit, S. *et al.*
Xu, S. H. *See* Mirchandani, P. B. and Xu, S. H.
Xu, S. H. *et al.* 43

Yakir, B. 17, 22, 23
Yamazaki, G. *See* Shanthikumar, J. G. *et al.*
Yamazaki, G. and Sakasegawa, H. 49
Yamazaki, G. *et al.* 52
Yang, K. -K. *See* Demeulemeester, E. *et al.*
Yang, T. 111
Yang, T. *See* Gerasoulis, A. and Yang, T.

Yang, T. and Gerasoulis, A.   112, 128, 129, 130, 132, 133, 136
Yannakakis, M. *See* Garey, M. R. *et al.*; Papadimitriou, C. and Yannakakis, M.; Papadimitriou, C. H. and Yannakakis, M.
Yao, D. *See* Chang, C. -S. and Yao, D.; Shanthikumar, J. G. and Yao, D.
Yao, D. D. *See* Chen, H. and Yao, D. D.; Shanthikumar, J. G. and Yao, D. D.
Yao, D. J. *See* Shanthikumar, J. G. and Yao, D. J.
Yechiali, U. *See* Browne, S. and Yechiali, U.
Yor, M. *See* Barlow, M. *et al.*; Pitman, J. and Yor, M.
Young, N. E. *et al.*   253
Yue, M.   3

Zahorjan, J. *See* Eager, D. L. *et al.*
Zawack, D. *See* Adams, J. *et al.*
Zdrzalka, S. *See* Grabowski, J. *et al.*
Zeghmi, A. H. *See* Graves, S. C. *et al.*
Zijm, W. H. M. *See* Adan, I. J. B. F. *et al.*

*Index compiled by Geoffrey C. Jones*

# Subject Index

Active schedule generation  288
Activity period  173–5, 188
Acyclic scheduling problems  241–58
Agreeability condition  318
Allocation functions  222
   load balancing  155–9
Allocation strategy  339
Approximate optimality  36, 46–7
Approximation algorithms  4, 5, 7, 71–2, 80–5, 289–90
Arcsine law  345
Ascendant sets  284–5
Associated schedules  216

Backtracking  265, 268
Bandit processes  41, 337–47
   branching  54–5
   control theory  340–2
   future directions of research  340–6
   Lévy  342
   Markovian  341, 342
   multi-armed  337, 339–40
   open problems  340–6
   prior research  338–9
   restless bandits  56–8
   special bandits  341
   switched processes  344–6
   with switching costs  343
Basic cyclic scheduling problem (BCS)  194, 195–8
   cyclic versions  209–12
   with deadlines  205
   with linear precedence constraints  205–6
   with resource constraints  208–12
Batch jobs  37–8
Bin packing problem  3
Bipartite graph  229, 231, 232, 235–6
Blocking  314–15
Borel mappings  305
Bottom-up level  125

Bounds  171–6
Bowl-shape phenomenon  315
   literature review  322
Branch-and-bound algorithms  217–18, 259–76
Branch-and-bound methods, elimination rules  283
Branching bandit processes  54–5
Branching rule  218
Branching schemes  288–9
Brownian approximations  59
Brownian control problem  59
Brownian model  329–32
Brownian motion  25, 26, 329, 341, 345
   nonlinearly-skewed  344–5
Busy period  173

Capacity constraints, relaxation  286
Central limit theorems (CLTs)  24–6, 162
Circular allocation functions  216
Classical list scheduling heuristic  125–6
Clique of disjunctions  282
CLIQUE problem  6, 69, 70
Cluster
   definition  118
   merging  130–1
      Sarkar algorithm  133–5
Clustering
   definitions  118–20
   linear  118, 120–2
   nonlinear  118, 120–1
   using dominant sequence algorithm (DSC)  128–30
Combinatorial optimization problem  4, 26
Communication delays  65–90
   problem definition  66–7
   with limited number of processors  85–6
   with no task duplication  73–86
   with task duplication  66–72
   with unlimited number of processors  67–72, 73–85

# SUBJECT INDEX

Communication network with real-time constraints 182–8
Connex notions 281
Constant profiles, optimal algorithms for 105
Convex task 114
Convexity 315
  constraint 114
  literature review 322
Core schedules 250–2
c.p.i. (cyclic pseudointerval) 236–7
CPM-graph 82–3
CPM-schedule 82–4
CPOSS 238
Critical circuits 196–7
Critical path (CP) priority list 125
Critical path diagrams 313–14
Critical path method (CPM) 194
  rearrangement 288–9
Critical subgraph 68
Cut constraints 330
Cyclic compact schedules 236–7
Cyclic dependence graph 247
Cyclic policy 308
Cyclic scheduling on identical processors (CSIP) 209, 211
Cyclic scheduling problems
  graph coloring models for 227–39
  modeling 202–3
  on parallel processors 193–226
  transforming into acyclic scheduling problems 241–58
Cyclic tasks 241, 245–7

DAG (directed acyclic program task graph) 111
DAG (directed acyclic weighted task graph) 112
  computing weights for 117
Data dependence graph 113–14
Data partitioning 116–17
Data structure 116
Data units 116
Deadline driven priority 168
Deadlines 205
Decreasing zigzag profile 104
Delaying alternatives 265, 268, 269, 271
Delaying point 268, 271
Delaying set 268, 271
Demeulemeester–Herroelen (DH) procedure 267–70
  GRCPSP 270–2

PRCPSP 272–4
Dependence distance 247
Depth-balanced schedule 156
Deterministic distributed program 147
Deterministic scheduling 33, 34
Dioids 201–4
Directed bipartite graph 79–80
Disjunctions 285
Disjunctive graph 281
Disjunctive programming problem 283
Distance between processors 117
Distributed systems 146
Doacross technique 208–9
DOALL 137
Dominance rules 269
Dominant sequence 120
DSC (dominant sequence algorithm) 82–4
  clustering using 128–30
Dual Jackson's preemptive schedule (DJPS) 283–4
Dual system 316
  literature review 322–3
Duality relaxations 287
Dynamic scheduling 198–208
  extensions 204–8

EDD (earliest due date) 36, 94, 99–100, 103, 323
Edge $k$-colorings 229
Eligible activity 272
Eligible set 270
Elimination rules 283
Elite theorem 95
ELS (extended list schedule) algorithm 87
Equitable edge $k$-colorings 232–6
ETF (earliest task first) heuristic 87
Euler's constant 21
Exact algorithms 217–18
Expected flowtime for i.i.d. jobs 44–5
Expected optimal schedule length 163
Expected weighted flowtime 46

FACTOR 34
Feasibility
  approaches to 168
  in real-time systems 167
  optimal access policy 183
  with bounds 186–8
  with trajectories 184–6
Feasibility graph 252–5
Feasibility tests 181–2

FFD (first fit decreasing) algorithm  3
FIFO (first in first out)  39, 170, 176, 177, 186, 188, 296
Finite buffers  314–15
First-come first-serve (FCFS)  303–6, 332–4
Flow shops  10–12, 26–8
  two-machine  49–50, 316–19
Fluid approximations  58
Folded graph  214
Forbidden sets  264
Forward induction  302
Free tasks  125
Fully polynomial approximation scheme  4
Functional central limit theorems (FCLTs)  24–6

Gantt chart  118, 119
General processing time distributions  35
Generalized cut constraints  330
Generalized resource-constrained project scheduling problem (GRCPSP)  260–2
  Demeulemeester–Herroelen (DH) procedure  270–2
Generic tasks  195–6, 199, 206, 222
Gittins index  36, 41, 48
Granularity
  coarse-grain  121–3
  fine-grain  121–3
  theory  120–4
Graph coloring  222–3
  models  227–39
Greedy policy, on-line  18–22

Half cuts  288
HARPOON problem  73–4
Harris-recurrent Markov chains  19
Heavy traffic analysis  329–32
Heuristics  35
  assessment of  35
  based on priority rules  36
HLF (highest level first) algorithm  94, 95, 103

IC algorithm  84–5
Identical processors, selection and mapping structure  214–16
i.i.d. jobs, expected flowtime for  44–5
ILR (increasing likelihood ratio)  44, 299

Immediate selection
  on ascendant sets  284–5
  on disjunctions  285
Implicit enumeration procedures  263
Impossibility theorem  7, 9, 11
In-forests  93, 103, 104, 106–7
Increasing failure rate (IFR)  299
Increasing likelihood ratio (ILR)  44, 299
Increasing zigzag profiles  93, 103, 104
Index field  340
Index function  341
Index policy  56–7
Index processes  339–40
Index strategies  339–40, 342–3
Indexing function  253–4
Induction hypothesis  187
Infinite time slicing (ITS)  168
Information process  339
Initial final task  92
Insertion  218–19
Integer linear programming procedures  263
Integer profiles  94–100
Interchange argument  302
Interchangeability of queues  319–21
  literature review  323
Internalization prepass  118
Interval chromatic index  232
Interval edge $k$-colorings  228
Interval order  93
  graphs  97, 105–6
ITS generalized (ITSG) policy  168

Jackson's preemptive schedule  283
Jackson's rule  279
Job shop scheduling problems  10–12, 277–93
  approximation algorithms  289–90
  optimization algorithms  280–9
  polynomially solvable particular cases  279–81
  problem formulation  281–3
  problem statement and complexity results  278–80
  with two jobs  279–81
  with two machines  279
Johnson's rule  49, 279

Klimov's model  36, 53–5
Klimov's problem  41
KNAPSACK problem  74
$k$-periodic schedules  210–11

Laplace limit theorem   160
Largest-first differencing method (LDM)   22, 24
Largest-processing time (LPT) policy   16, 49
Largest tree first (LTF)   105, 108
Last-come first-serve (LCFS)   303–6
Lebesgue measure   342
LEPT (longest expected processing time first)   43
Level scheduling   153–5
Lévy bandits   342
Lexicographic order schedules (LOS)   97, 98
LIFO (last in first out)   39, 296
Linear algebra   201–2
Linear constraints   205–6
Linear model   117
Linear recurrence system   204
LINPACK   138
List algorithms   91–110
List scheduling   85–6, 289–90
    makespan minimization by   95–9
    maximum minimization lateness by   99–100
Little's formula   39, 328
Load balancing   16, 133–5
    allocation function   155–9
Local breadth-first schedule   156, 158
Local search technique   290
Loop scheduling   207
Lower bounds   286–9
LRP (longest remaining path first) schedule   102–4

Machine model   242–4
Machines in parallel *See* Parallel machines
Machines in series *See* Series machines
Majorization   295–311
    and stochastic ordering   296–9
Makespan minimization
    by list scheduling   95–9
    on two machines   316–18, 323
Makespan scheduling problems   15
Manufacturing blocking   314–15
Markov chain   18–19
Markov decision processes   41, 296
Markovian bandit   341, 342
Markovian routing policies   309
Maximum minimization lateness by list scheduling   99–100
Merge theorem   96

Message size   117
MIMD architectures, scheduling algorithms for   125–35
Minimum resource violating sets (MRVS)   266
Mixed integer programming (MIP) formulation   280, 282–3, 287
Modified critical path (MCP) heuristic   124, 126–7
MONOTONE-NOT-ALL-EQUAL-3SAT   10
Most successors first (MSF) schedules   97
Multi-armed bandit model   339–40
Multi-armed bandit problem   337
Multiprocessor scheduling problem   5–9, 15
Multistep scheduling methods   127–33

Natural linear clustering   123–4
Near-optimal scheduling   1–14, 27
    general principle   2–5
Network load   117
Noncyclic problems   219–20
Non-idling (or work-conserving) policy   171
Nonpreemptive compact cyclic open shop scheduling (NCCOSS) problem   230–1
Nonpreemptive cyclic open shop scheduling (NCOSS) problem   229, 230
Nonpreemptive list algorithms and preemptive priority algorithms   102–3
Nonpreemptive OSS (NOSS)   228, 231, 238
Nonpreemptive priorities   178–82
Nonpreemptive scheduling   94
Number partition problem   3

Occupancy sequence   160
Occurrence vector   219
Off-line policies   22–4
One-step scheduling methods   125–31
On-line greedy policy   18–22
Open queueing networks *See* Queueing networks
Open shops   10–12, 227–9
OPT   34
Optimal algorithms
    for constant profiles   105
    for variable profiles   105–8
Optimal circular schedule   216
Optimal navigation problem   338

Optimal policy 27
Optimal preallocated schedule (OPS) 151–3
Optimal preemption of jobs 40
Optimal scheduling 35, 151–3
Optimal stopping problem 338
Optimization algorithms 280–9
Optional increasing path 339
Out-forests 93, 104, 107–8

PARAFRASE-2 system 136
Parallel channel access policy 183, 184
Parallel computing 221
Parallel machines 42–8
  additional optimality results for stochastic scheduling 43–4
  deterministic scheduling 42
  exact results for minimizing flowtime 43
  preemptive scheduling of stochastic jobs 48
  Smith's rule heuristic for minimization of weighted flowtime 44–8
Parallel processors 193–226
  algorithms 217–20
Parallel programs 145–65
Partial schedules 267–8, 270, 273
PARTITION 73
Partitioning algorithms 114–16
Pathwise solution 332, 333
Periodic arrivals 170–6
Periodic job shop, selection and uniform constraints 212–14
Periodic scheduling 197–8, 211–12, 247–50
  algorithms 217–20
  properties of 212–16, 249
Periodic tasks 167–91
Periodicity 171–3
Permutation flow shop scheduling 26
Permutation schedule 215, 216
PERT algorithms 194
Petri nets 208
Physical mapping 131
Poisson process 20, 38, 305
Policy-free error asymptotics 24–6
Polling systems 39–40
Polyhedral techniques 287
Polynomial approximation scheme 4, 5
Polynomial problems 75–80
Polynomial-time algorithms 5
Polynomial-time approximation algorithms 3, 11–12

Polynomial-time 2-approximation algorithm 6
Preallocation 145–65
Preallocation function 147, 156
Precedence constraints, relaxation 286–7
Precedence-based lower bound 265, 271
Preemption 170–6
Preemption allowed 286–7
Preemptive cyclic open shop scheduling problem (PCOSS) 234–6, 238
Preemptive OSS (POSS) 228, 232
Preemptive priorities 177
Preemptive priority algorithms and nonpreemptive list algorithms 102–3
Preemptive profile scheduling 100–4
Preemptive resource-constrained project scheduling problem (PRCPSP) 262
  Demeulemeester–Herroelen (DH) procedure 272–4
Primal Jackson's preemptive schedule (PJPS) 284
Priority index 40
Priority scheduling algorithms 100–2
Probabilistic analysis of schedule makespans 15–31
Processor component 117
Profile scheduling 91–110
  nonpreemptive 94–100
  notation 92–4
  preemptive 100–4
  stochastic 104–8
Program partitioning 113
  bottom-up 116
  top-down 115–16
PYRROS 135–40
  demonstration of 138
  experiments 138–40
  multistep scheduling algorithms 128–33
  software system 112
  task graph language 137–8

Queueing disciplines, extremal properties 303–6
Queueing networks 34, 327–36
  control problems 295–311
  dynamic or state dependent control strategies 328–9
  with single server 52–6
Queues
  interchangeability of 319–21
  literature review 323

# SUBJECT INDEX

Random allocation function  159–63
Rate monotonic priority  168
Real-time constraints  182–8
Real-time scheduling  167–91
Relative urgency (RU)  168
Release times  71–2
Reservation tables  208
Resource conflict  265, 268
Resource-constrained project scheduling problems (RCPSP)  259–76
  branch-and-bound procedures  263–70
  optimal procedures  263–7
Resource constraints  208–12, 269
Resource pooling effect  329
Restless bandits  56–8
Reversibility  315
  literature review  322
Root activities  156
Rooted trees  77, 80–2
Routing policies  306–9, 333–35
Routing strategy, stable  329–30
RVST (resource violating set time)  266

Sarkar's cluster merging algorithms  133–5
Scaling property  3
SCHEDULER  137
Scheduling, definitions  118–20
Scheduling algorithm  93
Scheduling algorithms
  and their properties  169
  for MIMD architectures  125–35
SCT problem  67, 73, 76–80
SD algorithm  80–1
SDR algorithm  80–2, 334
Selection circuit  216, 218–19
Selection graphs  213, 215, 216, 218–19
SEND problem  75–6, 80–2
SEND–RECEIVE problem  73
Sequencing policies  333–35
Series machines  48–52
  choosing order of machines  50
  interchangeability of tandem queues  50–1
  optimal order of machines with blocking  52
  ordering machines to reduce waiting time  51–2
Series–parallel graphs  77–9
Service times same at each machine  315–16
Shifting bottleneck procedure  290
Shop scheduling  9–12
  *see also* Flow shops; Job shops; Open shops
Shortest expected processing time first (SEPT)  37, 39, 52, 316, 319
Shortest processing time first (SPT)  36, 37, 49
Shortest queue (SQ) policy  296, 306–10
Simulated annealing  290
Simultaneous service  36–7
Single pooled resource  332
Single-machine nonpreemptive scheduling  37–40
Single-machine preemptive scheduling  40–2
Single-server deterministic scheduling problems  36
Single-server system
  under hard real-time constraints  176
  with periodic arrivals and preemption  170–6
Smallest laxity first (SLF) algorithms  101
Smallest remaining processing time (SRPT) policy  295–6
  optimality  299–303
Smith's rule  42, 44–8
Snapshot relation  333
Software engineering  34
Span constraints  221–2
Stability condition  190
Stable system  173
Static macro-dataflow model of execution  113
Stationary system  173, 175
Stationary version  173
Stochastic ordering
  and majorization  296–9
  applications  318–19, 323
Stochastic scheduling  33–64, 104–8
Stochastic waves  345
Stream of jobs  38–9
Subclique of disjunctions  289
Sum–depth-balanced schedule  156–8
Superposition limit theorem  22
Switched processes  339
Synchronizing processes  150

Taboo search techniques  290
Tandem queues  313–25
  interchangeability  50–1
Task graph  91, 104
Task graph language  137–8
Task ordering  131–3

TASKGRAPHER 137
Tasks 245–7
  definition 113
T-DAG 115–16, 124, 137, 138
Temporal constraint 268
Terminal action 153
Three-dimensional matching problem 8–9
Trajectories 178–81
Transformed graph 232
Transmission delay component 117
Traveling salesman problem 3
Triangle cuts 288
Truncated branch and bound methods 290
Turnpike optimality 36, 47–8
Two-jobs cuts 287–8
Two-machine flow shops 49–50, 316–19
Two-machine job shops 279

U-DAG 115–16, 123–4
UET (unit execution time) tasks 94–100, 151
UET–UCT problem 85
Unfinished activity 272

Uniform graph 203, 215
Unitary components 206
Unitary graph 206

Valid schedule 245
Value function 339
Variable profiles, optimal algorithms for 105–8

Weak ice $w$-coloring problem 222–3
Weak interval cyclic edge $k$-colorings (or weak ice $k$-colorings) 230
Weak majorization 297–8
Weak submajorization 297
Whittle's conjecture on asymptotic optimality 57–8
Whittle's relaxed problem 56–7
Work-in-process, optimization 207–8
Work-preserving schedule 148, 156
Worst case scheduling 148–51

Zigzag profiles 93

*Index compiled by Geoffrey C. Jones*